T0239731

Molecular Modeling at the Atomic Scale

SERIES IN COMPUTATIONAL BIOPHYSICS

Nikolay V. Dokolyan, Series Editor

Molecular Modeling at the Atomic Scale: Methods and Applications in Quantitative Biology
Edited by Ruhong Zhou

Forthcoming titles

Multiscale Methods in Molecular Biophysics
Cecilia Clementi, Ed.

Coarse-Grained Modeling of Biomolecules
Garegin A. Papoian, Ed.

Computational Approaches to Protein Dynamics
Mónika Fuxreiter, Ed.

SERIES IN COMPUTATIONAL BIOPHYSICS

Nikolay V. Dokolyan, Series Editor

Molecular Modeling at the Atomic Scale

Methods and Applications in Quantitative Biology

Edited by

Ruhong Zhou

IBM T.J. Watson Research Center
Yorktown Heights, New York, USA

CRC Press is an imprint of the
Taylor & Francis Group, an **informa** business

CRC Press
Taylor & Francis Group
6000 Broken Sound Parkway NW, Suite 300
Boca Raton, FL 33487-2742

First issued in paperback 2020

ISBN 13: 978-0-367-57607-3 (pbk)
ISBN 13: 978-1-4665-6295-0 (hbk)

Library of Congress Cataloging-in-Publication Data

Molecular modeling at the atomic scale : methods and applications in quantitative
    biology / editor, Ruhong Zhou.
      p. ; cm.
    Includes bibliographical references and index.
    ISBN 978-1-4665-6295-0 (hardback : alk. paper)
    I. Zhou, Ruhong, editor.
    [DNLM: 1. Models, Molecular. 2. Systems Biology. 3. Computational Biology. QU
26.5]

QH324.2
570.285--dc23   2014022780

To my father Guomei and my mother Hanqing
To my wife Grace for her everlasting support

Contents

SECTION I   Advanced Simulation Techniques

SECTION II   Self-Assembly of Biomolecules

SECTION III   Biomolecular Interactions

SECTION IV More Applications in Molecular Biology

Series Preface

The 2013 Nobel Prize in Chemistry was awarded for the "development of multiscale models for complex chemical systems." This prize was particularly special to the whole computational community as the role that computation has played since the pioneering works of Lifson, Warshel, Levitt, Karplus, and many others was finally recognized.

This Series in Computational Biophysics has been conceived to reflect the tremendous impact of computational tools in the study and practice of biophysics and biochemistry today. The goal is to offer a collection of books that will introduce the principles and methods for computer simulation and modeling of biologically important macromolecules. The titles cover both fundamental concepts and state-of-the-art approaches, with specific examples highlighted to illustrate cutting-edge methodology. The series is designed to cover modeling approaches spanning multiple scales: atoms, molecules, cells, organs, organisms, and populations.

The series publishes advanced-level textbooks, laboratory manuals, and reference handbooks that meet the needs of students, researchers, and practitioners working at the interface of biophysics/biochemistry and computer science. The most important methodological aspects of molecular modeling and simulations as well as actual biological problems that have been addressed using these methods are presented throughout the series. Prominent leaders have been invited to edit each of the books, and in turn those editors select contributions from a roster of outstanding scientists.

The Series in Computational Biophysics would not be possible without the drive and support of the Taylor & Francis Group series manager Luna Han. All the editors, authors, and I are greatly appreciative of her support and grateful for the success of the series.

Nikolay V. Dokholyan
Series Editor
Chapel Hill, North Carolina

Preface

Molecular modeling has seen great progress in a wide range of biological applications in recent years due to the advancement of novel algorithms and high-performance computing. The gap between the timescales achievable by experiments and by computer simulations has been significantly reduced due to the concurrent advances in both experimental and theoretical techniques. Scientists can nowadays access microsecond-to-millisecond timescales with atomic detail, which is sufficient to characterize many important biological processes, such as the folding dynamics of some key proteins. This book captures the perspectives of leading experts on this transformation in theoretical molecular biology and illustrates how molecular modeling approaches are being applied to change our understanding of many important aspects of structural and molecular biology. Together, I hope these chapters provide a comprehensive overview of the fundamentals of molecular modeling and its applications in modern quantitative biology.

Although molecular modeling has been around for a while, the groundbreaking advancement of massively parallel supercomputers and novel algorithms for parallelization is shaping this field into an exciting new area. There is an increased interest in molecular modeling, as witnessed by many new conferences and newcomers, particularly from the emerging world. Until now progresses, particularly the development of novel algorithms and massively parallel computing techniques, have been led by a small group of risk-takers. With the maturation of these algorithms and simulation techniques, a broadly accessible book that lays out the fundamentals of molecular modeling will helpfully broaden these techniques to a much larger community.

This book covers both the advanced techniques of molecular modeling and the latest research advancements in biomolecular applications from leading experts, so the content should be of interest to academic and industry professionals in addition to graduate students. The book begins with a brief introduction (preface) of the major methods and applications that are presented in the chapters for molecular modeling and its applications in structural and molecular biology. Section I: Advanced Simulation Techniques (Chapters 1 through 3) covers the development of cutting-edge methods/algorithms, new polarizable force fields, and massively parallel computing techniques. Chapters 4 through 12 will describe how these novel techniques can be applied in various research areas in molecular biology. Section II: Self-Assembly of Biomolecules (Chapters 4 through 7) covers topics related to the self-assembly of biomacromolecules, including protein folding, RNA folding, amyloid peptide aggregation, and membrane lipid bilayer formation. Section III: Biomolecular Interactions (Chapters 8 through 10) focuses on the important biomolecular interactions, including protein interactions with DNA/RNA, membrane, ligands, and nanoparticles. Section IV: More Applications in Molecular Biology (Chapters 11 and 12) covers other emerging topics in biomolecular modeling, such as DNA sequencing with solid-state nanopores, biological water under nanoconfinement, and so on.

This book provides advanced techniques and real-world examples, which should be useful to senior undergraduate and graduate students, as well as established researchers in chemistry, molecular biology, computer science, and bioengineering. I hope courses such as computational chemistry, biophysical chemistry, biophysics, statistical mechanics, and computational molecular biology might find some of the contents useful as textbook or supplementary material. In addition, it might be of interest to academic and industry professionals in drug discovery and other biotechnologies.

Finally, I thank all the contributing authors for their excellent work and fruitful collaboration on this project. The very helpful and efficient technical assistance of Luna Han in putting together this book in a timely manner is greatly appreciated.

Ruhong Zhou

Editor

Ruhong Zhou is a research staff scientist and manager of Soft Matter Theory and Simulation Group at IBM Thomas J. Watson Research Center, as well as an adjunct professor in the Department of Chemistry, Columbia University. He is an elected fellow of the American Association of Advancement of Science (AAAS) and American Physical Society (APS). He received his PhD under the guidance of Prof. Bruce Berne in chemistry from Columbia University in 1997. Dr. Zhou joined IBM Research in 2000 after spending two and a half years working with Prof. Richard Friesner (Columbia University) and Prof. William Jorgensen (Yale University) on polarizable force fields. He has authored or coauthored more than 150 journal publications and 20 patents, delivered more than 150+ invited talks at major conferences and universities internationally, and chaired or cochaired many conferences in the fields of computational biology, computational chemistry, and biophysics. He is part of the IBM BlueGene team that won the 2009 National Medal on Technology and Innovation. He has also won the IBM Outstanding Technical Achievement Award (the highest technical award within IBM) in 2014, 2008 and 2005, the IBM Outstanding Innovation Award (OIA) in 2012, the IBM Research Division Award in 2005, the Columbia University Hammett Award in 1997 (for best graduates), and the American Chemical Society DEC Award on Computational Chemistry in 1995. His current research interests include the development of novel methods and algorithms for computational biology and bioinformatics, as well as large-scale simulations for protein folding, ligand–receptor binding, protein–protein interaction, and protein–nanoparticle interaction.

Contributors

Bruce J. Berne
Department of Chemistry
Columbia University
New York, New York

Deepak R. Canchi
Center for Biotechnology and
    Interdisciplinary Studies
and
Department of Physics, Applied
    Physics and Astronomy
Rensselaer Polytechnic Institute
Troy, New York

Qiang Cui
Department of Chemistry
Theoretical Chemistry Institute
University of Wisconsin-Madison
Madison, Wisconsin

Marcus Elstner
Institute of Physical Chemistry
Karlsruhe Institute of Technology
Karlsruhe, Germany

Charles English
Center for Biotechnology and
    Interdisciplinary Studies
and
Department of Physics, Applied
    Physics and Astronomy
Rensselaer Polytechnic Institute
Troy, New York

Angel E. Garcia
Center for Biotechnology and
    Interdisciplinary Studies
and
Department of Physics, Applied
    Physics and Astronomy
Rensselaer Polytechnic Institute
Troy, New York

Michael Gaus
Department of Chemistry
Theoretical Chemistry Institute
University of Wisconsin-Madison
Madison, Wisconsin

Puja Goyal
Department of Chemistry
Theoretical Chemistry Institute
University of Wisconsin-Madison
Madison, Wisconsin

Alan Grossfield
Department of Biochemistry &
    Biophysics
University of Rochester Medical Center
Rochester, New York

Guanhua Hou
Department of Chemistry
Theoretical Chemistry Institute
University of Wisconsin-Madison
Madison, Wisconsin

Xuhui Huang
Department of Chemistry
The Hong Kong University of Science
   and Technology
Kowloon, Hong Kong

Changbong Hyeon
School of Computational Sciences
Korea Institute for Advanced Study
Seoul, Republic of Korea

Camilo A. Jimenez-Cruz
Center for Biotechnology and
   Interdisciplinary Studies
and
Department of Physics, Applied
   Physics and Astronomy
Rensselaer Polytechnic Institute
Troy, New York

Seung-gu Kang
Computational Biology Center
IBM Thomas J. Watson Research
   Center
Yorktown Heights, New York

Luca Larini
Department of Chemistry and
   Biochemistry
and
Department Physics
University of California, Santa Barbara
Santa Barbara, California

Nicholas Leioatts
Department of Biochemistry &
   Biophysics
University of Rochester Medical Center
Rochester, New York

Xiya Lu
Department of Chemistry
Theoretical Chemistry Institute
University of Wisconsin-Madison
Madison, Wisconsin

Binquan Luan
Computational Biology Center
IBM Thomas J. Watson Research
   Center
Yorktown Heights, New York

J. Andrew McCammon
Center for Theoretical Biological
   Physics
and
Howard Hughes Medical Institute
and
Department of Chemistry and
   Biochemistry
and
Department of Pharmacology
University of California, San Diego
La Jolla, California

Xiaojia Mu
Department of Biomedical
   Engineering
The University of Texas at Austin
Austin, Texas

Xueqin Pang
Department of Chemistry
Theoretical Chemistry Institute
University of Wisconsin-Madison
Madison, Wisconsin

Pengyu Ren
Department of Biomedical
  Engineering
The University of Texas at Austin
Austin, Texas

Joan-Emma Shea
Department of Chemistry and
  Biochemistry
and
Department Physics
University of California, Santa Barbara
Santa Barbara, California

Devarajan (Dave) Thirumalai
Institute for Physical Science and
  Technology
University of Maryland
College Park, Maryland

Qiantao Wang
Department of Biomedical
  Engineering
The University of Texas at Austin
Austin, Texas

Yi Wang
Department of Physics
Chinese University of Hong Kong
Shatin, Hong Kong

Zhen Xia
Department of Biomedical
  Engineering
The University of Texas at Austin
Austin, Texas

and

Computational Biology Center
IBM Thomas J. Watson Research
  Center
Yorktown Heights, New York

Xin Xu
Department of Chemistry
Fudan University
Yangpu, Shanghai, People's Republic
  of China

Ruhong Zhou
Computational Biology Center
IBM Thomas J. Watson Research
  Center
Yorktown Heights, New York

and

Department of Chemistry
Columbia University
New York, New York

Jan Zienau
Department of Chemistry
Theoretical Chemistry Institute
University of Wisconsin-Madison
Madison, Wisconsin

Advanced Simulation Techniques

Chapter 1

Introduction to Molecular Dynamics and Enhanced Sampling Algorithms

Xuhui Huang, Ruhong Zhou, and Bruce J. Berne

CONTENTS

Molecular dynamics (MD) simulations have been widely applied in many biological applications in recent years due to the advancement of novel algorithms and high-performance computing. Both structural and functional roles of many biomolecules such as proteins have been revealed, thanks to the advances in molecular modeling techniques. In order to understand molecular functions, it is crucial to adequately sample their conformational space. MD simulation is a powerful tool to explore the conformational space, but when applied to complex systems such as proteins, they are often trapped in local minima of the rugged free energy landscape. Recently, extremely long MD simulations enabled by specific hardware such as Anton [1,2] and Blue Gene [3] or a large ensemble of individual simulations generated by the distributed computing platform [4–9] start to make it possible to study the folding of small proteins in the explicit solvent. However, it is still a difficult task for conventional MD simulations to satisfactorily sample the configuration space of proteins due to the trapping problem. Generalized ensemble (GE) sampling algorithms such as replica exchange method (REM) (or parallel tempering) [10–13] and simulated tempering (ST) [14,15] were developed to overcome the sampling problem by inducing a random walk in temperature space in the expanded ensemble. In these methods, conformations sampled at high temperatures may be exchanged to lower ones and thus help the system to escape the local energy minima, because energetic barriers can be easily crossed with larger thermal fluctuations at high temperatures. These temperature-based GE algorithms have been widely applied in simulating proteins and other biological macromolecules [16–23]. In this chapter, we first review the basic theory of MD simulations. Then we introduce various GE-enhanced sampling algorithms, followed by some examples. Next, we discuss some limitations of the GE algorithms and recent development to overcome them. We end with a review of a few Hamiltonian REMs. The development of GE-enhanced sampling algorithms is a fast-evolving field, and hence, it is not possible to exhaustively review all the variants of GE algorithms due to the limit of space in this chapter.

1.1 MD SIMULATIONS

In MD simulations, motions follow the Newtonian laws:

$$m_i \frac{\partial^2 r_i}{\partial t^2} = F_i \tag{1.1}$$

where m_i and r_i denote to the mass and position of the tth atom of the system, respectively. The force can then be computed as

$$F_i = \frac{\partial U\left(r_1,\ldots,r_n\right)}{\partial r_i} \tag{1.2}$$

where $U(r_1,\ldots,r_n)$ is the potential energy of the conformation $X(r_1,\ldots,r_n)$ and can be computed from a force field with predetermined parameters.

From the earlier description, it is clear to see that MD simulations are classical simulations that describe the motions of the atomic nuclei while neglecting electronic degrees of freedom. In MD simulations, one assumes that electrons instantaneously adjust their positions according to the change in nuclei positions (Born–Oppenheimer approximation) and always reside in the ground state. Therefore, MD simulations cannot be applied to investigate chemical bond breaking or formation, electronic state excitation, etc. On the other hand, MD simulation has its advantages over the first principle calculations. Strikingly, MD simulation can be applied to model very large systems (as large as 1 million atoms such as ribosome) compared to quantum mechanics (QM) calculations (mostly limited to around 100 atoms). In addition, modern MD simulations can model the dynamics of small proteins or peptides at microseconds or even milliseconds, while it is difficult to obtain time-dependent information using the first principle methods. In this section, we will introduce the basic theory of MD simulations.

1.1.1   Force Field

1.1.1.1   Energy Terms

In MD simulations, the force is derived from the potential energies (U) that are computed from force field or potential functions with predetermined parameters. A typical force field for MD simulations contains the following terms:

$$U = U_{stretch} + U_{bend} + U_{torsion} + U_{LJ} + U_{Coulomb} \tag{1.3}$$

where $U_{stretch}$, U_{bend}, and $U_{torsion}$ are used to describe the chemical bond stretching, angle bending, and torsion interactions, respectively. These three energy terms belong to the bonded interactions. The bond stretching and angel bending interactions are typically modeled as harmonic functions:

$$U_{stretch} = \sum_{bonds} K_r \left(r - r_{eq}\right)^2 \tag{1.4}$$

$$U_{bend} = \sum_{angles} K_\theta \left(\theta - \theta_{eq} \right)^2 \qquad (1.5)$$

where

K_r and K_θ are force constants for bond stretching and angle bending, respectively

r_{eq} and θ_{eq} denote to the equilibrium chemical bond length and angle between two connected chemical bonds, respectively

Sometimes the Morse potential is also used to model the bond stretching interactions.

Torsion interaction is applied on the dihedral angle between two planes determined by four atoms. These atoms $(i, j, k,$ and $l)$ are connected linearly through covalent chemical bonds. The dihedral angle ϕ is then defined as the angle between the ijk and jkl planes. A periodic potential based on cosine function is often used to model the torsion interactions:

$$U_{torsion} = \sum_{dihedrals} \sum_{i=1}^{3} V_i \left[1 + \cos \left(i\phi + f_i \right) \right] \qquad (1.6)$$

where V_i is the force constant. There also exists a special type of dihedral interaction, improper dihedral, to force certain atoms to remain in the same plane or to prevent transition to a mirror image conformation.

In the force field, two energy terms are applied to model the nonbonded interactions. In particular, the Lennard-Jones potential (U_{LJ}) is used to model the van der Waals (vdW) interaction:

$$U_{LJ} = \sum_{i<j} 4\varepsilon_{ij} \left[\left(\frac{\sigma_{ij}}{r_{ij}} \right)^{12} - \left(\frac{\sigma_{ij}}{r_{ij}} \right)^{6} \right] \qquad (1.7)$$

where

σ describes the size of the atoms

ε determines the magnitude of the attractive energy

With the 6th and 12th powers, the Lennard-Jones potential decays very fast with distance, and thus it is often treated by using the *cutoff* method, where the potential is truncated or smoothly switched to zero at a distance larger than a certain threshold (i.e., cutoff). Normally, σ_{ij} and ε_{ij} can be obtained from the combination rules

$$\sigma_{ij} = \frac{1}{2} \left(\sigma_{ii} + \sigma_{jj} \right); \quad \varepsilon_{ij} = \sqrt{\varepsilon_{ii} \varepsilon_{jj}} \qquad (1.8)$$

or

$$\sigma_{ij} = \sqrt{\sigma_{ii}\sigma_{jj}}; \quad \varepsilon_{ij} = \sqrt{\varepsilon_{ii}\varepsilon_{jj}} \tag{1.9}$$

where σ_{ii} and ε_{ii} denote the σ and ε of the ith atom, respectively. The earlier two combination rules are applied in different force field, for example, the latter one is used by the OPLS force field.

The Coulomb potential ($U_{Coulomb}$) is used to model the electrostatic interactions such as dipole–dipole, ion–dipole, and ion–ion interactions:

$$U_{Coulomb} = \sum_{i<j} \frac{q_i q_j}{r_{ij}} \tag{1.10}$$

where q_i denotes the fixed partial charge applied on the ith atom, and the total charge on individual atom should sum to the charge on the whole molecule. In the popular force fields with fixed atomic charges, both dipole-induced dipole and ion-induced dipole interactions are not taken into account. In comparison with the Lennard-Jones potential, the Coulomb potential is long range and decays very slow. Therefore, the *cutoff* method cannot be applied to compute the Coulomb interactions leaving the calculation of this potential rather expensive.

In the homogeneous systems, one can assume an environment with a uniform dielectric constant (ε_{rf}) beyond a certain distance cutoff (r) from the solute. Under this condition, the Coulomb interaction can be modified and only the pairwise interactions within r need to be explicitly calculated. This method is called the *reaction field* and it can greatly reduce the computational cost for the Coulomb interactions.

Another popular method to compute the long-range electrostatic interactions is the Ewald summation or particle-mesh Ewald (PME). In the Ewald summation, the slow-decaying Coulomb interaction is decomposed into a couple of quickly converging terms: one can be directly computed in the Cartesian space, while the other one is computed in the reciprocal space upon the Fourier transformation. The reciprocal sum is over an infinite number of periodic images, so that the Ewald summations were originally designed to compute the long-range interactions of crystals. Due to the periodic boundary conditions, the Ewald summations can now be widely applied in MD simulations. The Ewald summation is computationally expensive because the cost of the reciprocal sum increases with $O(N^2)$, where N is the number of particles in the system. Therefore, the application of the Ewald summation is limited to small systems. In order to improve the performance of the reciprocal sum, the PME method is developed where the charges are assigned to grids rather than performing a direct summation of wave vectors. The computational cost of the PME method only scales with $N \log (N)$, and thus it has been widely used in simulating complex systems such as proteins.

The energy terms introduced earlier ($U_{stretch}$, U_{bend}, $U_{torsion}$, U_{LJ}, and $U_{Coulomb}$) describe fundamental intra- and intermolecular interactions in chemical and biological systems. The combination of these terms can sufficiently model more complicated interactions such as hydrogen bonding. Hydrogen bonding can be modeled by the combination of vdW and electrostatic interactions in a force field. For example, let us consider a hydrogen bond formed by a carbonyl oxygen approaching the hydrogen atom attached to another nitrogen atom: N–H...O=C. Since both O and N are highly electronegative, they will carry large negative partial charges, while H and C will then be assigned significant positive charges in the force field. Therefore, the accumulated Coulomb attractions (e.g., N...C and H...O) will produce strong attraction with a minimum of around a few kcal/mol at a N–O distance of about 2.8 Å, which is an order of magnitude higher than the vdW interactions alone. On the other hand, the vdW interactions will prevent the O to approach too close to H. Furthermore, vdW interactions also make the hydrogen bond directional, that is, the N, H, and O atoms in the hydrogen bond are almost linear. Otherwise, there will be strong steric repulsions from the vdW potential.

1.1.1.2   Derivation of Force Field Parameters

Since there exist millions of chemical compounds and a large variety of biological macromolecules, it is impossible to develop force field parameters for every single atom in these systems. On the other hand, one cannot simply assign identical parameters to all the atoms belonging to the same element, since the chemical surroundings can greatly influence the force field parameters. Therefore, the idea of the force field is to create a set of atom types that are sufficient to describe the majority of the molecules, and each atom type may represent one type of atoms in a particular chemical environment, for example, O atom in the C=O group. The number of atom types is thus not huge in typical force fields. For example, the OPLS/AA force field contains around 170 atom types. Next, we will discuss how the force field parameters of these atom types are derived.

For stretching and bending potentials, the equilibrium bond length (r_{eq}) and angle (θ_{eq}) are usually taken from experimental data such as x-ray structures. The force constants K_r and K_θ are in general obtained by fitting to the vibrational frequency data from spectroscopy experiments. For the torsion potential, the dihedral parameters are normally obtained by fitting to QM calculations by minimizing the difference of the energy profile computed from QM and the force field.

The partial charges (q_i) are also obtained by fitting to QM calculations. In particular, a set of atomic charges is found by reproducing the electrostatic potential (ESP) generated by the nuclei and electron wave functions. For example, in the AMBER charge fitting programs, the ESP energies are evaluated at a large number of points around the molecule of interest, and charges are then derived to reproduce the QM energies at these points. The ESP fitting method is the most popular one to determine atomic charges for the force field. However, the derived charges using this method may depend on molecular conformations, which is particularly problematic for molecules with multiple stable conformations. In addition, charges for buried atoms may be poorly determined, since the ESP is mainly determined by atoms close to the surface. The Lennard-Jones potential parameters (σ_{ii} and ε_{ii}) are adjusted to match certain experimental thermodynamic properties such as densities and heat of vaporization.

1.1.1.3 Limitation of the Force Field

As we discussed before, energies obtained from force field parameters only consider the ground electronic state; therefore, the force field could not be used to study chemical bond breaking or formation, electronic state excitation, etc. Furthermore, all the potential energy terms are pairwise additive, and thus many-body effects are ignored in the force field calculations. In general, force field parameters perform the best under the same conditions they were parameterized, and thus they are recommended to be used at this condition. Finally, the temperature dependence is currently not taken into account when deriving the force field parameters, and therefore, they should not predict the temperature-dependent properties accurately.

1.1.2 MD Simulation Theory

In MD simulations, one solves Newton's equations of motion (see Equations 1.1 and 1.2). All the atoms move together and forces between atoms change with time. For complex systems containing a large number of atoms, analytical solutions of Equations 1.1 and 1.2 for the positions and velocities are impossible, while the numerical solution is trivial. Therefore, one solves Newton's equations of motion approximately using numerical solutions when performing MD simulations. In this section, we will introduce the basic theory of MD simulations.

1.1.2.1 MD Simulation Procedure

The procedure of running MD simulations is as follows:

- *Initial values for the atomic positions.* One can start from experimental structures such as x-ray crystal structure and NMR structures; one can also start from modeled structure or even some random positions.

- *Initial values for the atomic velocities.* The initial velocities (v_i) are often selected from a Maxwell–Boltzmann distribution at a given temperature T.
- *Iterate: position $\Rightarrow$ forces $\Rightarrow$ new position.* The forces are computed from the potential energies. The configuration (positions and velocities) is updated according to Newton's equations of motion.

1.1.2.2 Integrators
As we discussed before, solving Newton's equations of motion analytically is impossible due to the complex form of the potential energy function $U(r_1,...,r_n)$ that depends on the positions of all the particles $(r_1,r_2,...,r_n)$ in the system. Therefore, a number of numerical integration algorithms have been developed to solve Newton's equations of motion. All these algorithms are based on the Taylor expansions of positions.

1.1.2.2.1 Verlet Integrator
In this integrator developed by Verlet in 1967 [24], the Taylor expansion of $r(t + \Delta t)$ at $\Delta t = 0$ is first obtained:

$$r(t+\Delta t)=r(t)+v(t)\Delta t+\frac{1}{2}a(t)\Delta t^2 +\frac{1}{6}b(t)\Delta t^3 +O\left(\Delta t^4\right) \tag{1.11}$$

where
$v(t)$ is the velocity
$a(t)$ is the acceleration
$b(t) = r^{(3)}(t)$

The Taylor expansion of $r(t - \Delta t)$ at $\Delta t = 0$ is

$$r(t-\Delta t)=r(t)-v(t)\Delta t+\frac{1}{2}a(t)\Delta t^2 -\frac{1}{6}b(t)\Delta t^3 +O\left(\Delta t^4\right). \tag{1.12}$$

By adding Equations 1.11 and 1.12 together, one can obtain the relationship of positions at $t + \Delta t$, t, and $t - \Delta t$:

$$r(t+\Delta t)=2r(t)+a(t)\Delta t^2 +O\left(\Delta t^4\right). \tag{1.13}$$

The velocities do not appear in Equation 1.13, but can be obtained by simply subtracting Equation 1.12 from Equation 1.11:

$$v(t)=\frac{r(t+\Delta t)-r(t-\Delta t)}{2\Delta t}. \tag{1.14}$$

1.1.2.2.2   Velocity Verlet Integrator

An improvement of the Verlet algorithm was later proposed by Swope et al. [25] to handle the velocities better:

$$r(t+\Delta t)=r(t)+v(t)\Delta t+\frac{1}{2}a(t)\Delta t^2 \tag{1.15}$$

$$v\left(t+\frac{\Delta t}{2}\right)=v(t)+\frac{1}{2}a(t)\Delta t \tag{1.16}$$

$$a(t+\Delta t)=-\left(\frac{1}{m}\right)\nabla U\big(r(t+\Delta t)\big) \tag{1.17}$$

$$v(t+\Delta t)=v\left(t+\frac{\Delta t}{2}\right)+\frac{1}{2}a(t+\Delta t)\Delta t. \tag{1.18}$$

1.1.2.2.3   Leapfrog Integrator

Leapfrog is another popular integrator [26] where the position and velocity are updated in a way like frogs leaping over each other's back:

$$v\left(t+\frac{\Delta t}{2}\right)=v\left(t-\frac{\Delta t}{2}\right)-\left(\frac{1}{m}\right)\nabla U\big(r(t)\big)\Delta t \tag{1.19}$$

$$r(t+\Delta t)=r(t)+v\left(t+\frac{\Delta t}{2}\right)\Delta t. \tag{1.20}$$

1.1.2.3   Periodic Boundary Conditions

In most situations, we need to simulate bulk systems such as a solid crystal or protein solution. Periodic boundary conditions are then used to minimize edge effects in a finite simulation box and allow us to simulate bulk systems. In the periodic boundary conditions, particles are placed in a simulation box, which is surrounded by translated copies of itself. The simulation box and its replicas or images will occupy the whole space and thus mimic the bulk systems. In this way, there are no boundaries of the system, and the boundary of the isolated simulation box is replaced by the periodic conditions. For example, under the periodic boundary conditions with 2D rectangular boxes, an atom leaving the box on the left will reenter from the right.

The periodic boundary conditions are desired for simulating a crystal, where the simulation box may precisely represent each unit cell. However, for liquids or solutions, the periodic boundary conditions will introduce errors. For example, in protein simulations, errors may arise due to interactions between the protein and its images; thus, the box size has to be sufficiently large to avoid significant image interactions. As we discussed before, the Ewald summation or PME methods can be used in conjunction with the periodic boundary conditions to compute the long-range electrostatic interactions in the system.

The simulation box can be designed in various different shapes. The cubic box is mostly widely used; however, it may not be ideal for simulating globular molecules such as proteins. Other box shapes such as the rhombic dodecahedron and truncated octahedron may be more suitable for the simulations of the spherical macromolecules, since these two shapes are closer to a sphere and thus require a smaller number of solvent molecules.

1.1.2.4 Neighbor List

With periodic boundary conditions, one can mimic the infinite system using a finite simulation box. For certain configurations, one would need to calculate the vdW and electrostatic interactions between infinite numbers of pairs of atoms, which is impossible. Since the nonbonded interactions (U_{LJ} and $U_{Coulomb}$) decay fast with the distance, a cutoff (r_c, so-called neighbor list) can be defined and only interactions between pairs of atoms within the cutoff are computed.

1.1.2.5 Temperature and Pressure Coupling

Direct application of the integrators introduced in Section 1.1.2.2 will produce simulations in the microcanonical ensemble (constants N, V, and E). However, in many cases, we wish to simulate the system at constant temperature condition or canonical ensemble (constants N, V, and T). Various temperature coupling algorithms have been developed to simulate the constant temperature such as the Andersen thermostat [27], the Berendsen thermostat [28], and the Nosé–Hoover algorithm [29,30].

1.1.2.5.1 Berendsen's Thermostat

In this algorithm, the deviation from the targeted temperature T_t is corrected through first-order kinetics:

$$\frac{dT}{dt} = \frac{T_t - T}{\phi} \tag{1.21}$$

where ϕ is a time constant. The Berendsen thermostat cannot produce a proper canonical ensemble since it suppresses the fluctuations of the kinetic energy. However, it has the advantage of easy tuning of the coupling strength. The Berendsen thermostat is thus recommended to be used in the equilibration phase of the simulations. A later study fixes the earlier issue of the Berendsen thermostat by introducing an additional stochastic term that ensures correct fluctuations of the kinetic energy. This improved Berendsen scheme is also called the velocity-rescaling thermostat [31].

1.1.2.5.2 Nosé–Hoover's Thermostat

This algorithm can rigorously produce the canonical ensemble distributions by introducing an additional degree of freedom (s) for the heat bath. The Hamiltonian can then be extended as

$$H = U\left(r_1, r_2, \ldots, r_N\right) + \sum_i \frac{p_i^2}{2ms^2} + \left(3N+1\right)kT\ln s + \frac{p_s^2}{2Q} \tag{1.22}$$

where

p_s is the momentum for the variable s
Q is the mass-like parameter for s

In this way, the system (with N particles) is coupled to a heat bath that is described by the variable s and its momentum p_s. Nosé [29] proved that the extended Hamiltonian in Equation 1.22 can generate canonical ensemble distributions independent of the values selected for s and Q. It is worth noting that it normally takes a bit longer for the system to relax to the targeted temperature using the Nosé–Hoover thermostat compared to the Berendsen thermostat. Therefore, the Nosé–Hoover thermostat is normally used for the production phase of the MD simulations.

MD simulations may also need to be performed under a constant pressure (constants N, P, and T) in some situations. Various algorithms have also been developed to couple the system with a pressure bath such as the Berendsen algorithm [28] and the Parrinello–Rahman approach [32]. In this chapter, we will not discuss these pressure coupling algorithms in detail and would like to refer the readers to the original references [28,32,33].

1.1.2.6 Constraints

In MD simulations, the time step has to be extremely small, 1 fs (or 10^{-15} s), which makes the performing of the long timescale simulations difficult. Chemical bond stretching ($U_{stretch}$) is the fastest motion among all potential energy terms and thus determines the time step of the MD simulations. In order to accelerate the MD simulations, one often fixes the length of the covalent chemical bonds, and the time step can then be increased to 2 fs, 4 fs, and even 6 fs in some force fields. Several algorithms such as SHAKE [34], RATTLE [35], and LINCS [36] have been developed to satisfy the constraints of the bond length in MD simulations. In this chapter, we will focus on the SHAKE algorithm.

SHAKE [34] is one of the most popular constraint algorithms, which can convert unconstrained atomic positions to the ones that satisfy a list of distance constraints. Lagrange multipliers are applied in SHAKE to solve the equations of motion subject to a set of constraints. Let us consider a system containing a list of K constraints: $\{d_1, d_2, \ldots, d_K\}$. For each constraint of the bond length, d_i,

$$\vec{r}_{ij} \cdot \vec{r}_{ij} - d_i^2 = 0. \tag{1.23}$$

The force can then be rewritten as

$$F_i = -\frac{\partial}{\partial r_i} \left(U(r_1, r_2, \ldots, r_N) + \sum_{k=1}^{k=K} \lambda_k \sigma_k \right) \tag{1.24}$$

where the second part of the earlier equation is sometimes called the constraint force. λ_k are the Lagrange multipliers that satisfy the constraint equations, and they are solved iteratively in SHAKE.

1.2   THEORY OF GE-ENHANCED SAMPLING ALGORITHMS

1.2.1   Replica Exchange MD

In replica exchange molecular dynamics (REMD), independent replicas of MD simulations are performed at different temperatures in parallel. At a certain interval, an exchange of conformations at two replicas is proposed:

$$(X_i, T_i) \rightarrow (X_j, T_i); \quad (X_j, T_j) \rightarrow (X_i, T_j) \tag{1.25}$$

where X_i and T_i are the conformation and the temperature of the ith replica, respectively. The exchanges are normally attempted at two closest neighbors in the temperature space, and the transition probabilities are defined as follows to satisfy the detailed balance:

$$P_i(X_i, p_i) P_j(X_j, p_j) P(i \rightarrow j) = P_j(X_i, p_i') \, P_i(X_j, p_j') \, P(j \rightarrow i) \tag{1.26}$$

where
  $P(i \rightarrow j)$ is the transition probability for the exchange $i \rightarrow j$
  $P_i(X_j, p_j)$ is the probability of the conformation X_j with momenta p_j in the canonical ensemble at temperature T_i

$$P_i(X_i, p_i) = \frac{1}{Q_i} \exp(-\beta_i E(X_i, p_i)) \tag{1.27}$$

where
  $\beta_i = 1/(k_B T_i)$
  p_i is the momenta
  Q_i is the partition function

The Hamiltonian $E_i(X_i, p_i)$ contains the contributions from both kinetic energy (K) and potential energy (U):

$$E_i(X_i, p_i) = K(p_i) + U(X_i). \qquad (1.28)$$

The momenta (p_i) are normally rescaled immediately after the exchange [12] so that the kinetic energy term can be cancelled out in Equation 1.27. This sometimes is also achieved by redrawing velocities according to the distributions at the new temperature if the Andersen thermostat is applied. The transition probabilities can then be rewritten as

$$\frac{P(i \rightarrow j)}{P(j \rightarrow i)} = \exp(-\Delta_{ij}) \qquad (1.29)$$

where

$$\Delta_{ij} = (\beta_j - \beta_i)(U(X_j) - U(X_i)). \qquad (1.30)$$

Applying the Metropolis criteria, the transition probabilities can be obtained as

$$P(i \rightarrow j) = min\{1, e_{ij}^{-\Delta}\}; \quad P(j \rightarrow i) = min\{1, e_{ij}^{\Delta}\}. \qquad (1.31)$$

1.2.2   Simulated Tempering

In ST, individual simulations are performed in an expanded canonical ensemble, where the canonical ensemble is attached with different weights at different temperatures [14,15]. In ST, we define the Hamiltonian $\Xi_i(X, p)$:

$$\Xi_i(X, p) = \beta_i E(X, p) - g_i \qquad (1.32)$$

where
   $\Xi_i(X, p)$ denotes the Hamiltonian of the system
   $\beta_i = 1/(k_B T_n)$
   g_i is the a priori determined weights assigned to the canonical ensemble at temperature T_i

The procedure of ST is as follows: a single MD simulation is started at a particular temperature T_i, and an attempt is made after a certain interval to exchange the conformation to another temperature T_j:

$$(X_i, T_i) \rightarrow (X_i, T_j) \qquad (1.33)$$

where X_i is the conformation of the system.

The probabilities for the exchange have to satisfy the detailed balance:

$$P_i(X_i, p_i)P(i \rightarrow j) = P_j(X_i, p_i')P(j \rightarrow i) \tag{1.34}$$

where the probability of conformation X_i in the expanded ensemble is

$$P_i(X_i, p_i) = \frac{1}{Q} \exp(-\Xi_i(X_i, p_i)) \tag{1.35}$$

where

Q is the partition function for the expanded ensemble
p_i is the momentum

As in REM, p_i is typically rescaled after the exchange, so that it will be canceled in Equation 1.34. After applying the Metropolis criteria, the probabilities for the exchange can then be written as

$$P(i \rightarrow j) = min\{1, \exp(-(\beta_j - \beta_i)U(X_i) + (g_j - g_i))\}$$
$$\tag{1.36}$$
$$P(j \rightarrow i) = min\{1, \exp(-(\beta_i - \beta_j)U(X_i) + (g_i - g_j))\}$$

where $U(X_i)$ is the potential energy of the conformation X_i and remains the same before and immediately after the exchange.

Different from REM, ST is a serial algorithm that does not require multiple replicas running in parallel. The efficiency of ST largely depends on the selection of the weights g_i, which are a set of constants determined a priori. Without proper reweighing, ST simulations may only be able to explore a subset of the temperature space. It is generally believed that the optimal weights will induce the ST simulations to perform a random walk in the temperature space and thus result in a uniform sampling across the temperature space. We will show in the next paragraph that this random walk can be achieved if g_i equals the Helmholtz free energy of the canonical ensemble at T_i [14,15,37–40].

In the expanded ensemble in ST, the probability for the temperature T_i to be visited is

$$P_i = \frac{Q_i e^{g_i}}{Q} \tag{1.37}$$

where Q_i and Q are partition functions of the canonical ensemble (T_i) and the expanded ensemble, respectively. The partition function Q_i and the Helmholtz free energy have the following relationship:

$$F_i = -\frac{1}{\beta_i} \ln Q_i. \tag{1.38}$$

When $g_i = \beta_i F_i$, all the temperatures have the identical probabilities to be visited in the expanded ensemble:

$$P_1 = P_2 = \cdots = P_n = \frac{1}{Q}. \qquad (1.39)$$

Since $\beta_i F_i$ corresponds to the unitless free energies, the earlier set of optimal weights is also referred to as the *free energy* weights [14,15].

It is a difficult task to determine these *free energy* weights a priori. In practice, these free energies can be obtained from the weighted histogram analysis method (WHAM) from pilot simulations [40–42]. To obtain a set of converged weights, an interactive procedure and relatively expensive pilot simulations are often required [43].

It is noted that the *free energy* weights given earlier must yield the same overall acceptance ratios for both forward and backward transitions between two temperatures [44]:

$$\langle P(i \to j) \rangle = \langle P(j \to i) \rangle. \qquad (1.40)$$

If the potential energy distribution functions $P(U_i)$ at T_i are known, one can also derive the weights g_i by imposing Equation 1.40, where

$$\langle P(i \to j) \rangle = e^{g_j - g_i} \int_{-\infty}^{(g_i - g_j)/(\beta_i - \beta_j)} e^{-(\beta_j - \beta_i)U_i} P(U_i) dU_i + \int_{(g_i - g_j)/(\beta_i - \beta_j)}^{\infty} P(U_i) dU_i$$

$$\langle P(j \to i) \rangle = \int_{-\infty}^{(g_i - g_j)/(\beta_i - \beta_j)} P(U_j) dU_j + e^{g_i - g_j} \int_{(g_i - g_j)/(\beta_i - \beta_j)}^{\infty} e^{-(\beta_i - \beta_j)U_j} P(U_j) dU_j$$

$$(1.41)$$

where $P(U_i)$ and $P(U_j)$ can be estimated from the pilot MD simulations.

A set of *free energy* weights will theoretically lead to the uniform sampling in equilibrium. However, the time it takes to reach the uniform sampling also depends on the acceptance ratios of the attempts to exchange. The acceptance ratios diminish quickly with the increase in the temperature gap ΔT_{ij}. Therefore, the attempted exchanges are often made between neighboring temperatures and all the temperature gaps have to be sufficiently small to yield significant acceptance ratios.

1.3   CONVERGENCE OF THE GE ALGORITHMS

In order to make use of the data produced by the GE algorithms, one has to make sure that certain convergence has been reached. Rao and Caflisch [45] have proposed that observing the reversible folding (multiple folding and unfolding events in the same replica) is sufficient for determining a dataset to have reached equilibrium. In GE algorithms such as REMD, the same temperature is constantly visited by different replicas due to the exchanges; therefore, trajectories collected at a single temperature may display artificially rapid transitions between the folded and unfolded states. If one follows a single replica that performs a random walk in the temperature space, it is rather difficult to observe the reversible folding because the protein is easy to unfold when visiting high temperatures but difficult to refold (see Section 1.5 for more discussions). Indeed, reversible folding is shown to be readily achieved for helical proteins with fast relaxation rates even in explicit solvent [44]. However, it is extremely difficult to attain reversible folding in explicit solvent for more complicated proteins [45].

An alternative to reversible folding is to run multiple independent sets of simulations started from very different initial configurations [44,46]. If each set of simulations converges to the same result, then they may be considered to have reached the true equilibrium distribution. Unfortunately, it is extremely difficult to compare two datasets directly because of the high dimensionality of the underlying free energy landscapes. Thus, researchers often test convergence by comparing projections onto one or two order parameters [17,18,44,47]. Common order parameters for protein folding include the number of native contacts, RMSD, radius of gyration (Rg), fraction folded, and secondary structure component. For example, Huang et al. [44] have performed two independent sets of ST simulations starting from the helical and random coil states of a 21-residue Fs-peptide on a distributed computing environment [48]. They show that these two sets of ST simulations have reached the convergence over a wide range of temperatures by comparing the helical content of the peptide.

One alternative to these projections is to group conformations into clusters (or states) and use these to judge the convergence. For example, Daura et al. [49] have plotted the number of clusters for two independent datasets as a function of time and proposed that convergence of this value is an adequate metric for converged sampling. However, this method fails to take into account cluster populations, which may differ drastically between datasets. More recently, Lyman and Zuckerman [50] proposed that two datasets should have the same number of clusters and that these clusters must have the same populations in order to be considered converged.

1.4   APPLICATIONS OF GE ALGORITHMS IN PROTEIN FOLDING

Since the introduction of the GE algorithms in simulating protein folding in 1990s, they have been widely adopted to study the thermodynamics of protein folding and other conformational changes in biological macromolecules

[5,11,12,16–18,43,44,47,51–53]. In 1993, Hansmann and Okamoto introduced the multicanonical method, a GE algorithm, in simulating a simple pentapeptide: Met-enkephalin [51]. In this study, they have successfully predicted the lowest-energy conformation and calculated various thermodynamic properties. A few years later, Hansmann [11] has applied the REMD or parallel tempering, a GE algorithm that originated from the spin glass field [54,55], to model the same pentapeptide. In this work, REMD was shown to greatly enhance the conformational sampling compared to conventional MD simulations. Sugita and Okamoto [12] have simulated the same pentapeptide using the REMD algorithm. In this study, he has calculated thermodynamic properties such as distributions of dihedral angles over a wide range of temperatures. In a later REMD work, Garcia and Sanbonmatsu [20] have identified four metastable states for Met-enkephalin separated by low free energy barriers and with comparable populations. They further suggested that this structural flexibility allows Met-enkephalin to bind to several different receptors.

In the early 2000s, Zhou, Berne, and coworkers have published a series of seminal work [17,18,53] on applying REMD simulations to simulate the folding of various peptides, which have provided great insight into the mechanisms of protein folding. In particular, Zhou et al. [17] have investigated the folding free energy landscape of the C-terminal beta hairpin of protein G using REMD simulations in the explicit solvent. Projections of the free energy landscape at low temperatures on various order parameters such as the number of hydrogen bonds, RMSD, and Rg all display a rugged landscape. Furthermore, their results indicated that the hydrophobic core and the beta strand hydrogen bond form simultaneously. These REMD simulations predicted the beta hairpin component that is in reasonable agreement with the experiments at physiological temperature, though the predicted melting temperature is substantially higher than the experimental value. In a follow-up study, Zhou et al. [17] have simulated the same peptide in the surface-generalized Born implicit solvent. To their surprise, the implicit solvent failed to generate a reasonable free energy landscape even with exhausted sampling. Specifically, the lowest free energy state in the implicit solvent model is not a β-hairpin conformation, and some non-native states with significant populations have also been observed. Zhou [18] has also performed REMD simulations to investigate the folding of a 20-residue miniprotein Trp-cage. A metastable intermediate state was observed with the structural features of two partially formed hydrophobic cores separated by a salt bridge between Asp8 and Arg16. Zhou proposed that this intermediate state may be crucial for the superfast folding of this miniprotein. Again, the REMD simulations predicted a higher melting temperature (>400 K) than the experimental value (315 K), indicating that the force field parameterized at room temperature may not be sufficient to predict temperature-dependent thermodynamic properties. The same conclusion has also been made in later

REMD studies on other protein systems including Trpzip2 [56] and 1BBL [57]. In the study of 1BBL by Pitera et al. [57], they have observed two distinct structural transitions in the folding of this protein: breaking of the protein's tertiary structure at low temperatures and complete loss of secondary structure and hydrophobic core at high temperatures. These results suggested a more complex folding process than a simple two-state folding for 1BBL.

REMD simulations have also been applied to predict the folded structure of the proteins. Mohanty and Hansmann [58] have compared the folding thermodynamics of three small proteins with distinct folded structures: helical protein (1RIJ), all sheet (beta3s), and mixed helix sheet (BBA5). They have successfully identified the native structures of these proteins as the global minima of the free energy landscapes computed from REMD simulations. However, the folding process strongly differs. Im and Brooks III [59] have performed de novo folding simulations using REMD for the major pVIII coat protein from filamentous fd bacteriophage with a membrane generalized Born model. Their predicted native structure is in reasonable agreement with the solid-state NMR in lipid bilayers and solution-phase NMR on the protein in micelles. They further suggested that REMD simulations may serve as an emerging technique for structural characterizations of the membrane proteins as a complementary approach to the experimental techniques.

1.5 LIMITATION OF TEMPERATURE-BASED GE ALGORITHMS

GE algorithms show great improvement compared to conventional MD for small rapidly equilibrating systems in implicit solvent, but less improvement for more complexed peptides or proteins in explicit solvent [20,44,45,60,61]. For example, Zhang et al. [60] studied the Fs-peptide (a 21-residue helical peptide that folds in a few hundred ns) in implicit solvent. They conclude from their study that REMD simulations are 14–72 times more efficient than conventional MD simulations depending on the temperatures used. However, a number of other studies have demonstrated that implicit solvent may not be sufficient for generating accurate folding free energy landscapes [47,62], so despite the huge efficiency increase, the degree to which one can draw physical insight from such results is unclear. On the other hand, the more accurate free energy landscapes gained with explicit solvent do come with a cost: much less efficient sampling. This loss in efficiency results from the fact that the energy is no longer directly related to the solute configuration as solvent–solvent interactions dominate the potential energy. In their study of a pentapeptide in explicit solvent, Garcia and Sanbonmatsu [20] found that REMD is only about five times more efficient than MD. In another study of a β-heptapeptide in explicit solvent, Periole and Mark [61] found that REMD simulations are more than four times efficient than multiple independent MD simulations at 300 K. Huang et al. [44] recently investigated the folding of the same Fs-peptide as Zhang et al. [60], but in explicit solvent using a serial version of REMD (SREMD) [63]. It was found that REMD is around three times more efficient than conventional MD for this system.

Efficiency of the temperature-based GE algorithms is limited by the non-Arrhenius kinetics in protein folding, that is, while high temperatures may facilitate the crossing of energetic barriers, entropic barriers will be more difficult to cross. Zuckerman and Lyman [64,65] investigated the maximum gain in efficiency of REMD relative to conventional MD by assuming that barrier crossing is the rate limiting factor and that the barrier crossing rates follow the Arrhenius equation. They suggested that REMD gives approximately an order of magnitude greater efficiency for sampling at physiological temperatures when only moderate energetic barriers are presented (a few kcal/mol). Nymeyer [66] later reached similar conclusions using an analytical approach assuming a two-state folding and the Arrhenius behavior. However, Rhee and Pande [46] have suggested that the relative efficiency of REMD is reduced when the dominant barriers are entropic because this will result in non-Arrhenius kinetics, and in many protein systems, the entropic barriers may be dominant at high temperatures. This idea has recently been explored in more detail by Zheng et al. [23,67] using a kinetic network model to simulate REMD simulations. They show that there is an optimal temperature for the fastest refolding kinetics, and visiting temperatures higher than this optimum reduces the efficiency of REMD. In fact, this kinetic network model by Zheng et al. [23,67] may be oversimplified since their model assumes a two-state folding. If this condition does not apply, the efficiency advantage of REMD will be reduced further.

Some effort has been made to improve the effectiveness of REMD by optimizing either the temperature spacing [68,69] or the exchange frequency [70,71]. However, these studies have not explored the effect of the underlying folding kinetics and its temperature dependence on the efficiency of REMD. For example, Hansmann and coworkers [68,69] have developed an iterative algorithm to optimize the temperature spacing by maximizing the diffusion rate of individual replica in the temperature space. For a 36-residue protein HP-36, their algorithm has shown that the bottleneck of the replica diffusion locates at the transition temperatures, where smaller temperature spacing is required. This algorithm is particularly useful for the implicit solvent REMD simulations, where the effect energy (i.e., potential energy $U(X_i)$ in the temperature-based REMD) is highly correlated with the protein conformations. However, in the explicit solvent, since water–water interactions dominate the effect energy, the replica diffusion rate in the temperature space may barely correlate with the protein conformational changes and thus provide little help on crossing the entropic barriers. On the other hand, Roitberg and coworkers [71] suggested to exchange more often. Using 5 peptides with length ranging from 1 to 21 residues, they show that the efficiency of sampling increases with increasing exchange-attempt frequency, which is contradictory to the general view that the system needs sufficient time (e.g., 1 or a few ps) to relax between two exchanges.

As we discussed earlier, energetic barriers may be crossed more easily at high temperatures in GE simulations; however, entropic barriers will become more significant. This poses a problem because the dominant barriers to conformational change are entropic for many biological systems, such as hairpins [72]. In order to overcome the entropic barriers for protein folding, Huang et al. [72] developed an efficient algorithm called the adaptive seeding method (ASM) that uses nonequilibrium GE simulations to identify the meta-stable states and seeds short simulations at constant temperature from each of them to determine their equilibrium populations. In ASM, the equilibrium properties are predicted by the Markov state model (MSM) [73–77]. MSMs are a powerful algorithm that can be used to extract equilibrium thermodynamic properties from a dataset that only reaches the local equilibration. In MSM, the conformational space is divided into a number of metastable states and the fast protein motions can be integrated out by coarse graining in time (Δt). If Δt is longer than the intrastate relaxation time, we can obtain a Markovian model, where the probability of a given state at time $t + \Delta t$ depends only on the state at current time t. The system's dynamics can then be modeled by a first-order master equation:

$$P(n\Delta t) = \left[T(\Delta t) \right]^{n} P(0) \tag{1.42}$$

where

$P(n\Delta t)$ denotes the populations of metastable states at time $n\Delta t$

$T(\Delta t)$ is the transition probability matrix

MSMs are usually used to study dynamics, but it is also easy to obtain the thermodynamic properties from MSMs. In particular, the equilibrium-state populations can be obtained as the normalized first eigenvector components of $T(\Delta t)$. The ASM [72] takes advantage of the broad sampling potential with GE algorithms but may cross the entropic barriers more efficiently during the seeding simulations at low temperature. Using an RNA hairpin system as the example, Huang et al. [72] show that only local equilibrium is neces-sary for ASM, so very short seeding simulations may be used. They further suggested that the ASM may be used to recover equilibrium properties from existing datasets that failed to converge and is well suited to run on modern computer clusters.

1.6 HAMILTONIAN REPLICA EXCHANGE ALGORITHMS

The applications of the temperature-based REMD are restricted to relatively small peptides and proteins since the required number of replicas increases dramatically with the number of degrees of freedom (f) of the system, roughly as $O(f^{1/2})$ [78]. To alleviate the problem of the poor scaling of REMD with the system size, Hamiltonian replica exchange algorithms have been introduced. In these algorithms, all the replicas are typically run at the same temperature, but one or multiple parts of Hamiltonian are modified at each replica. The effective energies that determine the acceptance ratio of the exchange will thus only depend on this subset of the Hamiltonian. Therefore, the number of replicas required in Hamiltonian replica exchange algorithms is scaled as $O(f_p^{1/2})$, where f_p is the degrees of freedom corresponding to this subset of the Hamiltonian and $f_p \ll f$. Therefore, Hamiltonian replica exchange algorithms hold great potential to gain speedup against REMD in terms of obtaining the converged thermodynamic properties. However, it is not an easy task to determine which parts of Hamiltonian to modify and the selections are often system dependent. In this section, we will review a few Hamiltonian replica exchange algorithms.

1.6.1 Hydrophobic-Aided Replica Exchange

Hydrophobic in aided replica exchange (HAREM) [79] is a variant of Hamiltonian REM, which is inspired by the observation that the increase in the protein hydrophobicity can greatly accelerate the protein hydrophobic collapse [80–82]. Previous studies show that protein hydrophobicity can be tuned by scaling the protein–water attractive interactions [80–82]. For example, Zhou et al. [81] observed that the collapse of the two domains of the BphC enzyme can be accelerated by around 10 times when the electrostatic protein–water interactions (E_{elec}) are turned off.

In HAREM, different replicas contain different Hamiltonian with scaled protein–water interactions:

$$E_i = E_i^{other} + \lambda_i \left(U_{elec}^{pw} + E_{LJ6}^{pw} \right) \qquad (1.43)$$

where

λ_i is the scaling factor for protein–water interactions (pw stands for protein–water) including the van der Waals attractive potential and ESP

E_i^{other} contains all other energy terms in the Hamiltonian for the ith replica

$$U_{elec} = \sum_{i<j} \frac{q_i q_j}{r_{ij}} \qquad (1.44)$$

$$U_{LJ6} = \sum_{i<j} 4\varepsilon_{ij} \left(\frac{\sigma_{ij}}{r_{ij}} \right)^6 \tag{1.45}$$

where

q_i is the partial change of atom i

r_{ij} is the distance between a pair of atoms i and j

Now, we consider the exchange of configurations between the replica i and j:

$$\left(X_i, E_i(X_i) \right) \rightarrow \left(X_j, E_i(X_j) \right)$$
$$\left(X_j, E_j(X_j) \right) \rightarrow \left(X_i, E_j(X_i) \right) \tag{1.46}$$

where X_i, $E_i(X_i)$ corresponds to the configuration and the potential energy of the replica i, respectively. The equilibrium probability for this state is then

$$P_i = \frac{1}{Z_i} \exp\left(-\beta E_i(X_i) \right) \tag{1.47}$$

where

Z_i is the partition function at replica i

$\beta = 1/(k_B T)$ is a constant

We can then obtain the following equation by applying the detailed balance condition on the exchange:

$$P_i(X_i) P_j(X_j) P(i \rightarrow j) = P_j(X_i) P_i(X_j) P(j \rightarrow i) \tag{1.48}$$

and the ratio of the transition probabilities is

$$\frac{P(i \rightarrow j)}{P(j \rightarrow i)} - \exp\left(-\Delta_{ij} \right) \tag{1.49}$$

where

$$\Delta_{ij} = \beta \left[E_i(X_i) + E_j(X_j) - E_i(X_j) - E_j(X_i) \right]. \tag{1.50}$$

If we further apply the Metropolis criteria, the acceptance probability can be obtained by

$$P(i \rightarrow j) = min\left\{ 1, e^{-\Delta_{ij}} \right\}; \quad P(j \rightarrow i) = min\left\{ 1, e^{\Delta_{ij}} \right\} \tag{1.51}$$

which is essentially the same as Equation 1.31 in normal REMD method. Following the earlier scheme, different replicas have different protein–water attractive interactions and thus different strengths of protein hydrophobicity. When the scaling factor $\lambda < 1$, the hydrophobic interaction will be scaled up to facilitate the protein folding because the hydrophobic effect is one of the main driving forces for protein folding. Another advantage of HAREM is that the number of replicas required is much smaller than the temperature-based REMD because only part of the system's Hamiltonian is involved in computing the acceptance ratio of the exchanges. Liu et al. [79] have applied HAREM to study the folding of an α-helical peptide in the explicit solvent and show that HAREM is an order of magnitude more efficient than REMD. They also pointed out that HAREM should be particularly useful for proteins where the hydrophobic effect is the main driving force for the folding, for example, globular proteins with large hydrophobic cores.

1.6.2 Replica Exchange with Solute Tempering

The physical motivation of the replica exchange with solute tempering (REST) algorithm is to only heat the protein while keeping the solvent at low temperature as the simulation climbs the ladder of the replicas [83]. This algorithm is particularly useful for explicit solvent simulations because the number of replicas required in REST is only scaled with the protein's degrees of freedom but not the number of water molecules. In REST, the deformation of the Hamilton is smartly designed so that the acceptance probability for the exchange does not depend on water self-interactions. In particular, the potential energy of protein simulations can be partitioned into three terms:

$$E(X)=E_{pp}(X)+E_{pw}(X)+E_{ww}(X) \tag{1.52}$$

where E_{pp}, E_{pw}, and E_{ww} are protein–protein, protein–water, and water–water interactions, respectively. In REST, the potential energy surface, $E(X)$, is then modified according to

$$E_i(X)=E_{pp}(X)+\left[\frac{\beta_0+\beta_i}{2\beta_i}\right]E_{pw}(X)+\left[\frac{\beta_0}{\beta_i}\right]E_{ww}(X) \tag{1.53}$$

where $\beta_i = 1/(kT_i)$ and $\beta_0 = 1/(kT_0)$. T_0 is the lowest temperature replica, while T_i is the temperature of the ith replica. When $T_i = T_0$, $E_0(X) = E_{pp}(X) + E_{pw}(X) + E_{ww}(X)$ and the original potential energy surface is recovered. In this scheme, the protein is defined as the central group. After applying the detailed balanced condition for the exchange, the ratio of the transition probabilities is

$$\frac{P(i \rightarrow j)}{P(j \rightarrow i)}=\exp\left(-\Delta_{ij}\right) \tag{1.54}$$

where

$$\Delta_{ij} = \left(\beta_j - \beta_i\right)\left[\left(E_{pp}\left(X_i\right) + \frac{1}{2}E_{pw}\left(X_i\right)\right) - \left(E_{pp}\left(X_j\right) + \frac{1}{2}E_{pw}\left(X_j\right)\right)\right] \quad (1.55)$$

where water–water interaction energy (E_{ww}) vanishes from the acceptance criterion. It is obvious that with the same temperature spacing, the acceptance probability of REST will be much higher than that of REMD, because the effective energy in REST is much smaller than REMD:

$$\left|E_{pp}(X) + \frac{1}{2}E_{pw}(X)\right| \ll \left|E_{pp}(X) + E_{pw}(X) + E_{ww}(X)\right|. \quad (1.56)$$

Berne and coworkers [83] have compared REST with REMD for an alanine dipeptide molecule in the explicit solvent. The comparisons show that REST is around eight times more efficient than REMD in terms of calculating the thermodynamic properties for the alanine dipeptide system. In this work, the authors also proposed a variant of REST by including the solvation shells of water with the protein as the central group. However, this may impose some overhead to the calculations because a Metropolis criterion needs to be added to properly treat the water that moves in or out of the solvation shell between the two exchanges.

In a follow-up study, the researchers from the same group [84] have applied REST to an α-helical, a β-*hairpin*, and the Trp-cage peptide in the explicit solvent. To their surprise, they found that REST is sometimes less efficient than REMD. Further investigations show that in REST, exchanges between folded and unfolded protein conformations become extremely difficult because the difference of the effective energy is large, $\Delta(E_{pp}(X) + 1/2E_{pw}(X)) \gg 0$. On the other hand, the REMD method does not suffer this problem because the water self-interaction energy dominates the potential energy, and its fluctuations are sufficient to overshadow the energy difference caused by the protein conformational change. Based on this observation, they suggested that REST should only be applied to proteins where energy gap between the folded and unfolded states is not large. In the same paper, they have also proposed a new scheme by including a number of random water molecules into the central group and show that this scheme can greatly alleviate the problem that caused the poor performance of REST.

More recently, Berne and coworkers [85] have developed a new version of the REST algorithm, which they call REST2. Compared to REST, only a small change is made in the potential energy surface in REST2:

$$E_i(X)^{\text{REST2}} = \left[\frac{\beta_i}{\beta_0}\right]E_{pp}(X) + \sqrt{\frac{\beta_i}{\beta_0}}E_{pw}(X) + E_{ww}(X). \quad (1.57)$$

In REST2, all the replicas are run at the sample temperature T_0. The protein–protein interaction energy is scaled by β_i/β_0. Since T_0 is selected as the lowest temperature, $\beta_i/\beta_0 < 1$. Therefore, the energy barriers separating different protein conformations are lower. In this way, REST2 can greatly alleviate the problem encountered by REST due to the large energy difference between folded and unfolded protein conformations. At the same time, water self-interaction energy, E_{ww}, still vanishes in the effect energy; therefore, RESTs can also bypass the poor scaling with system size of REMD and largely reduce the number of replicas. Using a β-hairpin peptide, they show that REST2 greatly outperform both REST and REM [85].

1.6.3 Resolution Replica Exchange

Coarse-grained simulations have become a popular approach to alleviate the conformational sampling problem because the number of degrees of freedom of the system is greatly reduced in coarse-grained simulations by using a single site to represent a group of atoms [86]. Although coarse-grained simulations can make the computation more efficient, the atomistic degrees of freedom play crucial roles in many biological systems. Therefore, it is necessary to combine both atomistic and coarse-grained degrees of freedom in many cases.

REM provides an ideal framework to mix the simulations at different resolutions. Lwin and Luo have developed a dual-resolution REM to couple an accurate high-resolution simulation with an efficient low-resolution one [87]. This dual-resolution REM is a variant of Hamiltonian replica exchange where the exchanges of two Hamiltonians (one at high resolution and one at low resolution) are attempted periodically and the acceptance is subjected to the detailed balanced criterion. The authors show that this method greatly outperformed REMD in reproducing the correct canonical distributions when applying to a β-hairpin peptide. They further suggest that the low-resolution model can greatly help overcome the entropic barriers that exist in the free energy landscape of the high-resolution model for protein folding. A year after Lwin and Luo's work, Lyman et al. [88] have developed another resolution REM where a number of high-resolution and low-resolution models are coupled to enhance the conformational sampling.

One challenge associated with these resolution REMs is to properly reconstruct the high-resolution conformations from the low-resolution ones. Recently, a number of methods have been developed to address this challenge [89,90]. In a study by Christen and van Gunsteren [89], the coarse-grained particles are presented as virtual sites in the atomistic model to facilitate the reconstruction process. They have further selected a number of intermediate replicas between the coarse-grained model and atomistic model to ensure a reasonable acceptance ratio between the neighboring replicas. Liu and Voth [90] adopted a rather different approach using only two replicas in their method: one at atomistic resolution and the other at coarse-grained resolution. They allow the atomistic conformation constructed from the coarse-grained resolution replica to be fully relaxed before attempting the exchange by applying the smart walking algorithm developed by Zhou and Berne [91]. This new algorithm can greatly enhance the efficiency of the conformational sampling, even though only an approximated canonical distribution can be reproduced at either resolution.

1.7   SUMMARY AND FUTURE PERSPECTIVE

Despite the extensive studies of MD algorithms and methods in the last decades, the groundbreaking advancement of massively parallel supercomputers and novel algorithms for enhanced sampling is shaping this field into an exciting new area. There is an increasing interest in MD simulations, as witnessed by the recent Nobel Prize in chemistry and the ever-increasing numbers of publications on MD method, particularly from the emerging world such as China and India. Until now, modern progresses, particularly the development of novel algorithms and massively parallel computing techniques, have been led by only a small group of risk takers. With the maturation of these algorithms and simulation techniques, we hope this chapter on the fundamentals of MD and enhanced sampling will help broaden these techniques to a much larger community.

REFERENCES

1. Shaw, D. E.; Dror, R. O.; Salmon, J. K.; Grossman, J.; Mackenzie, K. M.; Bank, J. A.; Young, C. et al. *Proceedings of the Conference on High Performance Computing, Networking, Storage and Analysis* (*SC09*), Portland, OR, 2009, pp. 39:1–39:11.
2. Grossman, J. P.; Kuskin, J. S.; Bank, J. A.; Theobald, M.; Dror, R. O.; Ierardi, D. J.; Larson, R. H. et al. *Proceedings of the Eighteenth International Conference on Architectural Support for Programming Languages and Operating Systems* (*ASPLOS'13*), Houston, TX, 2013, pp. 549–560.
3. Fitch, B. G.; Germain, R. S.; Mendell, M.; Pitera, J.; Pitman, M.; Rayshubskiy, A.; Sham, Y. et al. *J. Parallel Distrib. Comput.* 2003, *63*, 759–773.
4. Snow, C.; Nguyen, H.; Pande, V. S.; Gruebele, M. *Nature* 2002, *420*, 102–106.
5. Rhee, Y.; Sorin, E.; Jayachandran, G.; Lindahl, E.; Pande, V. S. *Proc. Natl. Acad. Sci. USA* 2004, *101*, 6456–6461.
6. Sorin, E.; Pande, V. S. *Biophys. J.* 2005, *88*, 2472–2493.
7. Jayachandran, G.; Vishal, V.; Garcia, A. E.; Pande, V. S. *J. Struct. Biol.* 2007, *157*, 491–499.
8. Jayachandran, G.; Vishal, V.; Pande, V. S. *J. Chem. Phys.* 2006, *124*, 164902.

9. Sorin, E. J.; Pande, V. S. *J. Am. Chem. Soc.* 2006, *128*, 6316–6317.
10. Geyer, C. J.; Thompson, E. A. *J. Am. Stat. Assoc.* 1995, *90*, 909–920.
11. Hansmann, U. H. E. *Chem. Phys. Lett.* 1997, *281*, 140–150.
12. Sugita, Y.; Okamoto, Y. *Chem. Phys. Lett.* 1999, *314*, 141–151.
13. Swendsen, R. H.; Wang, J. S. *Phys. Rev. Lett.* 1986, *57*, 2607–2609.
14. Lyubartsev, A. P.; Martsinovski, A. A.; Shevkunov, S. V.; Vorontsov-Velyainov, P. N. *J. Chem. Phys.* 1992, *96*, 1776–1789.
15. Marinari, E.; Parisi, G. *Europhys. Lett.* 1992, *19*, 451–458.
16. Hansmann, U.; Okamoto, Y. *Curr. Opin. Struct. Biol.* 1999, *9*, 177–183.
17. Zhou, R.; Berne, B. J.; Germain, R. *Proc. Natl. Acad. Sci. USA* 2001, *98*, 14931–14936.
18. Zhou, R. *Proc. Natl. Acad. Sci. USA* 2003, *100*, 13280–13285.
19. Simmerling, C.; Strockbine, B.; Roitberg, A. E. *J. Am. Chem. Soc.* 2002, *124*, 11258–11259.
20. Garcia, A. E.; Sanbonmatsu, K. Y. *Proc. Natl. Acad. Sci. USA* 2001, *99*, 2782–2787.
21. Lin, C.; Hu, C.; Hansmann, U. *Proteins* 2003, *52*, 436–445.
22. Schug, A.; Herges, T.; Wenzel, W. *Proteins* 2004, *57*, 792–798.
23. Zheng, W.; Andrec, M.; Gallicchio, E.; Levy, R. *Proc. Natl. Acad. Sci. USA* 2007, *104*, 15340–15345.
24. Verlet, L. *Phys. Rev.* 1967, *159*, 98.
25. Swope, W. C.; Andersen, H. C.; Berens, P. H.; Wilson, K. R. *J. Chem. Phys.* 1982, *76*, 637.
26. Hockney, R.; Goel, S.; Eastwood, J. *J. Comput. Phys.* 1974, *14*, 148.
27. Andersen, H. C. *J. Chem. Phys.* 1980, *72*, 2384.
28. Berendsen, H. J. C.; Postma, J. P. M.; DiNola, A.; Haak, J. R. *J. Chem. Phys.* 1984, *81*, 3684.
29. Nosé, S. *Mol. Phys.* 1984, *52*, 255.
30. Hoover, W. *Phys. Rev. A* 1985, *31*, 1695.
31. Bussi, G.; Donadio, D.; Parrinello, M. *J. Chem. Phys.* 2007, *126*, 014101.
32. Parrinello, M.; Rahman, A. *J. Appl. Phys.* 1981, *52*, 7182.
33. Martyna, G.; Tuckerman, M. E.; Tobias, D.; Klein, M. *Mol. Phys.* 1996, *87*, 1117.
34. Ryckaert, J. P.; Ciccotti, G.; Berendsen, H. J. C. *J. Comput. Phys.* 1977, *23*, 327.
35. Andersen, H. *J. Comput. Phys.* 1983, *52*, 24.
36. Hess, B.; Bekker, H.; Berendsen, H. J. C.; Fraaije, J. G. E. M. *J. Comput. Chem.* 1997, *18*, 1463–1472.
37. Park, S.; Pande, V. S. *Phys. Rev. E* 2007, *76*, 016703.
38. Park, S. *Phys. Rev. E* 2008, *77*, 016709.
39. Park, S.; Ensign, D. L.; Pande, V. S. *Phys. Rev. E* 2006, *74*, 066703.
40. Bartels, C.; Karplus, M. *J. Comput. Chem.* 1997, *12*, 1450–1462.
41. Kumar, S.; Bouzida, D.; Swendsen, R. H.; Kollman, P. A.; Rosenberg, J. M. *J. Comput. Chem.* 1992, *13*, 1011–1021.
42. Chodera, J. D.; Swope, W. C.; Pitera, J. W.; Seok, C.; Dill, K. A. *J. Chem. Theory Comput.* 2007, *3*, 26–41.
43. Mitsutake, A.; Okamoto, Y. *Chem. Phys. Lett.* 2000, *332*, 131–138.
44. Huang, X.; Bowman, G.; Pande, V. *J. Chem. Phys.* 2008, *128*, 205106.
45. Rao, F.; Caflish, A. *J. Chem. Phys.* 2003, *119*(7), 4035–4042.
46. Rhee, Y.; Pande, V. *Biophys. J.* 2003, *84*, 775–786.
47. Nymeyer, H.; Garcia, A. *Proc. Natl. Acad. Sci. USA* 2003, *100*, 13934–13939.

48. Shirts, M.; Pande, V. S. *Science* 2000, *290*, 1903–1904.
49. Daura, X.; van Gunsteren, W.; Mark, A. *Proteins* 1999, *34*, 269–280.
50. Lyman, E.; Zuckerman, D. *Biophys. J.* 2006, *91*, 164–172.
51. Hansmann, U.; Okamoto, Y. *J. Comput. Chem.* 1993, *14*, 1333.
52. Hukushima, K.; Nemoto, K. *J. Phys. Soc. Jpn.* 1996, *65*, 1604–1608.
53. Zhou, R. *J. Mol. Graph. Model.* 2004, *22*(5), 451–463.
54. Geyer, G. *Stat. Sci.* 1992, *7*, 437.
55. Tesi, M.; van Rensburg, E.; Orlandini, E.; Whittington, S. *J. Stat. Phys.* 1996, *82*, 155.
56. Zhuang, W.; Cui, R. Z.; Silva, D.-A.; Huang, X. *J. Phys. Chem. B* 2011, *115*(18), 5415–5424.
57. Pitera, J.; Swope, W.; Abraham, F. *Biophys. J.* 2008, *94*(12), 4837–4846.
58. Mohanty, S.; Hansmann, U. *Biophys. J.* 2006, *91*(10), 3573–3578.
59. Im, W.; Brooks III, C. L. *J. Mol. Biol.* 2004, *337*(3), 513–519.
60. Zhang, W.; Wu, C.; Duan, Y. *J. Chem. Phys.* 2005, *123*, 154105.
61. Periole, X.; Mark, A. E. *J. Chem. Phys.* 2007, *126*, 014903.
62. Zhou, R.; Berne, B. J. *Proc. Natl. Acad. Sci. USA* 2002, *99*(20), 12777–12782.
63. Hagen, M.; Kim, B.; Liu, P.; Friesner, R. A.; Berne, B. J. *J. Phys. Chem. B* 2007, *111*, 1416–1423.
64. Zuckerman, D. M.; Lyman, E. *J. Chem. Theory Comput.* 2006, *2*(4), 1200–1202.
65. Zuckerman, D. M.; Lyman, E. *J. Chem. Theory Comput.* 2006, *2*, 1693.
66. Nymeyer, H. *J. Chem. Theory Comput.* 2008, *4*(4), 626–636.
67. Zheng, W.; Andrec, M.; Gallicchio, E.; Levy, R. M. *J. Phys. Chem. B* 2008, *112*(19), 6083–6093, PMID: 18251533.
68. Nadler, W.; Hansmann, U. H. E. *Phys. Rev. E* 2007, *76*, 065701.
69. Trebst, S.; Troyer, M.; Hansmann, U. *J. Chem. Phys.* 2006, *124*, 174903.
70. Abraham, M. J.; Gready, J. E. *J. Chem. Theory Comput.* 2008, *4*(7), 1119–1128.
71. Sindhikara, D.; Meng, Y.; Roitberg, A. *J. Chem. Phys.* 2008, *128*, 024103.
72. Huang, X.; Bowman, G.; Bacallado, S.; Pande, V. *Proc. Natl. Acad. Sci. USA* 2009, *106*, 19765–19769.
73. Noe, F.; Fischer, S. *Curr. Opin. Struct. Biol.* 2008, *18*(2), 154–162.
74. Chodera, J. D.; Singhal, N.; Pande, V. S.; Dill, K. A.; Swope, W. C. *J. Chem. Phys.* 2007, *126*(15), 155101.
75. Buchete, N. V.; Hummer, G. *J. Phys. Chem. B* 2008, *112*(19), 6057–6069.
76. Swope, W. C.; Pitera, J. W.; Suits, F. *J. Phys. Chem. B* 2004, *108*(21), 6571–6581.
77. Swope, W. C.; Pitera, J. W.; Suits, F.; Pitman, M.; Eleftheriou, M.; Fitch, B. G.; Germain, R. S. et al. *J. Phys. Chem. B* 2004, *108*(21), 6582–6594.
78. Fukunishi, H.; Watanabe, O.; Takada, S. *J. Chem. Phys.* 2002, *116*, 9058–9067.
79. Liu, P.; Huang, X.; Zhou, R.; Berne, B. J. *J. Phys. Chem. B* 2006, *110*(38), 19018–19022, PMID: 16986898.
80. Huang, X.; Margulis, C.; Berne, B. *Proc. Natl. Acad. Sci. USA* 2003, *100*, 11953.
81. Zhou, R.; Huang, X.; Margulius, C.; Berne, B. *Science* 2004, *305*, 1605–1609.
82. Liu, P.; Huang, X.; Zhou, R.; Berne, B. J. *Nature*, 2005, *437*, 159–162.
83. Liu, P.; Kim, B.; Friesner, R. A.; Berne, B. J. *Proc. Natl. Acad. Sci. USA* 2005, *102*, 13749–13754.
84. Huang, X.; Hagen, M.; Kim, B.; Friesner, R. A.; Zhou, R.; Berne, B. J. *J. Phys. Chem. B* 2007, *111*(19), 5405–5410, PMID: 17439169.

85. Wang, L.; Friesner, R. A.; Berne, B. J. *J. Phys. Chem. B* 2011, *115*(30), 9431–9438.
86. Ayton, G. S.; Noid, W. G.; Voth, G. A. *Curr. Opin. Struct. Biol.* 2007, *17*(2), 192–198.
87. Lwin, T.; Luo, R. *J. Chem. Phys.* 2005, *123*, 194904.
88. Lyman, E.; Ytreberg, F.; Zuckerman, D. *Phys. Rev. Lett.* 2006, *96*, 028105.
89. Christen, M.; van Gunsteren, W. *J. Chem. Phys.* 2006, *124*, 154106.
90. Liu, P.; Voth, G. *J. Chem. Phys.* 2007, *126*, 045106.
91. Zhou, R. H.; Berne, B. J. *J. Chem. Phys.* 1997, *107*, 9185.

Chapter 2

Toward Quantitative Analysis of Metalloenzyme Function Using MM and Hybrid QM/MM Methods

Challenges, Methods, and Recent Applications

Michael Gaus, Puja Goyal, Guanhua Hou, Xiya Lu, Xueqin Pang, Jan Zienau, Xin Xu, Marcus Elstner, and Qiang Cui

CONTENTS

2.1   INTRODUCTION

Transition metal ions play essential structural and catalytic roles in metallo-enzymes.[1,2] For example, a recent survey[2] indicated that, among 1371 different enzymes for which 3D structures are available, ~47% contain metal ions with 41% hosting metals at the catalytic site. Since many transition metal ions are redox active, most transition metal enzymes catalyze complex chemical transformations that require mechanisms beyond relatively simple general acid/base pathways and involve unusual chemical intermediates; although the nature of these intermediates can often be probed with various spectroscopic techniques, the interpretation of the data at a molecular scale is often not straightforward.[3,4] Moreover, although many mechanistic studies focus on the coordination chemistry of metal ions, the importance of second-sphere effects and other long-range (allosteric) contributions has been increasingly recognized[5,6]; the relative contributions of various factors, however, are often difficult to untangle. To effectively complement experimental investigations in tackling these mechanistic issues, it is essential to develop quantitative computational approaches for structural, reactive (energetic), and spectroscopic properties of metalloenzymes.

For transition metal sites in proteins that are structurally rigid and well shielded from bulk solvent, a battery of computational techniques have been established. Both carefully constructed cluster models[7,8] and hybrid quantum mechanical/molecular mechanical (QM/MM) models[9–13] have been shown to provide novel mechanistic insights. QM/MM studies, mostly with a density functional theory (DFT) QM method,[14] have been done with reaction path type of calculations,[10–13] approximate treatment of protein fluctuations,[15–17] or sub-nanosecond direct dynamics.[18–21] For systems that feature a high degree of structural flexibility and/or solvent accessibility, however, existing methods are not yet robust. In these systems, the coordination environment of the metal ion may fluctuate significantly, and therefore the applicability of cluster models and decoupled treatment of fluctuations for QM and MM regions is more limited.

The general goal of our research in the area of metalloenzymes is to develop effective computational methods that target systems that are particularly flexible and therefore not easily studied by methods already available in the literature. Depending on the specific problem of interest, this involves the development of MM or QM/MM models for the metal site, although most of our developments have been in the general area of QM/MM methods. In this chapter, we discuss some of our recent developments, with illustrative applications; in addition to mechanistic insights gleaned by simulation studies, we also discuss the limitations of these methods and briefly comment on further developments that will help alleviate the limitations. Finally, we conclude by a short summary regarding some of the remaining technical challenges and research opportunities for computational studies of metalloenzymes.

2.2 BASIC METHODOLOGIES: CLASSICAL (MM) AND HYBRID (QM/MM) MODELS

Considering the complexity of biomolecular systems, the key strategy is to properly balance computational accuracy and efficiency for the particular problem in hand. Depending on the specific problem of interest, either an MM or a QM/MM model is appropriate. For probing catalytic pathways, a QM/MM model is usually required. Even in this case, an MM model might be used for more thoroughly equilibrating the structure, such as the level of solvation of the active site at different stages of the catalytic cycle. For structural and ligand binding properties, an MM model might suffice, although the functional form of the MM model might be more complex than typically used for organic molecules; for binding affinity of transition metal ions to proteins, a QM/MM model is likely most reliable. In the following, we first briefly summarize strategies for constructing an MM model for metal sites and then discuss QM/MM methods for studying metalloenzymes, focusing mainly on those developed in our group.

2.2.1 MM Models

To describe metal–ligand interactions for a metal site in a protein, either a bonded or nonbonded MM model can be used. As the nomenclature implies, a bonded model uses explicit bonded terms to describe metal–ligand coordination[22]; the simplest bonded terms include the standard bond stretch, angle bending, and dihedral/improper terms,[23] although more sophisticated models have been introduced to describe the anharmonic nature of metal–ligand bonds, polarization,[24–26] or charge transfers between the metal ion and ligands.[27] Other effects unique to transition metals such as ligand field[28,29] and *trans* effects[30] have also been formulated and implemented. An interesting model worth mentioning is the VALBOND model developed by Landis and coworkers[31-34] based on a valence bond framework for describing metal complexes.[35] The model has been found effective even for transition metal compounds with fairly unusual geometries such as the distorted trigonal prism reported for $[W(CH_3)_6]$.[36] In VALBOND, the standard harmonic angular bending potential is replaced by terms related to orbital overlap and therefore more appropriate for describing large-amplitude motions. VALBOND also supports hypervalent compounds using a 3-center 4-electron bonding model.[32,33] For transition metals, VALBOND traditionally considers only sd hybrids. Therefore, compounds that count more than 12 electrons in their valence orbitals are considered hypervalent and 3-center 4-electron bonds are considered instead of the p-orbitals. The VALBOND model has been recently introduced into the biomolecular simulation package CHARMM,[30] laying the ground work for describing metal centers in biomolecules.

Alternatively, especially when the metal ion is closed shell, such as Mg^{2+} and Zn^{2+}, a nonbonded model is likely sufficient for many applications. In the simplest form, the metal ion is represented as a point charge with the Lennard-Jones parameters; this has been found (surprisingly) effective for structural properties.[37] For reliable energetics, more sophisticated treatments of polarization[38,39] and potentially charge transfer are required. The advantage of the nonbonded models compared to the bonded ones is that multiple coordination modes can be simultaneously described; for example, carboxylate (in Asp or Glu) has been found to switch between monodentate and bidentate binding modes to Zn^{2+}, Mg^{2+}, and Ca^{2+}, and it was proposed that this flexibility has functional implications.[40-44] A reliable description of the weights for different coordination modes likely requires going beyond the simple coulombic functional form for electrostatics. For example, Zhang and coworkers found that a point charge model for Zn^{2+} with coulombic electrostatics leads to incorrect coordination numbers in some protein simulations[41,45]; accordingly, they proposed a revised functional form for the short-range behavior of metal–ligand interactions.[41] The model was shown to give more reliable description of zinc sites in multiple proteins, including several systems that feature bimetallic zinc sites, while allowing a flexible coordination of carboxylate groups and water molecules. In our recent work,[46] we tested simple point charge models of Fe(II) for the nonheme iron site in AlkB family enzymes (see Section 2.3.1). QM calculations for active site models and a statistical survey of Fe(II) sites in the PDB suggest that carboxylate (Asp/Glu) prefers to adopt the monodentate binding mode when the iron is also coordinated with 1–3 His residues. Therefore, a coulombic electrostatic model for the Fe(II) with a harmonic restraint that controls the binding mode of Asp/Glu was found adequate for structural properties. For binding energetics, however, the point charge model was found only semiquantitative as compared to QM calculations. See Section 2.3.1 for additional discussions.

For the parameterization of the MM model, several strategies can be adopted. For bonded models, many of the key parameters for the bonded terms can be determined rather uniquely by comparing to QM results for relevant active site models, such as hessian matrices.[22,47] For nonbonded parameters, strategies involve MD sampling of condensed-phase models and matching specific quantities to either high-level QM or experimental data, such as force on atoms,[41] solvation free energy,[48,49] and osmotic pressure.[50]

2.2.2 QM/MM Models

2.2.2.1 QM Methods: Ab Initio vs. Approximate Approaches

As mentioned in Section 2.1, ab initio QM/MM calculations have been found valuable for mechanistic studies of metalloenzymes.[7-13] Nevertheless, balancing accuracy and sampling remains a topic that deserves further efforts. To this end, we target our QM/MM developments to leverage our extensive experience in the development and application of an approximate DFT (SCC-DFTB[51-53]), which has also been successfully applied by many other groups to a broad set of enzyme systems.[54-59] In the original form, however, the method is most successful for geometry and has limited accuracy in energetics when applied to complex chemistry in metalloenzymes,[60] such as phosphoryl transfers catalyzed by zinc enzymes, and its applicability to open-shell metals has not been thoroughly explored. Therefore, we have further developed the next generation of SCC-DFTB, which we termed DFTB3[61]; we found that including third-order contributions substantially improves the performance of SCC-DFTB in many applications. In addition to parameterizing DFTB3 for main group elements of biological interest, we have also initiated efforts on systematically improving DFTB3 for open-shell metal ions, such as Cu^{2+} and $Fe^{2+/3+}$ in proteins.

The original formulation of SCC-DFTB[51] starts with a second-order expansion of the total energy in the DFT framework:

$$E = \sum_{i}^{occ} \left\langle \Psi_i | \hat{H}^0 | \Psi_i \right\rangle + \frac{1}{2} \iint' \left(\frac{1}{|\vec{r}-\vec{r}'|} + \left. \frac{\delta^2 E_{xc}}{\delta\rho\delta\rho'} \right|_{\rho_0} \right) \delta\rho\delta\rho'$$

$$- \frac{1}{2} \iint' \frac{\rho_0'\rho_0}{|\vec{r}-\vec{r}'|} + E_{xc}[\rho_0] - \int V_{xc}[\rho_0]\rho_0 + E_{cc}. \tag{2.1}$$

With the monopole approximation and spherical charges,

$$\delta\rho \approx \sum_{\alpha} \delta\rho_{\alpha} \approx \sum_{\alpha} \Delta q_{\alpha} F_{00}^{\alpha} Y_{00}, \tag{2.2}$$

the second-order term becomes

$$E^{\text{second}} = \frac{1}{2}\sum_{\alpha\beta}\Delta q_\alpha \Delta q_\beta \iint' \left(\frac{1}{|\vec{r}-\vec{r}'|} + \left.\frac{\delta^2 E_{xc}}{\delta\rho\delta\rho'}\right|_{\rho 0} \right) F_{00}^\alpha F_{00}^\beta Y_{00}^2 \approx \frac{1}{2}\sum_{\alpha\beta}\Delta q_\alpha \Delta q_\beta \gamma_{\alpha\beta}. \quad (2.3)$$

To obtain explicit expressions for $\gamma_{\alpha\beta}$, Elstner et al.[51] used Slater-like charge distributions to evaluate the approximate expression of the second-order term

$$F_{00}^\alpha = \frac{\tau_\alpha}{8\pi}\exp\left(-\tau_\alpha|\vec{r}-\vec{R}_\alpha|\right), \quad (2.4)$$

in which the exponent is related to the Hubbard parameter (U_α):

$$\tau_\alpha = \frac{16}{5}\gamma_{\alpha\alpha} = \frac{16}{5}U_\alpha = \frac{32}{5}\eta_\alpha, \quad (2.5)$$

where η_α is the chemical hardness parameter for element α. In other words, the Hubbard parameter is closely related to the effective size of the atom (charge distribution).

In recent studies,[61–63] it was shown that for proton affinity (PA) calculations, it is important to include charge dependence of the Hubbard parameter. Physically, this is because as the molecule deprotonates, the chemical hardness and size of the titration site are expected to change; similar argument can be made for other chemical processes that involve a major change in atomic charge. Numerically, this can be accomplished by extending the energy expansion around the reference density to third order. For example, including the diagonal (on-site) third-order terms involves the following contributions:

$$E^{\text{third-diagonal}} = \frac{1}{6}\sum_\alpha \Delta q_\alpha^3 \frac{\partial U_\alpha}{\partial q_\alpha}, \quad (2.6)$$

in which the charge derivative of the Hubbard parameter, $\partial U_\alpha/\partial q_\alpha$, can be either computed based on atomic calculations or fitted based on observables of interest, such as PA. More recently, the complete third-order terms (i.e., including off-diagonal terms) have been implemented by Gaus et al.[61] Although the off-diagonal terms make relatively minor contributions to the PAs of *simple* organic compounds, such as amino acid side-chain analogs, they were shown to be important to PA of phosphate compounds, whose chemical transformations in biology are catalyzed by various metalloenzymes. For example, as shown in Table 2.1, with only the diagonal third-order terms, it is difficult to describe PAs of phosphate compounds despite good results for nonphosphate species[63]; with complete third orders, the errors relative to high-level QM calculations (G3B3) are much reduced. We have published

TABLE 2.1   PROTON AFFINITIES FOR PHOSPHOROUS CONTAINING MOLECULES (IN KCAL/MOL) RELATIVE TO G3B3[a]

Molecule[b]	G3B3	DFTB2	DFTB3-diag[c] Calc	DFTB3-diag[c] Fit	DFTB3[c] Calc	DFTB3[c] Fit
H_3PO_4	334.0	+17.1	+23.9	+18.5	+18.3	+5.5
$H_2PO_2^-$	464.5	+26.2	+26.6	+20.0	+17.2	−4.3
Dimethyl hydrogen phosphate	336.3	+9.6	+20.3	+14.9	+15.9	+4.8
$P(O)(OH)(OH)(OCH_3)$	336.7	+12.0	+20.9	+15.4	+15.8	+3.8
$P(O)(O)(OH)(OCH_3)-$	460.5	+21.5	+26.3	+19.9	+18.3	−1.2
PH_3OH^+	201.6	−8.6	+7.1	+2.7	+4.8	−0.0
PH_2OHOH^+	201.6	−2.8	+10.8	+6.3	+8.2	+2.1
$PHOHOHOH^+$	200.8	+4.5	+16.3	+11.7	+13.6	+6.2
$PH_2(OH)=O$	336.6	+3.0	+15.7	+10.4	+12.2	+3.3
$PH(OH)(OH)=O$	334.7	+10.7	+20.4	+15.0	+16.0	+5.3
$P(O)(OH)(-O-CH_2CH_2-O-)$	336.3	+7.2	+17.7	+12.3	+13.4	+2.4
$P(OH)(OH)(-O-CH_2CH_2-O-)(OH^*)$	359.0	−3.3	+18.6	+12.8	+12.9	+0.3
$P(OH^*)(OH)(-O-CH_2CH_2-O-)(OH)$	350.4	+6.7	+15.9	+10.6	+11.0	−0.4
$P(OH^*)(OH)(-O-CH_2CH_2-O-)(OCH_3)$	351.2	+1.8	+12.8	+7.4	+8.9	−1.4
$P(OH)(OCH_3)(-O-CH_2CH_2-O-)(OH^*)$	359.6	−8.3	+7.9	+2.1	+3.3	−7.2
$P(OH^*)(OCH_3)(-O-CH_2CH_2-O-)(OH)$	352.9	+3.6	+14.5	+9.2	+10.1	−0.5
$P(OH)(OH)(OH)(OH^*)(OH)-ax$	357.3	+4.0	+21.8	+15.9	+14.2	−1.2
$P(OH)(OH)(OH)(OH^*)(OH)\_eq^d$	347.0	+14.0	—	—	—	−0.0
MUE		9.2	17.5	12.1	12.6	2.8
MSE		+6.6	+17.5	+12.1	+12.6	+1.0
MAX		26.2	26.6	20.0	18.3	7.2

[a] The PA is computed with the potential energies at 0 K without any zero-point energy correction. For the DFTB methods, the deviation is given as the difference to the G3B3 method ($E^{method} - E^{G3B3}$).

[b] The molecules are given in the protonated form.

[c] DFTB3-diag includes only diagonal third-order contributions.

[d] The molecule $P(OH)(OH)(OH)(OH^*)(OH)\_eq$ dissociates forming H_2O for DFTB3-diag and DFTB3/calc. Depending on the basis set, this dissociation also occurs for PBE and B3LYP: dissociation with 6311G(2d,2p), no dissociation with cc-pVTZ.

DFTB3 parameters for O, N, C, and H along with extensive tests for organic molecules[61,64]; parameters for P, S, Na, Mg, and Zn are also being finalized and will be published in the near future.

To describe open-shell metal ions, the key innovation in our methodology is to enhance the flexibility of DFTB3 by allowing the Hubbard parameters and their charge derivatives to depend on orbital angular momentum (l). This accounts for the different localization of $3d$ electrons and $4s/4p$ electrons and was briefly explored at the second order for iron clusters.[65] The energy expression including the third-order contribution with spin polarization is

$$E^{\text{DFTB3}} = \sum_{iab}\sum_{\mu \in a}\sum_{\nu \in b} n_i c_{\mu i} c_{\nu i} H^0_{\mu\nu} + \sum_{ab}\sum_{l_a l_b} \Delta q_{l_a} \Delta q_{l_b} \left(\frac{1}{2}\gamma_{l_a l_b} + \frac{1}{3}\Delta q_a \Gamma_{l_a l_b} \right)$$

$$+ \frac{1}{2}\sum_a \sum_{l a l'_a} p_{al_a} p_{al'_a} W_{al_a al'_a} + E^{\text{rep}}, \tag{2.7}$$

in which the angular momentum dependence of γ_{ab}, Γ_{ab}, and $W_{al_a l'_a}$[65-67] is made explicit. For example, $\Gamma_{l_a l_b} = (\partial \gamma_{l_a l_b} / \partial U_{l_a})(\partial U_{l_a})(\partial U_{l_a} / \partial q_a)$. While the rigorous derivation requires different derivatives with respect to the charge of every l_a shell, our approximation requires only derivatives with respect to the total atomic charge ($\partial U_{l_a} / \partial q_a$).

As an example, we show some preliminary results for the parameterization of the l-dependent DFTB3 for Cu ions. The parameterization has been carried out using B3LYP/aug-cc-pVTZ as the reference, although we recognize that for transition metal ions, it is generally less obvious what QM methods are systematically reliable,[14,68-72] especially regarding different DFT methods. For example, very recently,[72] a fairly thorough set of benchmark calculations for transition metal thermochemistry suggests that the B97 family of functionals is superior in performance than B3LYP for energetics. Therefore, we will also explore using B97 as the reference for another parameterization of DFTB3 in the future. As shown in Figure 2.1a and b, the preliminary parameterization is very successful for geometry; for both Cu(I) and Cu(II) complexes, including those that exhibit the Jahn–Teller distortions, the RMSD values in bond distance and bond angles in comparison to B3LYP/aug-cc-pVTZ results are very small and represent major improvements over existing semiempirical methods such as PM6.[73,74] We note that this level of agreement would be *impossible* to accomplish, especially for the Cu(II) complexes, without the l-dependence of the Hubbard parameters. We have also compared to 26 crystal structures in the Cambridge structure database, and the performance is also rather impressive, especially considering that some of the structural differences are due to crystal packing effects on floppy aromatic ligands that can easily rotate. In Figure 2.1c and d, the overlap

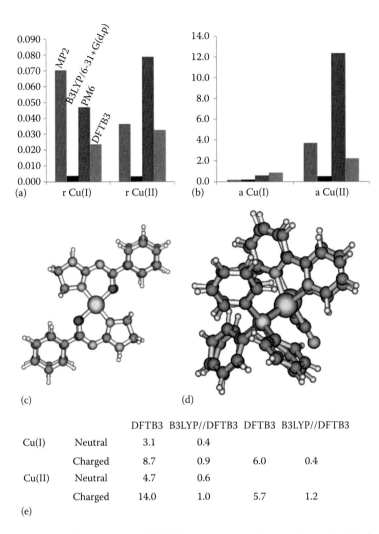

		DFTB3	B3LYP//DFTB3	DFTB3	B3LYP//DFTB3
Cu(I)	Neutral	3.1	0.4		
	Charged	8.7	0.9	6.0	0.4
Cu(II)	Neutral	4.7	0.6		
	Charged	14.0	1.0	5.7	1.2
(e)					

Figure 2.1 A preliminary set of DFTB3 parameterization for Cu with orbital angular momentum-dependent Hubbard parameters. Mean absolute deviations for (a) bond lengths (Å) and (b) bond angles (°) of 36 Cu(I) and 46 Cu(II) species with ligands such as H_2O, NH_3, PH_3, SH_2, their deprotonated analogs, CO and CH_3^- in comparison to B3LYP/aug-cc-pVTZ optimized geometries. For PM6, all Cu–S bonds are excluded due to exceptionally large errors. Basis set for MP2 is aug-cc-pVTZ. (c) Overlap of DFTB3 and crystal structure from the Cambridge Structural Database[300] (entry AJILIO); the RMSD is 0.26 Å. (d) In analogy for entry AWEMAQ; the RMSD is 0.50 Å. (e) Mean absolute deviations (kcal/mol) of sequential bond dissociation energies (sBDEs) and PAs in comparison to B3LYP/aug-cc-pVTZ energies and geometries. The values shown are calculated using DFTB3 and B3LYP/aug-cc-pVTZ single-point energies at DFTB3 optimized geometries.

of DFTB3 optimized and crystal structures is shown for two cases with small and typical RMSD values, supporting the encouraging quality of the current parameterization on structures.

Regarding energetics (Figure 2.1e), the performance is satisfactory for the binding energy of charge-neutral ligands (e.g., water, ammonia, and H_2S) to both Cu(I) and Cu(II), with an RMSD value of a few kcal/mol in comparison to the B3LYP reference. The performance is decidedly worse when the ligands are negatively charged (e.g., thiolate), with RMSD on the order of 10 kcal/mol, which is not entirely unexpected since DFTB3 is a minimal-basis approach; it should be borne in mind, however, the binding energies for these anionic ligands are on the order of >200 kcal/mol in the gas phase! Most importantly, single-point energy calculations at the DFTB3 structures give binding energies, for *all* cases, *very close* to the B3LYP reference, again reflecting the high quality of the DFTB3 structures. This applies also to the calculated PA of copper ligands. Therefore, DFTB3 is promising as a low-level method in a multilevel QM/MM approach[75–77] in which the conformational space is sampled at the low level, while energetics are evaluated based on free energy perturbation between the low- and high-level QM/MM potential functions. The semiquantitative nature of DFTB3 energetics is significant in this context because a good overlap between the potential energy surfaces at different QM/MM levels is essential to the convergence of multilevel QM/MM methods.[16,76,78]

2.2.2.2 QM/MM Interface and Boundary Conditions

In addition to the accuracy of the QM method, at least two other issues also contribute to the reliability of QM/MM simulations. The first concerns the interaction between QM and MM atoms, while the second issue concerns the boundary condition of QM/MM simulations; another topic that has been thoroughly discussed in the literature is the treatment of QM/MM boundary/frontier[9,18,79–81] (i.e., when the partitioning between QM and MM occurs across covalent bonds) and is considered well understood.

For the interaction between QM and MM atoms, it is common to include both electrostatic and the van der Waals contributions; bonded terms are also included when the partitioning is across covalent bonds. The QM/MM van der Waals terms usually take the standard Lennard-Jones form, with parameters predetermined based on model systems.[82,83] This is not ideal when the QM region undergoes significant changes in charge distribution and polarizability during the process under study; thus, more sophisticated models in which the QM/MM van der Waals terms are dependent on the QM charge distribution have been formulated.[84]

In most condensed-phase applications, electrostatics tend to dominate and therefore it is essential that electrostatic interactions between QM and MM atoms are properly described before more sophisticated treatment for the QM/MM van der Waals interactions is incorporated. For SCC-DFTB, the QM/MM electrostatic interaction is approximately calculated in the original implementation[85] as the coulombic interaction between the QM Mulliken charge (Δq_a) and MM point charge (Q_I). The error due to this approximation can be significant when QM and MM atoms approach each other where charge penetration effect becomes important. As a result, reactions that involve highly charged solute/substrate are difficult to study with the original SCC-DFTB/MM Hamiltonian.[86] The problem can be partially solved by enlarging the QM region, but this introduces not only additional cost but also technical complications for cases that involve highly mobile solvents, such as the need of changing QM/MM partitioning on the fly.[87,88]

In recent work,[89] motivated by the Klopman–Ohno (KO) expression for the two-center two-electron integrals in semiempirical QM methods,[90] which also inspired the development of the γ_{ab} kernel in the original SCC-DFTB,[51] we have implemented a different Hamiltonian for the SCC-DFTB/MM electrostatics. It takes the form

$$H_{\text{elec,KO}}^{\text{QM/MM}} = \sum_{\alpha \in \text{QM}} \sum_{A \in \text{MM}} \frac{\Delta q_\alpha Q_A}{\sqrt{R_{\alpha A}^2 + a_\alpha (U_\alpha^{-1}(\Delta q_\alpha) + U_A^{-1})^2 \exp(-b_\alpha R_{\alpha I})}} = \sum_{\alpha \in \text{QM}} \sum_{A \in \text{MM}} \gamma_{\text{KO}} \Delta q_\alpha Q_A$$

(2.8)

in which a_α and b_α are element type–dependent parameters. Together with the van der Waals parameters in the QM/MM Hamiltonian, there are four QM/MM parameters for each element type; using a minimal number of parameters that depend on element type rather than atom type is essential to the transferability of the model. To be consistent with the third-order formulation of SCC-DFTB,[61] the Hubbard parameter in the KO functional is dependent on the QM charge. As a result, the effective size of the QM charge distribution naturally adjusts as the QM region undergoes chemical transformations, making the KO-based QM/MM scheme particularly attractive for describing chemical reactions in the condensed phase.

In our proof of concept study,[89,91] the QM/MM parameters were optimized for O, C, H, and P for a specific version of SCC-DFTB (SCC-DFTBPR[86]) based on *microsolvation* clusters of 23 solutes that mimic different charge states of amino acid side chains and phosphate species. For the training set, optimized KO scheme gives slightly better results compared to the optimized point-charge-based QM/MM model; the mean unsigned error (MUE) is 3.3 and 4.8 kcal/mol, respectively, compared to full QM calculations (note that the errors are for *total solute–solvent interactions*, which are often >100 kcal/mol; thus, the error is typically less than 5%!). However, the KO scheme is much more transferable. For 16 stable structures and 24 transition states in the QCRNA database, the

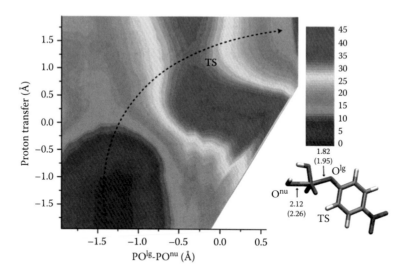

Figure 2.2   Potential of mean force (PMF) of pNPP^{2-} hydrolysis in water calculated by SCC-DFTBPR/MM simulation with the QM/MM-KO scheme. Also included is the transition state structure; numbers without parentheses are calculated with QM/MM; those with parentheses are calculated with SCC-DFTBPR and a PB model with charge-dependent atomic radii. (From Hou, G. et al., *J. Chem. Theory Comput.*, 8, 4293, 2012; Hou, G. et al., *J. Chem. Theory Comput.*, 6, 2303, 2010.)

MUE is 4.3 kcal/mol for the KO scheme but 16.2 kcal/mol for the point-charge-based QM/MM model. Clearly, the KO model is much more robust! As another example, the hydrolysis of phosphate monoesters in a solution was studied.[89] With point-charge-based QM/MM Hamiltonian, the hydrolysis barrier for pNPP^{2-} (see Figure 2.2) is grossly overestimated (~11 kcal/mol) with SCC-DFTBPR/MM simulations. With the KO scheme, the computed barrier is in close agreement (within 2 kcal/mol) with available experimental data. In the future, we will develop the QM/MM-KO parameters for all essential element types based on DFTB3. The parameterization will be carried out following the same *microsolvation* cluster model as in Ref. 89, but the full QM reference will be based on high-level QM calculations rather than SCC-DFTB. This is expected to be essential for the accuracy of the multilevel QM/MM approach as discussed earlier.

Regarding the boundary condition of QM/MM simulations, although the Ewald approach has been implemented by several authors for semiempirical QM/MM potential functions,[92–94] periodic boundary condition (PBC)-based QM/MM simulations remain somewhat expensive. This is partly because most semiempirical methods do not benefit well from highly parallel computations unless the QM region becomes very large[95]; as a result, for the typical application that employs a QM region of 100 atoms or so, the QM part of the computation substantially limits the efficiency of PBC-based QM/MM simulations. As an alternative, when the process of interest is localized, we have adopted a *multiscale* model based on combining QM/MM with the generalized solvent boundary potential (QM/MM-GSBP[53,96]) approach of Roux and coworkers.[97] The system is partitioned into *inner* and *outer* regions (e.g., see Figure 2.3a), which are modeled using microscopic and continuum electrostatic models, respectively. The inner region contains both the QM region (e.g., those explicitly involved in catalysis) and nearby MM atoms (e.g., those within 25 Å from the QM region), and they undergo thermal fluctuations following standard mixed Newtonian/Langevin dynamics sampling.[98] The outer region contains part of the protein and solvent far away from the QM region and, in the case of transmembrane proteins, the bulk membrane represented as dielectric slabs. The outer region has fixed geometry, which makes it straightforward to compute its continuum electrostatic contribution (second term on the rhs of Equation 2.9). The electrostatic component of the GSBP approach is formally equivalent to solving the Poisson–Boltzmann (PB) equation at each MD step; numerically, this is avoided by expanding Green's function of the linearized PB equation in terms of a basis set, which is spherical harmonics for a spherical inner region and Legendre polynomials for a rectangular inner region. This leads to a reaction field matrix (M) that needs to be computed *once* prior to the MD simulation; during the simulation, the instantaneous charge distribution of the inner region (q_α) is projected onto the same basis set and the electrostatic (reaction field) component of the energy is given by the product of the precomputed reaction field matrix and the projected charge distributions ($\mathbf{Q}$):

$$\Delta W_{\text{elec}}^{(ii)}(X) + \Delta W_{\text{elec}}^{(io)}(X) + U_{\text{elec}}^{(io)}(X) = \frac{1}{2} Q^{\mathrm{T}} M Q + \sum_{\alpha \in (i)} q_\alpha \phi_s^{(o)}(r_\alpha). \tag{2.9}$$

In this way, the GSBP approach is computationally efficient and allows one to focus the sampling on the region of interest.

The QM/MM-GSBP framework has been carefully tested with a number of solution and enzyme systems.[92,96,99,100] For example, using the zinc-enzyme carbonic anhydrase (CA), we have compared structural properties and water distribution of the active site from GSBP and Ewald simulations[99]; with a spherical inner region of 25 Å centered at the site of interest (zinc ion), the GSBP and Ewald results were found to be in good agreement. By contrast, significantly different behaviors were observed when the bulk solvents were ignored or when different cutoff schemes were used for QM/MM and MM/MM interactions.[96] For energetics, we have focused on microscopic pK_a calculations for both globular soluble proteins[99,101] and transmembrane proteins, such as

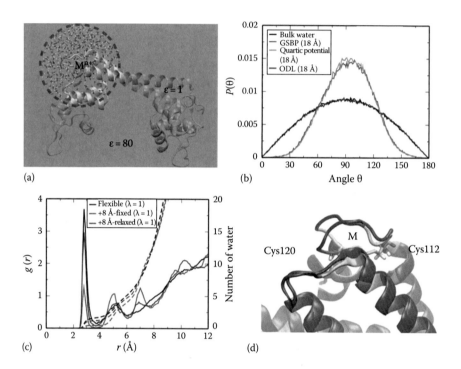

Figure 2.3   (See color insert.) Properties of PBC and GSBP simulations with a partially or fully flexible CueR. (a) A snapshot that illustrates the GSBP (20 Å) setup; protein atoms beyond 20 Å of M are fixed (cyan), those within 20 Å of M are flexible (yellow), and the bulk continuum is indicated with a gray background. (b) The orientation distribution of water molecules in bulk (from PBC calculations), an orientationally disordered liquid (ODL) ensemble, and two other water droplet simulations; for the last three cases, orientations for water molecules within 1.5 Å from the droplet surface are shown. The orientation is defined as the angle between the water dipole and the normal vector of the local surface. (c) Radial distribution function of water oxygen around M (oxidized state, $\lambda = 1$) in PBC simulations with different parts of the CueR protein held fixed. The integrated RDFs are shown as dashed lines. (d) Overlay of +8 Å-fixed and +8 Å-relaxed structures for the oxidize state after 1 ns production run. The flexible protein region from the +8 Å-fixed structure is shown in yellow, and that in the +8 Å-relaxed structure is shown in purple; the fixed protein region is colored in blue. Clearly, the metal binding loop responds to the metal oxidation and therefore should be allowed to relax for reliable oxidation free energy calculations (see Table 2.2 and text for discussion).

bacteriorhodopsin[102] and cytochrome c oxidase.[103] These are nontrivial test cases because structural response of the protein/solvent environment to titration, especially for buried residues, may implicate multiple processes, such as water penetration and structural transitions at both side-chain and main-chain levels.[104] Some of these structural responses might be difficult to describe with GSBP that involves a small inner region due to the frozen nature of outer region atoms.

In a recent study,[105] to understand the quantitative impact of various approximations in GSBP, we compared charging free energies in both solution and protein systems calculated with GSBP and PBC simulations. The solution model involved charging a cation or anion in water, while the protein model dealt with the oxidation free energy of a metal site in the copper efflux regulator (CueR) protein (see in Section 2.3.4), which does *not* have redox biological function and was chosen because its structural features (in terms of size and flexibility, see Figure 2.3a) make it an appropriate system for comparing GSBP and PBC simulations.

For simple ions in solution, we found good agreements between GSBP and PBC charging free energies, once the relevant correction terms are taken into consideration. For PBC simulations with PME for long-range electrostatics, the contribution ($\Delta G_{\text{P-M}}$) due to the use of a particle rather than molecule-based summation scheme in real space was found to be significant (~10 kcal/mol), as was first pointed out by Kastenholz and Hünenberger.[106,107] For GSBP, when the inner region is close to be charge neutral, the key correction is the over polarization of water molecules at the inner/outer dielectric boundary (for an illustration, see Figure 2.3b); the magnitude of the correction ($\Delta G_{\text{s-pol}}$), however, is relatively minor (~2 kcal/mol with TIP3P).

For charging (oxidation) free energy in proteins, the situation is more complex although good agreement between GSBP and PBC can still be obtained when care is exercised. For example, we observed that the smooth dielectric boundary approximation inherent to GSBP (which is what makes it possible to use a fixed reaction field matrix as in Equation 2.9 in MD simulations) tends to make significant errors when the inner region is featured with a high net charge. However, the error can be corrected with PB calculations using snapshots from GSBP simulations in a straightforward and robust manner.[108] Due to the more complex charge and solvent distributions, the magnitudes of $G_{\text{P-M}}$ and $G_{\text{s-pol}}$ in protein simulations appeared to be different from those derived for solution simulations. Therefore, it remains somewhat challenging to directly compare absolute charging free energies in PBC and GSBP simulations for protein systems; the relative charging/oxidation free energies, however, are more robust and PBC/GSBP results agree much better even for GSBP simulations with a fairly small inner region (Table 2.2).

TABLE 2.2   OXIDATION FREE ENERGY (IN KCAL/MOL) OF A METAL SITE M IN CUER CALCULATED USING A LINEAR RESPONSE MODEL WITH PBC/GSBP AND DIFFERENT PARTS OF PROTEIN FIXED

System	$\Delta G(\lambda = 1 - \lambda = 0)^a$	$\Delta G(\lambda = 2 - \lambda = 1)^a$	$\Delta\Delta G_{12}$
PBC (fully flexible)	−112.7 (0.8[b])	−255.8 (0.7)	−143.1 (1.1)
PBC (+20 Å-fixed[c])	−113.3 (1.3)	−254.7 (1.1)	−141.4 (1.7)
PBC (+8 Å-fixed[d])	−90.1 (0.7)	—	—
PBC (+8 Å-relaxed[e])	−115.4 (0.6)	—	—
GSBP (20 Å)[f,g]	−132.8/−119.0 (0.7)	−271.4/−259.2 (0.7)	−138.6/−140.2 (1.0)
GSBP (8 Å)[f,h]	−128.9 (0.9)	−273.1 (0.8)	−144.2 (1.2)

[a] $\lambda = 2$ is the *doubly oxidized* state of M with a partial charge of +2.1158 e, $\lambda = 1$ is the oxidized state with a partial charge of +1.1158 e, and $\lambda = 0$ is the reduced state with a partial charge of +0.1158 e.

[b] The values in parentheses are statistical errors based on block average based on ~5 ns of simulation.

[c] Fix protein beyond 20 Å of M and all water molecules are mobile.

[d] Fix protein beyond 8 Å of M and all water molecules are mobile.

[e] An equilibrated structure from the oxidized state of +20 Å-fixed simulation is used as the initial structure (see Figure 2.3d).

[f] The dielectric constant of protein in the outer region is set to 1 for all GSBP simulations.

[g] For each entry, the numbers after the slash contain PB correction for the smooth dielectric boundary.

[h] The initial structures are equilibrated structures from PBC simulations with the corresponding charge state of M and a fully flexible protein.

For the specific protein system (CueR) studied, the use of a low-dielectric *constant* for the outer region was found not to introduce any major deviation of the oxidation free energy, especially when the inner region radius is as large as 20 Å; even with an 8 Å radius inner region, using a dielectric constant of 1 for the outer region led to a relative oxidation free energy that agreed fairly well with PBC and GSBP (20 Å) simulations (Table 2.2). These results are likely due to the fact that the site of interest is fairly solvent accessible (Figure 2.3a) and therefore well screened from the outer region. For a deeply buried site, a more sophisticated treatment for the dielectrics for the outer region[109,110] might be needed; more sophisticated enhanced sampling will also benefit the analysis of the fully flexible system.

The effect of freezing the protein structure in the outer region of a GSBP setup was found to be small, unless a very small (8 Å) inner region was used; even in the latter case, the result became substantially improved when the nearby metal binding loop was allowed to respond to metal oxidation by adopting the structure from a PBC simulation as the starting structure of GSBP simulations (Table 2.2); this impacts the level of solvation of the metal site (Figure 2.3c). In other words, while it is important to leave flexibility in the model so that rearrangement of nearby structural motifs in response to metal oxidation occurs without artificial hindrance, it is less important to sample the thermal fluctuation of atoms beyond 8 Å. These observations are informative to ab initio QM/MM calculations. On one hand, they highlight the importance of carrying out adequate sampling rather than limited structural minimizations. On the other hand, they support a practical approach in which one equilibrates the system using MM or SCC-DFTB/MM with either PBC or large-inner-region GSBP simulations, then carries out ab initio QM/MM calculations using GSBP or SMBP[111,112] but with a smaller inner region. SMBP is an approach similar in spirit to GSBP but designed for geometry/path optimizations; it effectively solves for the Poisson(–Boltzmann) equation for each geometry optimization step and is useful when the number of geometry optimization step is small.[111] SMBP is also useful when localized vibrations are computed for interpreting infrared spectra of biomolecular systems.[112,113]

2.3 ILLUSTRATIVE EXAMPLES: VALUE AND LIMITATIONS OF COMPUTATIONAL MODELS

In the following, we briefly discuss several recent studies of metalloenzymes by our group. They are selected to illustrate both the value and limitations of various models and methodologies developed in the group. For each example, we also provide brief discussion of the general background to put the computational studies into context.

2.3.1 MM Model: Substrate Selectivity of AlkB Enzymes

2.3.1.1 General Background

Aberrant methylation of cellular DNA can result in cytotoxic or mutagenic effects and therefore pose a serious threat to organisms. To repair such damages, several strategies exist within cells, one of the simplest being direct DNA repair. One recently discovered set of repair proteins is the AlkB family enzymes,[114–117] which are a-ketoglutarate (2-KG) and Fe(II)-dependent dioxygenase type of enzymes.[118,119] These mononuclear nonheme iron enzymes employ a unique repair mechanism by oxidative demethylation. Nine human homologues of AlkB have been identified initially through sequence alignment,[120] and they are implicated in various serious human diseases. For example, ABH3[121] has been identified as a prostate cancer marker that is overexpressed and involved in apoptosis of prostate cancer cells. ABH8[122] is thought to be involved in human bladder cancer. Several genome-wide association studies have independently identified FTO as an obesity susceptibility gene that affects body mass index.[123–129] Recent studies in the He lab pointed to N6-methyladenosine (6meA) in nuclear RNA as a major substrate for FTO,[130] highlighting the potential significance of RNA epigenetic regulation.[131]

Members of the AlkB family[120,132] have distinct repair efficiencies for different alkylated bases. *Escherichia coli* AlkB,[114] human ABH3,[133] and FTO[129] prefer ssDNA/RNA (ssNA) as substrate, while ABH2[134] is markedly more active on dsDNA than on ssDNA. Regarding the repair specificity, AlkB is most efficient at repairing 1-methyladenine (1meA) and 3-methylcytosine (3meC)[114–116] and has notable efficiency toward the repair of 3-methylthymine (3meT),[135] 1-methylguanine (1meG),[136,137] and 1,N6-ethenoadenine (εA).[138] By contrast, FTO is most active toward 3meT and much less active toward 1meA, 3meC, and 1meG.[129] Although structural studies[139–141] have provided key clues regarding the mechanistic origin of these selectivities, the DNA structure is often highly distorted in the crystal structures, raising questions regarding the impact of crystal packing and robustness of mechanistic hypothesis based on structural analysis alone. Whether the chemical steps contribute to the selectivity of the repair specificity also remains to be clarified.[142] Along this line, by using a distal cross-linking protocol, the He lab has, for the first time, trapped oxidation demethylation repair intermediates inside the crystals of AlkB–dsDNA complexes, which opened up the exciting possibility of scrutinizing the repair mechanism with atomic level of details.[143]

Our goal is to use computational studies to complement experimental analysis to tackle some of these mechanistic questions. One major challenge is that the active site of AlkB and related enzymes contains a mononuclear Fe(II). Therefore, an effective model needs to be developed for this cofactor to allow efficient computational studies. For an analysis of the catalytic cycle, the active site needs to be treated at a QM level, as done in several recent studies.[143–145] For a proper equilibration of the structure, and for an analysis of protein–DNA binding using free energy simulations, a pure MM model would be preferable due to the higher computational efficiency. Accordingly, in our recent work,[46] we have tested a simple point charge model for Fe(II) in the AlkB enzymes; we have selected the nonbonded MM model due to its simplicity and flexibility in coordination modes. The model was first tested based on QM calculations for active site models, focusing mainly on the structural features; particular attention is paid to the possibility that carboxylate may adopt multiple binding modes to the iron, a situation discussed rather extensively for Zn^{2+}, Mg^{2+}, and Ca^{2+}.[40–44] Then, the model was used in ~150 ns MD simulations for AlkB–dsDNA and ABH2–dsDNA systems based on their crystal structures[140]; the two systems were selected because AlkB and ABH2 prefer ss- and ds-DNA as their natural substrates, respectively. We are interested in whether the simple point charge model is able to properly describe the active site features and whether classical MD simulations are able to provide additional insights regarding the substrate selectivity of the two AlkB enzymes.

2.3.1.2   Structural Features of AlkB–dsDNA Complexes

As shown in Figure 2.4, the simple MM model is able to preserve the structural integrity of Fe(II) sites in AlkB and ABH2 on the ~100 ns scale relative to the crystal structures; as discussed earlier in Section 2.2.1, this is possible only when a harmonic restraint was added to maintain the monodentate coordination mode of Asp (Figure 2.4a and b) to the iron ion. The striking difference between AlkB–dsDNA and ABH2–dsDNA simulations is that the dsDNA structure and protein–DNA contacts remain close to those observed in the crystal structure in the ABH2 system, while much more significant structural relaxations in the dsDNA and protein–DNA interface are apparent for the AlkB system. For example, the RMSD for the dsDNA is below 3 Å in the ABH2 simulations (Figure 2.4d), while it quickly increases to beyond 4 Å in the AlkB simulations (Figure 2.4c). Moreover, as illustrated in Figure 2.4e and f, the AlkB–dsDNA complex becomes substantially less compact after ~100 ns of simulations while the integrity of the complex remains largely unaffected for the ABH2–dsDNA complex. These features are qualitatively consistent with the fact that dsDNA is the preferred substrate for ABH2, which forms extensive interactions with both strands of the DNA. For AlkB, which prefers ss-DNA as the substrate, very few interactions exist between the protein and chain B of the dsDNA. Even for the AlkB-chain A interface, many of the interactions observed in the crystal structure are not stable,[46] again highlighting strains in the crystal structure due to crystal packing effects.

One way to identify regions under strain in the DNA is to analyze its backbone structural features, such as the distribution of α/χ backbone torsional angles. Previous studies indicate that except for Z-DNA, the α/γ torsional angles in free DNA exclusively sampled canonical g^-/g^+ conformations in both crystal structure and MD simulations, while they often exhibit noncanonical values in protein–DNA complexes; the DNA sequence does not significantly influence the distribution.[146] As a representative example, the GpC base step was studied for which the potential energy surface as a function of the α and γ torsional angles was mapped.[147] In addition to the canonical state, these analyses suggested five other stable or metastable noncanonical substates; the lowest barrier between the canonical state and these five substates is ~7 kcal/mol. This potential map also suggested some *forbidden* conformations that feature a strain energy of more than 25 kcal/mol.

Using tools available in the 3DNA package,[148] we mapped the crystal structures of AlkB–dsDNA (PDBID:3O1M) and ABH2–dsDNA (PDBID:3RZJ) as well as the MD structures onto the α–γ map established in previous studies (Figure 2.5a); regions with the strain energy of more than 25 kcal/mol are

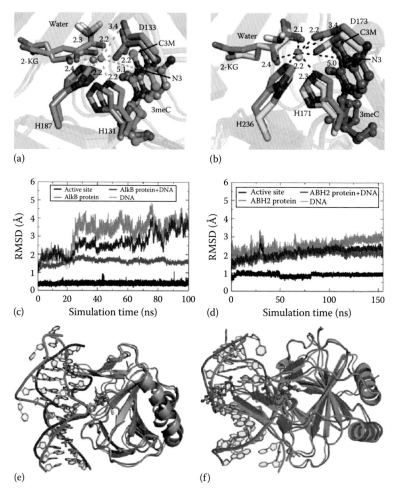

Figure 2.4   (See color insert.) Results for AlkB–dsDNA (a,c,e) and ABH2–dsDNA (b,d,f) simulations with a simple MM model for the iron site. (a, b) Overlay of the active site between crystal structure and the snapshot at the end of the ~100 ns MD simulation. For AlkB, the crystal structure is colored in cyan, and the MD snapshot is colored in pink. Residues in the active site are shown in sticks, lesion bases (3meC) in sticks and spheres, and the distances in the crystal structure (in Å) are shown in black. For ABH2, the crystal structure is colored in green while the lesion base colored in marine, and the snapshot from MD simulation is colored in yellow with the lesion base colored in magenta. (c, d) RMSD of the active site, protein, DNA, and the entire protein–DNA complex between the MD simulation and crystal structure. All nonhydrogen atoms are included in the calculation. (e, f) Superposition of the AlkB–dsDNA/ABH2–dsDNA crystal structure and the snapshot after 100 ns of MD simulation to illustrate conformational changes in the dsDNA and protein–DNA interactions; much larger changes are observed for AlkB.

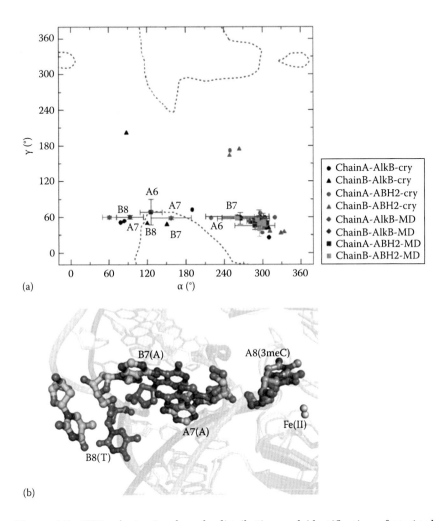

Figure 2.5 DNA α/γ torsional angle distribution and identification of strained regions in DNA in the AlkB–dsDNA/ABH2–dsDNA complexes. (a) α/γ torsional angle distributions from crystal structures and MD simulations; the dashed lines encompass the high-energy backbone regions identified in previous potential energy scan of a DNA dodecamer.[147] (b) The conformation of strained bases, A7(Ade), B7(Ade), and B8(Thy), in the AlkB–dsDNA complex according to the analysis of α/γ torsional angle distributions.

circled by dashed lines. Clearly, although two bases in chain B of the AlkB–dsDNA complex are located in the high-energy region in the crystal structure, relaxation of the DNA structure leads these bases into lower-energy regions during the MD simulation. However, A7(Ade), the 5′ direction neighbor of the lesion, remains in the high-energy conformation (Figure 2.5b), which explains explicitly why dsDNA adjusts its conformation and binds weakly to AlkB. For the ABH2–dsDNA complex, only A6(3meC) in the crystal structure is close to the high-energy area, and it remains near the high-energy region during the MD simulation; all other bases are in low-energy regions in both the crystal structure and during the MD simulation. These features are consistent with the selectivity of AlkB and ABH2 toward different DNA strand types and help pinpoint regions under strain.

2.3.1.3   Energetic Features

Regarding the energetic features of AlkB enzymes, our analysis suggests that the simple point charge MM model for the iron site provides a semiquantitative description. For example, a preliminary application of the model[149] on FTO found that it was able to distinguish different binding affinities of 3meT and 3meC to FTO; while 3meT remained stable in the active site of FTO for more than 10 ns, 3meC quickly dissociated from the active site at 3–6 ns scale. In another analysis,[46] we compared the interaction energy between a model base and the iron site of AlkB calculated with QM and MM models. The structures were collected from the MD simulations of AlkB–dsDNA simulations with a lesion (3meC) flipped into the active site; 2600 snapshots were taken and then clustered into 40 groups based on the geometry of the iron site. Representative structures in these groups were reduced to contain only the iron, its immediate ligands, and the base; the interaction between the iron site and the base is calculated using either DFT (B3LYP and M06) or an MM model. The 3meC was also modified to a normal cytosine (Cyt), allowing us to compare the relevant interactions of 3meC/Cyt to the iron site with different models. The absolute interaction values for Cyt and 3meC differ between B3LYP and M06 calculations, although the computed relative affinities are close, both being ~23 kcal/mol. With the MM model, compared to the QM results, the interaction with Cyt is fairly close, while the interaction with 3meC is underestimated. As a result, the relative binding of Cyt/3meC to the iron site is ~16 kcal/mol, which is lower than the QM result by ~7 kcal/mol; this is significant compared to the magnitude of fluctuation in the relative interaction energy (~4 kcal/mol) among different model structures.

In short, we found that a relatively simple modeling strategy in which the iron is represented by a simple +2 point charge with the Lennard-Jones parameters provides a satisfactory description of structural properties of the Fe(II) site in AlkB enzymes; an essential aspect of the modeling strategy is to apply a restraining potential to control the binding mode of carboxylate (Asp) to the iron. The MD simulations highlight that crystal packing may have a significant impact on the structure of protein–DNA complexes. Therefore, for systems in which the protein/DNA may undergo significant changes from their conformations in isolation, combining MD simulations and structural studies is a valuable approach for better understanding protein–DNA interactions and their functional implications. For energetic properties, however, the simple point charge model is only semiquantitative in nature, emphasizing the importance of developing more sophisticated MM[24,27,29] and possibly semiempirical QM[60,73,150,151] models for iron sites in quantitative applications.

2.3.2   SCC-DFTB/MM for Metalloenzymes: Catalytic Promiscuity of Enzymes in the AP Superfamily

2.3.2.1   General Background

Although a high level of catalytic specificity has been regarded as an important hallmark of enzymes, it is increasingly recognized that many enzymes have promiscuous catalytic activities.[152–156] The discovery of catalytic promiscuity challenged the traditional notion that efficient enzymes are also highly specific. It has also been proposed that catalytic promiscuity plays an important role in enzyme evolution since it can give an enzyme an evolutionary *head start*, providing a modest rate enhancement that is (nearly) sufficient to provide a selective advantage.[157,158] For applications, catalytic promiscuity can be taken advantage of in enzyme engineering for the development of new catalytic reactions.[159–164]

The AP superfamily represents striking examples of catalytic promiscuity.[165,166] AP catalyzes the hydrolytic reaction of phosphate monoesters for its physiological function but also exhibits promiscuous activity for the hydrolysis of phosphate diesters and sulfate esters. Similarly, although the main function of NPP is to hydrolyze phosphate diesters, it can also cleave phosphate monoesters and sulfate esters with considerable acceleration over solution cases. Recent studies of two other members of the AP superfamily, AS and PMH, have revealed that their rates of native and promiscuous activities cover a relatively narrow range,[167,168] especially concerning the sulfatase activity.

The observed promiscuity is remarkable considering the differences among the substrates: their charges range between 0 and –2, the reactions involve transition states of a different nature (dissociative vs. associative), attack at two different reaction centers (P and S), and diverse intrinsic reactivities (with half-lives between 20 and 85,000 years under near-neutral conditions[156]). For a mechanistic understanding, Herschlag and coworkers have characterized the nature of the TS in AP enzymes using linear free energy relation (LFER) and kinetic isotope (KIE) analyses[169–173]; they suggested that AP is able to stabilize

TSs of different nature for different substrates (a synchronous TS for phosphate diesters and loose TS for phosphate monoesters), although factors that govern the stabilization of multiple TSs are not clearly understood.[174] By contrast, recent QM/MM calculations found very loose TSs for both types of substrates in AP and the closely related NPP,[175–178] citing potential uncertainty in the interpretation of LFER/KIE data.[179,180] A notable feature emerged from those calculations is that the bimetallic zinc site undergoes significant structural changes during the reaction, helping to stabilize the expanded structure of a loose TS. Although the scale of the changes observed in those studies is likely overestimated, computational analyses for other metalloenzymes[181,182] collectively highlight possible structural plasticity in multimetallic sites[183–187] that is not visible from *ground state structural data*. Considering the intense interest in the connection between enzyme motion and catalysis,[188–190] this is an important but underexplored subject.

Motivated by these considerations, we have initiated systematic efforts in understanding the catalytic mechanism and nature of phosphoryl transition states in the AP enzymes. From a technical perspective, the AP enzymes are ideal systems that benefit from recent developments[61,89] of SCC-DFTB/MM because (1) the active sites of AP enzymes are readily accessible to solvents (an important reason that they are able to bind rather different substrates and have a high degree of catalytic promiscuity), thus a meaningful study requires extensive sampling; (2) the bimetallic zinc site and phosphoryl transfer chemistry are difficult to treat with other semiempirical methods; and (3) the substrates, especially monoesters, are highly charged thus a reliable description of QM/MM electrostatic interactions is essential. As discussed earlier (Section 2.2.2.2), reliable results were obtained for monoester reactions with SCC-DFTB/MM simulations only when the QM/MM-KO scheme[89] is used.

2.3.2.2 Diester Substrates

In Ref. 191, we discussed extensively the results of SCC-DFTBPR/MM simulations for phosphate diesters in AP and NPP; reference reactions in solution and enzyme simulations with thio-substituted substrates were also carried out to probe the reliability of the SCC-DFTBPR/MM methodology. We note that the original QM/MM electrostatic Hamiltonian[85] was used in these simulations. As shown in Figure 2.6a, the most striking difference between our results and the AM1(d)/PhoT/MM results[175–178] is that the phosphoryl transfer transition state is synchronous in both AP and NPP with SCC-DFTBPR/MM, while they are much looser in nature in the AM1(d)/PhoT/MM simulations. This is best illustrated by the value of the tightness coordinate (TC) in the transition state, which is the sum of distances between P and oxygen atoms in the nucleophile (O^{nu}) and leaving group (O^{lg}), that is, $PO^{nu} + PO^{lg}$. In the SCC-DFTBPR/MM simulations, the TC value is about 3.9 Å for the reaction involving MpNpp$^-$ in both

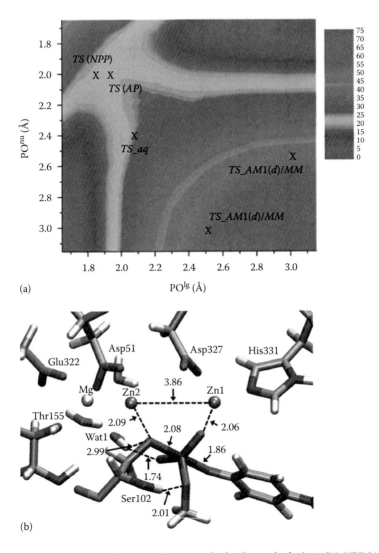

Figure 2.6   SCC-DFTBPR/MM simulation results for diester hydrolysis (MpNPP⁻) in the AP enzymes. (a) Two-dimensional PMF for the phosphoryl transfer reaction; the crosses indicate the location of transition state from the SCC-DFTBPR/MM simulations[191] and AM1(d)/PhoT/MM studies[175–178]; note that the nature of the transition states is much looser from the AM1(d)/PhoT/MM studies, due likely to the much expanded Zn^{2+}–Zn^{2+} observed in those studies (see text). (b) The average structures for the transition state region in PMF simulations for R166S AP; distances are in Å. Note the binding mode of the leaving group.

(continued)

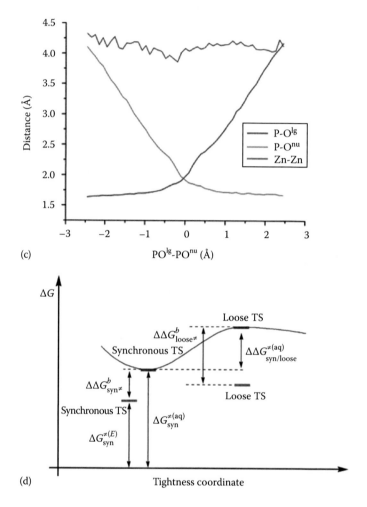

Figure 2.6 (continued)   SCC-DFTBPR/MM simulation results for diester hydrolysis (MpNPP$^-$) in the AP enzymes. (c) Key geometrical properties during the PMF simulations. (d) A scheme that illustrates how relative energetics of synchronous and loose transition states in the enzyme (in red) compare to those in solution (in blue). $\Delta G_{sun}^{\neq(aq/E)}s$ gives the free energy barrier (relative to infinitely separated substrate and nucleophile) in solution/enzyme; $\Delta\Delta G_{syn/loose^{\neq}}^{b}$ gives the binding free energy of a syn/loose TS structure to the enzyme; $\Delta\Delta G_{syn/loose}^{\neq(aq)}$ is the free energy difference between the synchronous and loose transition structures in solution. For the enzyme to shift the nature of TS from synchronous to loose, $\Delta\Delta G_{loose^{\neq}}^{b}$ needs to be larger than $\Delta\Delta G_{syn^{\neq}}^{b} + \Delta\Delta G_{syn/loose}^{\neq(aq)}$, which we argued is unlikely for AP and diesters (see text for discussions).

AP (Figure 2.6b) and NPP; by contrast, the value is 5.66 (NPP)/5.00 (AP) in the AM1(d)/PhoT/MM simulations. Regarding the possible origin of the difference, we note that the AM1(d)/PhoT/MM simulations found rather large variations in the Zn^{2+}–Zn^{2+} distance relative to the crystal structure; the value reached about 6–7 Å as compared to the value of ~4 Å in both crystal structures and from EXAFS analyses.[192] In the SCC-DFTBPR/MM simulations, the Zn^{2+}–Zn^{2+} distance did exhibit thermal fluctuations and systematic difference between the ground state and transition state (Figure 2.6c), but the deviation from the crystal structure is much smaller (see Figure 2.6b). To explicitly illustrate the impact of the Zn^{2+}–Zn^{2+} distance on the nature of phosphoryl transfer transition state, we also carried out simulations in which the Zn^{2+}–Zn^{2+} distance was restrained to specific values ranging from 3.6 to 4.6 Å. The results clearly indicated that the nature of the transition state becomes increasingly looser as the Zn^{2+}–Zn^{2+} distance elongates. These studies were not meant to dwell on criticism of previous work[175–178]; rather, we emphasize the importance of carefully benchmarking the reliability of QM/MM models based on multiple types of data, especially when the interpretation of results appears to contrast those from experimental studies.

Since the cognate substrates of AP and NPP are monoesters and diesters, respectively, it is worthwhile considering whether the observation that features a synchronous phosphoryl transfer transition state for diesters, similar to the uncatalyzed reaction in solution, is consistent with available experimental data. We do so with the scheme outlined in Figure 2.6d, which is qualitatively similar to that used by Herschlag and coworkers.[172,173,193] For the enzyme to shift the nature of TS from synchronous to loose, the driving force needs to be large enough to overcome the binding energy for the synchronous TS ($\Delta\Delta G^{b}_{syn^{\neq}}$) plus the energy gap between these two kinds of structures in solution ($\Delta\Delta G^{\neq(aq)}_{syn/loose}$). The latter, although not measurable directly from experiments, can be estimated based on calculations; our calculations[191] gave a value ~8 kcal/mol (the *loose* structure is taken to have a TC of ~5.7 Å as found for the TS in NPP in previous QM/MM calculations[177]). In other words, the enzyme needs to bind to the loose TS by more than 16 kcal/mol ($\Delta\Delta G^{b}_{syn^{\neq}} + \Delta\Delta G^{\neq(aq)}_{syn/loose} = 8 + 8$) to make it more favorable in the enzyme than a synchronous TS; the value of $\Delta\Delta G^{b}_{syn^{\neq}}$ was estimated based on experimental values for aryl phosphate diester hydrolysis in solution and R166S AP.

Considering what we know about the activity of AP toward phosphate monoesters, however, we argued[191] that such a strong binding is unlikely for phosphate diesters. For a phosphate monoester related to the diesters studied, such as pNPP^{2-} (*p*-nitrophenylphosphate), the solution barrier is about 32 kcal/mol[193] and the barrier in R166S AP is 10.6 kcal/mol.[194] Since LFER data indicate that the nature of TS is loose in both AP and solution,[170,194] these results suggest that R166S AP binds a loose TS for monoesters by ~21 kcal/mol. Since diesters feature less charge and are promiscuous substrates of (R166S) AP, the binding energy of a loose TS for diesters to R166S AP is expected to be substantially lower than 21 kcal/mol. Therefore, R166S AP is not able to shift the TS for diesters to be much looser in nature, in agreement with our QM/MM calculations. Similar arguments can be made for NPP.[191] In other words, the binding of AP/NPP to diester substrates is simply not strong enough, presumably due to their lower charge compared to monoesters, to significantly shift the nature of the transition state from synchronous (preferred in solution) to loose.

2.3.2.3   Monoester Substrates

In more recent studies,[195] we have initiated the analysis of monoester reactions in the AP enzymes, taking advantage of the QM/MM-KO scheme[89] developed for reactions that involve highly charged species. As discussed earlier for the study of diesters, we found that the nature of the transition state correlates with the zinc–zinc distance. If such correlation also applies to the hydrolysis of phosphate monoesters, since only limited structural fluctuations in the bimetallic motif were observed in our (unconstrained) QM/MM simulations, one might predict that the transition state of monoester hydrolysis becomes substantially tighter in the AP enzymes than the solution counterpart. This prediction appears to be contradictory to both LFER/KIE analysis[172,173] and AM1(d)/PhoT/MM simulations,[176] which predict a loose transition state for the hydrolysis of monoesters in AP, similar to uncatalyzed reactions in a solution.

We explored the hydrolysis of pNPP^{2-} in R166S AP and wild-type (WT) NPP; pNPP^{2-} is a phosphate monoester whose hydrolysis has been widely studied in a solution[193] and AP/NPP,[172,173] and it is a natural counterpart to the phosphate diester (MpNPP$^-$) that we have studied.[191] With a simple correction scheme for the intrinsic errors of SCC-DFTBPR for phosphate monoesters,[89,191] our QM/MM simulations were in good agreement with available experimental data for both solution and enzyme cases. As shown in Figure 2.7, our calculations suggest that pNPP^{2-} hydrolysis proceeds through a loose transition state in both AP and NPP, despite the relatively rigid bimetallic zinc motif, and that the nature of the transition state in these enzymes is fairly close to that in solution (Figure 2.2). The difference in the TC values between a monoester and a comparable diester (e.g., pNPP^{2-} vs. MpNPP$^-$) is consistently about 0.6–0.7 Å for both R166S AP and NPP (compare Figures 2.6b and 2.7). Therefore, we have explicitly confirmed that a single active site is able to recognize and stabilize transition states of different nature. Comparisons of the transition state structures suggest fairly similar configurations for the bimetallic zinc motif for the hydrolysis of different substrates; the Zn–Zn distance is ~3.9 Å for diester TS[191] and slightly expanded to ~4.1 Å for monoester TS, while the magnitude of the thermal fluctuation is about 0.2 Å in both cases. Therefore, the *plasticity*

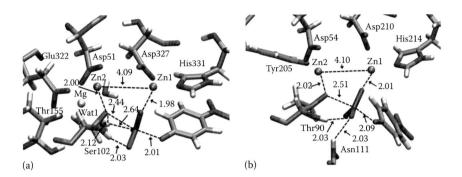

(a)   (b)

Figure 2.7 SCC-DFTBPR/MM simulation results for monoester (pNPP[2-]) hydrolysis in the AP enzymes; the QM/MM interactions use the KO scheme.[89] The average structures for the transition state region in PMF simulations are shown for (a) R166S AP and (b) WT NPP. Distances are in Å. Note the binding mode of the leaving group (see text for discussions). See Figure 2.2 for a comparison of the transition state in solution.

of the bimetallic zinc motif is only one of the reasons that the AP enzymes can recognize multiple types of transition state. Our calculations suggested that the significant degree of solvent accessibility of the active site might be another reason for the AP enzymes to accommodate transition states of different nature: with the leaving group interacting with the solvent molecules but still being fairly close to the zinc ions (Figure 2.7), a loose transition state is stabilized without the need of causing large structural distortions of the bimetallic site.

The latter observation leads to an interesting point regarding the use of vanadate as phosphoryl transition state analog. For reactions in the AP enzymes, crystal structures with a bound vanadate were used to suggest that the leaving group oxygen directly interacts with one zinc ion in the transition state.[192] In our diester studies,[191] this direct interaction was not observed and we speculated that the reason is mainly due to the different charges: a diester only bears −1 charge, while vanadate has a high charge of −3. In the monoester study, the phosphate monoester pNPP[2-] bears a −2 charge and therefore more similar to vanadate. However, the TC in the transition state was more than 4.5 Å, which is substantially longer than the corresponding value for vanadate (~3.6 Å) and the zinc–zinc distance in AP/NPP. Therefore, from a structural perspective, vanadate is also not an ideal analog, which might explain that the direct interaction between the leaving group and zinc ion is not observed in the pNPP[2-] hydrolysis transition state. Although more extensive analysis of zinc–ligand interactions at the SCC-DFTB(PR)/MM level should be carried out, our results do highlight the potential caveats of using vanadate to infer the binding mode of phosphoryl transfer transition state in enzymes, especially when the active site is flexible and/or solvent accessible.

In short, our SCC-DFTB/MM analyses[191,195] of diester and monoester hydrolysis in the AP enzymes generally support the hypothesis[172,173] that AP enzymes are able to recognize and stabilize multiple types of transition state in a single active site. In addition to interesting implications regarding the design/evolution strategies of natural enzymes, the results also provide indirect support for the use of LFER to infer the structural features of transition states in enzymes,[193] although a more conclusive statement is possible only after explicit comparison between computations and experimental observables, which include LFER, KIE, and infrared spectra.

2.3.3   DFTB3/MM for Open-Shell Systems: A Preliminary Study of Redox Potential in Blue Copper Proteins

As discussed earlier, we have recently initiated the development of DFTB3 for open-shell metal ions, such as Cu^{2+}. To illustrate the status of the development, we briefly discuss our preliminary application of the methodology to redox potentials of blue copper proteins.

Blue copper proteins (or cupredoxins) are fundamental electron transfer (ET) units in biology and therefore are featured with a rich experimental background.[5,196,197] Their copper site is typically coordinated with a cysteine thiolate, two histidines, and a methionine as the axial ligand. The reduction potential of natural cupredoxins spans >800 mV[198] and it is well appreciated that in addition to the covalent characters of first-shell ligands, hydrogen-bonding and other noncovalent interactions with second shell groups may contribute substantially to the tuning of redox potential.[15,199] For example, Lu and coworkers were able to modulate the reduction potential of azurin by 700 mV range using only a handful of inner- and outer-sphere mutations.[200]

Despite these advances, contributions from nearby solvent and other long-range contributions are less well understood. For example, *Trametes versicolor* laccase and Fet3p have reduction potentials of ~790 and ~430 mV, respectively, while crystal structures observed little variation in either the first or second coordination spheres.[201] Therefore, despite the general backbone rigidity of blue copper proteins,[202,203] protein and solvent dynamics[204,205] need to be considered for a thorough understanding of redox potential tuning,[5,197] especially regarding enthalpy–entropy compensation.[206] Cupredoxins have been analyzed with diverse mutations in both the first and second coordination shells as well as other nearby residues.[5,197,199,200] The impact of these mutations on the redox potential has been well documented, yet their interpretation has not always been straightforward; most of the discussions were based on a static structure,[5,199,207] while recent analysis of protein pK_a by Garcia-Moreno's group[104,208,209] highlighted the importance of diverse protein/solvent structural response to the titration of even *not* deeply buried sites.[210] It is likely that protein fluctuations and solvent response are also essential to the proper interpretation of mutation data on cupredoxins, as echoed by thermodynamic analysis[206,211] and recent simulation studies of other redox proteins.[212]

There is a notable body of computational work on cupredoxins,[199,213–219] but most employed limited sampling and, in particular, the issue of pH-redox coupling has not been deeply explored,[220,221] providing a nice opportunity for DFTB3-based QM/MM studies. From a biomedical perspective, the analysis of cupredoxins will add new insights into the understanding of regulatory mechanism of biological ET processes, which enhances our understanding of metabolism and respiration at the molecular level. New mechanistic information has medical relevance since defective (proton coupled) ET leads to production of the reactive oxygen species and free radicals that are associated with aging and many disease states, including certain mitochondrial myopathies[222] and degenerative diseases such as Alzheimer's disease (AD), Huntington's disease, muscular dystrophies, and the Leigh syndrome.[223–225]

As preliminary results, we present data on WT (poplar) plastocyanin (PCY) and (*Thiobacillus ferrooxidans*) rusticyanin (RCY) (Figure 2.8 and Table 2.3). As shown in Figure 2.8a, the average structure of the active site from a nanosecond simulation compares well with the crystal structure (PDB ID 5PCY, 1.80 Å resolution, Cu(I)), including for the Cu(II) state, providing further support to our preliminary parameterization; the Cys ligand has larger fluctuations in the Cu(I) state, consistent with weaker interaction than in the Cu(II) state. Figure 2.8b and c illustrate that the level of solvation of both the copper site and His87 in PCY changes notably upon copper reduction, in qualitative agreement with CPMD studies[216]; our simulation, however, is two orders of magnitude longer (~1 ns vs. ~10 ps) and therefore statistically much more robust.

Regarding the reduction potential[226] (Table 2.3), DFTB3/MM predicts the proper trends for PCY and RCY at different pH values. The magnitude of the corrections related to the smooth inner–outer boundary in the GSBP framework is correlated to the net charge of the inner region[105] (–8.75 for PCY, 1.37 for RCY at pH 7, and 7.69 for RCY at pH 2–3.2 in the Cu(I) state), and inclusion of these corrections brings the calculated E_{red} of PCY closer to the experimental value. Including B3LYP/6-31+G(d,p) single-point calculations via a second-order cumulant expansion as in our previous studies[103] further improves the quantitative agreement with experiments, supporting the applicability of the multilevel QM/MM framework; we will explore other high-level QM methods in the future. For PCY, in independent DFTB3/MM simulations in which the QM region geometry is constrained to be that from a QM/MM minimization of the crystal structure, a change in the reduction potential of almost 300 mV(!) is observed, highlighting again the importance of sampling. On the other hand, if the geometry of the QM region is fixed at the average structure from simulations, the effect on the reduction potential is quite modest. This suggests that

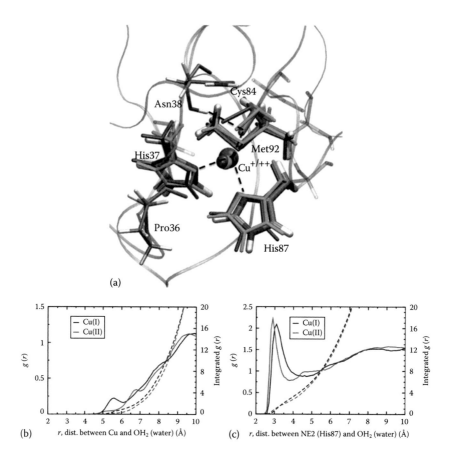

Figure 2.8 Preliminary calculations of reduction potential in the blue copper–protein PCY using DFTB3/MM-GSBP simulations. (a) Overlay of MD snapshots (purple: Cu(II); brown: Cu(I)) and crystal structure (PDB code: 5PCY, 1.80 Å; Cu(I)) for the active site. (b, c) Level of solvation of copper and H87(N ε2) from 1 ns MD simulations; both show clear dependence on the redox state of the copper ion.

snapshots from trajectories where the QM region is *frozen* at the average structure are good candidates for higher-level single-point calculations, possibly leading to better convergence of higher-level QM/MM corrections.

The reorganization energy calculated for PCY is also in good agreement with experiments (Table 2.3). It has been argued that including electronic polarization is essential to reliable reorganization energy,[16,215,227] thus the good agreement found here using a nonpolarizable force field (CHARMM22) might be fortuitous. We will examine this issue by using polarizable force fields[228–230] available in CHARMM.

TABLE 2.3 REDUCTION POTENTIAL AND REORGANIZATION ENERGY OF PLASTOCYANIN AND RUSTICYANIN CALCULATED WITH A PRELIMINARY PARAMETERIZATION OF DFTB3 FOR COPPER

		$\langle \Delta U \rangle_{Cu^{2+a}}$ (kcal/mol)	$\langle \Delta U \rangle_{Cu^{1+a}}$ (kcal/mol)	ΔG_{red} (LRA)[b] (kcal/mol)	E_{red} w.r.t. SHE[c] (mV)	λ^d (eV)
PCY (pH 7)	Traj. 1[e]	−68.7 (1.2)	−123.2 (1.0)	−95.1	−316/−30[f]/ 157[g]	1.21
	Traj. 2[e]	−65.5 (1.6)	−122.8 (1.5)			
	QM frozen (avg. str.)[h]	−73.1 (0.8)	−120.3 (1.4)	−96.7	−247	1.02
	QM frozen (mini. str.)[i]	−65.7 (0.8)	−112.1 (1.7)	−88.9	−585	1.01
RCY (pH 7)	Traj. 1[e]	−71.1 (1.4)	−133.9 (2.4)	−101.8	−28/−76[f]/ 293[g]	1.33
	Traj. 2[e]	−70.9 (1.8)	−131.0 (1.9)			
	QM frozen (avg. str.)[h]	−72.4 (1.5)	−129.8 (1.9)	−101.1	−56	1.24
RCY (pH 2–3.2)	Traj. 1[e]	−78.7 (1.5)	−144.4 (1.2)	−113.6	486/274[f]/ 555[g]	1.37
	Traj. 2[e]	−85.2 (1.8)	−146.1 (1.2)			

[a] The DFTB3/MM calculations are done with a GSBP setup with an inner region of 22 Å; the DFTB3 region contains the copper ion and its immediate ligands. The energy gap is defined as $\Delta U = \Delta U_{Cu^{1+}} + \Delta U_{Cu^{2+}}$. The subscript for $\langle \cdots \rangle$ indicates the copper oxidation state for the MD simulation. Numbers within parentheses are standard deviations.

[b] $\Delta G_{red} = \frac{1}{2}(\langle \Delta U \rangle_{Cu^{2+}} + \langle \Delta U \rangle_{Cu^{1+}})$.

[c] $E_{red} = -\Delta G_{red}/nF$, where F is Faraday's constant and $n = 1$. The reduction potential of the SHE is 4.44 ± 0.02 V at 25°C. The experimentally measured reduction potential values of PCY (pH 7),[218] RCY (pH 7),[296] and RCY (pH 2–3.2)[297,298] are 375, ~590, and 680 mV, respectively.

[d] $\lambda = \frac{1}{2}(\langle \Delta U \rangle_{Cu^{2+}} - \langle \Delta U \rangle_{Cu^{1+}})$. The experimentally measured λ of PCY is 1.296 eV.[299]

[e] Includes smooth boundary corrections related to GSBP,[105] estimated using 100 snapshots separated by 5 ps taken from the first 500 ps of Traj. 1.

[f] Includes B3LYP/6-31+G(d,p) single-point corrections, estimated using 100 snapshots separated by 5 ps taken from the first 500 ps of Traj. 1.

[g] 1 ns long.

[h] 500 ps long; starting snapshot is the snapshot from the first 500 ps of Traj. 1, which has the lowest RMSD w.r.t. QM heavy atoms compared to the average structure.

[i] 500 ps long; starting snapshot is the DFTB3/MM-minimized crystal structure for the respective Cu oxidation state.

Collectively, the quantitative comparison between calculations and experiments for cupredoxins will serve as a stringent benchmark to firmly establish the applicability of the DFTB3/MM-based framework to copper–protein interactions, setting the stage for other exciting applications in the near future.

2.3.4 Binding Selectivity of Metal Ions to Proteins: Initial Insights from Ab Initio QM/MM-PB Studies

2.3.4.1 General Background

Since the catalytic reactivities of different metals are often distinct,[1] it is important that the correct metal ion(s) is incorporated in the catalytic site; similarly, metal-sensing transcription factors are evolved to be activated by specific metal ions. Clearly, revealing the underlying mechanism for metal binding selectivity is a crucial step toward understanding how the function of metalloproteins is regulated in cells[231–233] as well as the design of novel proteins that recognize specific metal ions.[234,235] The most significant biomedical relevance of studying metal binding selectivity is to better understand the structure–function relations in proteins involved in *metal homeostasis*. With many metabolic processes requiring metals, dyshomeostasis is expected to feature widely in diseases.[236] For example, aberrant copper transport is the cause of the Menkes disease and Wilsons disease,[237] aberrant zinc transport is a cause of acrodermatitis enteropathica,[238] and aberrant iron transport is a cause of haemochromatosis.[239] Progressive accumulations of competitive and/or redox-active metals in the wrong locations have been proposed to underlie the links between metals and multiple neurological disorders.[240,241] Therefore, a more complete knowledge of metal homeostasis and, by implication, metal sensing is likely to precede an understanding of its aberrations. A better understanding of how these proteins in human discriminate between metals has the potential to aid the development of therapies, while the parallel studies of these proteins in bacteria will lead to novel strategies to battle infectious pathogens such as *Salmonella*.[242]

Indeed, understanding the mechanistic basis for metal binding selectivity in metalloproteins has emerged as one of the major frontiers in bioinorganic chemistry.[6,231–233,243] In vivo, metal-ion selectivity is believed to be the result of several factors that include structural properties of the active site (affinity), global conformations (allostery), as well as the metal availability in the cellular pool (access). Among those, the local structural properties are most clearly defined in physical terms and have been analyzed in experimental and computational studies. Some of the proteins have been characterized structurally using x-ray crystallography[244] or NMR,[245,246] which provided hints regarding the mechanism for metal selectivity. In most cases, the selectivity is *qualitatively* rationalized in terms of the *preorganized* coordination environment of the metal ion; for example, an unusual linear, two-coordinated geometry was invoked[247] to explain the high selectivity of the CueR in *E. coli* toward Cu[+] over Zn[2+]. However, it was also realized that many metal-sensing/trafficking proteins are structurally very flexible,[248] as required by their function, thus the importance of the coordination geometry might have been overestimated (see more discussions in the succeeding text).

From the computational side, a series of computational studies has been carried out by Dudev and Lim[249,250] to address binding selectivity of metal ions such as Zn[2+] vs. Hg[2+], [251] Mg[2+] vs. Ca[2+], [252] and Al[3+] vs. Ln[3+] [253] as well as Na[+] vs. K[+].[254,255] Useful insights have been obtained from these studies, which highlighted the importance of several ligand properties including charge, polarizability, charge-donating ability, and denticity. However, these studies[251–253] used relatively small active site models and replaced most of the protein environment by an implicit dielectric continuum. Therefore, only limited insights were available regarding contributions from residues beyond the first coordination shell of the metal ion; the role of protein flexibility and/or active site rigidity was also not possible to evaluate with these models. In the context of ion selectivity of ion channels, which has been largely analyzed with classical models,[256,257] studies start to emerge in which one employs a QM description for the ions and their ligands[258] and a classical description of the protein environment.[259] An accurate estimate of the binding free energy and selectivity, however, remains challenging due to the need of ample sampling[260] or approximations in the treatment of the ion environment.[257,261]

Clearly, new computational methods that analyze coordination chemistry of transition metal ions and nonlocal protein contributions on equal footing are sorely needed to complement experimental studies for a deeper understanding. The ability to rationally modulate the metal binding selectivity of proteins will facilitate engineering new metal sensors[262,263] for various chemical biology, biomedical, and biotechnology applications.

2.3.4.2 Ab Initio QM/MM-PB Calculations

As a proof of concept study,[264] we have selected to analyze the binding affinity and selectivity in the CueR protein. It is one of the best-studied metal-sensing transcription factors in terms of both structure and binding affinity/selectivity.[247,265] In vivo assays[247] showed that CueR responds to monovalent coinage metal ions (Cu[+], Ag[+], Au[+]) but not divalent metal ions like Zn[2+] and Hg[2+]. High-resolution x-ray structure of CueR revealed a linear, two-coordinated geometry of Cu[+] (see Figure 2.9a), which was used to explain the low affinity of Zn[2+], an ion that prefers a tetrahedral coordination.[1] On the other hand, the binding site is in a loop region (Figure 2.9a), which can be sufficiently flexible to accommodate an ion (e.g., Zn[2+]) that prefers a tetrahedron type of coordination, possibly involving water molecules as some of the ligands. The electrostatic interaction between

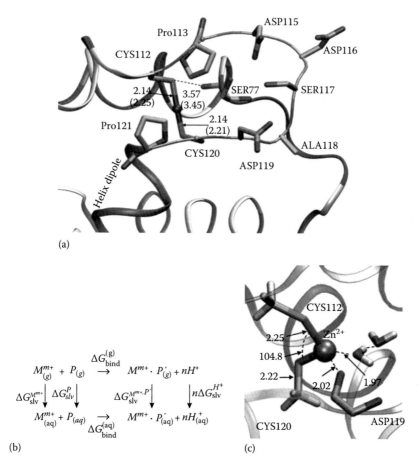

Figure 2.9 Analysis of binding selectivity of CueR toward different metal ions. (a) Active site of CueR based on the crystal structure[247] that contains a bound Cu^+, which is shown as a brown sphere bound to two Cys residues. Nearby amino acids and the helix dipole that has been proposed to stabilize the electrostatics of the binding site are indicated; a few distances from the two binding sites (with and without parentheses) in the crystal structure are also included. (b) Thermodynamic cycle for the binding of a metal ion (M^{m+}) to a protein (P); the notations highlight the fact that metal binding is likely coupled to changes in the protein structure (P′ vs. P) and protonation state of active site residues (release of nH^+). Nevertheless, since we are mainly interested in *relative* binding affinity of different metal ions to CueR, potentially large structural transitions in the active site during the binding process and entropic components of the binding free energy have not been considered in the calculations discussed here. (c) Optimized structure of the active site in which two water molecules are also included in the QM region to allow a tetrahedron coordination sphere around Zn^{2+}; the structure was optimized following 1 ns of MD simulation in which the zinc was described with an MM model. (From Rao, L. et al., *J. Am. Chem. Soc.*, 132, 18092, 2010.)

the conserved Lys81 and the metal binding site in CueR has been invoked[247] to explain the discrimination against Hg^{2+}, which also prefers a linear coordination; however, Lys81 is solvent exposed and therefore its role in electrostatic stabilization might be rather small. Finally, as typical for metal binding processes, the protonation state of active site residues may change upon metal coordination. For CueR, both ligands are Cys residues; while some proposed that both Cys are deprotonated in the metal-bound CueR,[247] others argued[266] that only one Cys is deprotonated. Therefore, CueR is ideally suited for computational analysis because there is both solid experimental background and ambiguities regarding the selectivity mechanism.

The QM/MM protocol we adopted[264] in an initial analysis of metal binding selectivity is based on the thermodynamic cycle outlined in Figure 2.9b, which clearly highlights the fact that metal binding is coupled to structural changes in the protein (P′ vs. P) and protonation state change of active site residues (release of nH[+]). The quantity of interest for each metal ion (M^{m+}) is the binding free energy in solution, $\Delta G_{\text{bind}}^{(\text{aq})}$, which, according to the thermodynamic cycle, can be written as the sum of several contributions:

$$\Delta G_{\text{bind}}^{(\text{aq})} = \Delta G_{\text{bind}}^{(\text{g})} + \Delta G_{\text{slv}}^{M^{m+}P'} + n\Delta G_{\text{slv}}^{H+} - \Delta G_{\text{slv}}^{M^{m+}} - \Delta G_{\text{slv}}^{P}. \tag{2.10}$$

As discussed in detail in Ref. 264, we compute these contributions individually using different methods: $\Delta G_{\text{bind}}^{(\text{g})}$ based on gas-phase QM (DFT) calculations, $\Delta G_{\text{slv}}^{M^{m+}}$ based on QM calculations of metal–water clusters with an implicit solvent model (IEF-PCM[267]), and $\Delta G_{\text{slv}}^{M^{m+}P'}$, $\Delta G_{\text{slv}}^{P}$ based on classical PB calculations[268]; for $\Delta G_{\text{slv}}^{H+}$, we take the experimental value of –272.54 kcal/mol as adjusted for pH 7.[269] As to the value of n (number of proton released during metal binding), we tested several possibilities: $n = 0$ (i.e., both Cys112 and 120 remain protonated after metal binding), $n = 1$ (either Cys112 or Cys120 becomes deprotonated after metal binding), and $n = 2$ (both Cys112 and Cys120 become deprotonated).

The key results are summarized in Table 2.4. Overall, the calculated binding affinities by the QM/MM-based models are consistent with transcription induction experiments in that the coinage metal ions are found to bind to CueR with significant affinity, while the divalent ions have weaker affinities to CueR than to other chelating ligands such as dithiothreitol (DTT) and glutathione (GSH). This supports the use of the computational methodology to analyze factors that dictate binding affinity and selectivity. By contrast, small active site models do not reproduce all trends, especially concerning the binding affinity of Hg^{2+} and how the protonation pattern of the Cys residues in the active site influences the binding affinity of the coinage metal ions.

We found that the following factors contribute to the binding selectivity and affinity of metal ions to CueR. For the coinage metals, the binding affinity to the

TABLE 2.4   CALCULATED BINDING AFFINITY (IN KCAL/MOL) OF METAL IONS TO CUER WITH ACTIVE SITE AND QM/MM MODELS

Metal Ion	Active-Site Model[a]	QM/MM Model[b]
Thiol-Thiolate ($n = 1$)		
Cu^+	−19.3(−20.7)[−21.7]	−15.8/−21.3
Ag^+	−24.0(−25.9)[−27.3]	−19.0/−24.4
Au^+	−35.0(−36.2)[−37.1]	−30.6/−38.5
Zn^{2+}	33.8(23.6)[16.5]	63.9/57.0
Hg^{2+} [c]	−27.7(−40.1)[−48.8]	11.5/4.2
Thiolate-Thiolate ($n = 2$)		
Cu^+	−12.5(−18.8)[−22.8]	−24.4
Ag^+	−15.0(−21.4)[−25.4]	−25.0
Au^+	−29.2(−35.3)[−39.2]	−39.3
Zn^{2+}	−2.4(−5.3)[−7.6]	10.3
Zn^{2+}-W[d]	6.6(3.8)[1.8]	18.8
Hg^{2+} [c]	−63.3(−68.3)[−72.1]	−42.7

[a] The three values are computed with the dielectric constant for the implicit protein environment set to 4(8)[20].
[b] For the case of $n = 1$, the numbers before and after slashes are for deprotonated Cys112 and Cys120 (see Figure 2.9a), respectively.
[c] For comparison (see text), the binding affinity of Hg^{2+} to DTT is −57.8 kcal/mol, to GSH is −62.6 kcal/mol.
[d] One water molecule is included in the binding site of zinc-bound form of the protein following MD simulation, which together with the backbone carbonyl of Asp 119 allows Zn^{2+} to adopt a tetrahedral coordination (see Figure 2.9c).

protein largely follows the intrinsic binding properties, or the softness, of the metal ions and their desolvation penalty, that is, $Au^+ > Ag^+ \sim Cu^+$. Under in vivo condition, the sensitivity of CueR to Cu^+ is likely due to the higher concentration of Cu^+ than Au^+ and Ag^+ (i.e., access[232,233]). Along this line, we note that the absolute binding affinity of hydrated Hg^{2+} to CueR is, in fact, larger than the coinage metals. However, Hg^{2+} binds more strongly to ligands such as DTT and GSH, which is the reason that its binding to CueR in buffer solution or in cells is unfavorable. This highlights the importance of considering the proper reference when discussing the binding affinity of metal ions to proteins.

The calculated binding affinity of Zn^{2+} was unfavorable with both active site and protein models, regardless of whether the coordination environment is linear (with two Cys residues) or tetrahedron (with additional water molecules, see Figure 2.9c). Therefore, it is likely that the weak binding of Zn^{2+} to CueR is due largely to the large desolvation penalty for Zn^{2+} compared to other ions (which is consistent with the large hardness value[270] of Zn^{2+}), while the previously discussed effect of coordination geometry[247] seemed much less important.

Finally, perturbative analysis of QM/MM calculations[264] indicated that the electrostatic environment of the protein is well tuned to favor the binding of coinage metal ions over divalent ions. In particular, charged residues and the helix dipole near the binding site (Figure 2.9a) contributed either favorable or small unfavorable interactions when the binding site with a coinage metal ion is charge neutral (one Cys deprotonated) or bears a negative charge (both Cys deprotonated). With a divalent ion such as Hg^{2+}, however, the interaction was found more unfavorable even when the binding site is charge neutral.

The application of our QM/MM-based approach to CueR encourages similar studies to other systems that exhibit interesting binding properties to transition metals, such as those that bind specifically to divalent ions (e.g., ZntR[247]) or more *exotic* ions such as uranyl.[235] A particularly interesting system is the closely related CupR,[271] which has a highly similar binding site as CueR but very different relative binding affinities toward Cu^+ and Au^+; a related system is GolS, another member of the MerR family transcription factors,[6,242] for which in vivo expression assay also indicated to be about 100-fold more responsive to Au^+ than to Cu^+.[272] From a methodology perspective, it is important to better understand the limitations of the QM/MM-PB-based methodology and therefore identify ways to improve its robustness. For example, the impact of structural fluctuations in the protein was only briefly explored by using different initial structures collected from molecular dynamics simulations[264]; a more rigorous test involves comparison to QM/MM-based free energy simulation,[15] which is more feasible with calibrated DFTB3 as QM.

2.4   CONCLUDING REMARKS: FUTURE CHALLENGES AND OPPORTUNITIES

As the examples discussed in this chapter and other studies[7–13,15–17,19–21] demonstrate, computational studies play a major role in better understanding the structure, dynamics, and therefore function of metalloenzymes. Due to the richness of chemistry of transition metals, quantitative computations are indispensable for interpreting experimental data and for disentangling molecular factors that dictate the reactivity of metalloenzymes. As computational methodologies and hardwares further improve, the impact of computational studies on the field of bioinorganic chemistry/enzymology will only grow.

In terms of future challenges and research opportunities, in addition to the topics discussed here, there are several areas that are particularly of interest from our perspective. First, although dramatic progress has been made in the area of rational enzyme design,[273-275] limited success has been reported for the design of novel enzymatic reactivity that relies on transition metal chemistry.[276] This is in stark contrast to the fraction of natural enzymes that employ transition metal ions for complex chemistry. Developing effective strategies for combining efficient QM(/MM) and protein structure modeling approaches[277] to rationally design metalloenzymes is an exciting research topic that clearly has broad range of potential applications. In the similar vein, rational design of proteins that recognize specific metal cofactors for chemical biology applications[234,235,262,263] or metal separation has emerged as an active topic of research, for which computational studies should be able to provide much guidance; it is particularly exciting if the protein motif can be engineered with switch characteristics.[278]

Another area for which novel computational methods are sorely needed concerns understanding interactions between various metal ions and proteins in the context of neurodegenerative diseases, such as copper binding to α-synuclein, prion, and Aβ proteins.[240,279-284] To properly deal with the likely existence of multiple binding modes/sites and to reveal the impact of metal binding on the structure, stability, and aggregation tendency of the relevant protein(s), existing computational tools[285-289] are not adequate and it is essential to develop new simulation techniques based on efficient QM/MM or sophisticated MM force fields.[25,29] In this context, considering the flurry of recent developments[61,89,290,291] that aim to systematically improve its applicability, some of which briefly discussed here, the DFTB3/MM approach holds great promise as an intermediate level of theory that is able to integrate state-of-the-art QM methods[292-294] and statistical mechanics techniques[87,295] to tackle complex metalloprotein problems.

ACKNOWLEDGMENT

This work is supported in part by NIH grant R01-GM084028 and NSF grant CHE-0957285. J. Zienau is indebted to the *Nationale Akademie der Wissenschaften Leopoldina* for a postdoctoral fellowship (LPDS 2009-39), and X. Pang acknowledges support from the Chinese scholarship council. We are in debt to the discussions with our experimental collaborators, especially Professors C. He and D. Herschlag on the relevant problems. Computational resources from the National Center for Supercomputing Applications at the University of Illinois and the Centre for High Throughput Computing (CHTC) at UW–Madison are greatly appreciated. Computations are also supported in part by the National Science Foundation through a major instrumentation grant (CHE-0840494).

REFERENCES

1. Lippard, S. J.; Berg, J. M. *Principles of Bioinorganic Chemistry*, University Science Books: Mill Valley, CA, 1994.
2. Andreini, C.; Bertini, I.; Cavallaro, G.; Holliday, G. L.; Thornton, J. M. *J. Biol. Inorg. Chem.* 2008, *13*, 1205–1218.
3. Solomon, E.; Brunold, T. C.; Davis, M. I.; Kemesley, J. N.; Lee, S. K.; Lehnert, N.; Neese, F.; Skulan, A. J.; Yang, Y. S.; Zhou, J. *Chem. Rev.* 2000, *100*, 235–349.
4. Neese, F. *Coord. Chem. Rev.* 2009, *253*, 526–563.
5. Warren, J. J.; Lancaster, K. M.; Richards, J. H.; Gray, H. B. *J. Inorg. Biochem.* 2012, *115*, 119–126.
6. Reyes-Caballero, H.; Campanello, G. C.; Giedroc, D. P. *Biophys. Chem.* 2011, *156*, 103–114.
7. Siegbahn, P. E. M.; Himo, F. *J. Biol. Inorg. Chem.* 2009, *14*, 643–651.
8. Noodleman, L.; Lovell, T.; Han, W. G.; Li, J.; Himo, F. *Chem. Rev.* 2004, *104*, 459–508.
9. Senn, H. M.; Thiel, W. *Angew. Chem. Int. Ed.* 2009, *48*, 1198–1229.
10. Shaik, S.; Cohen, S.; Wang, Y.; Chen, H.; Kumar, D.; Thiel, W. *Chem. Rev.* 2010, *110*, 949–1017.
11. Friesner, R. A.; Guallar, V. *Annu. Rev. Phys. Chem.* 2005, *56*, 389–427.
12. Monard, G.; Merz Jr., K. M. *Acc. Chem. Res.* 1999, *32*, 904–911.
13. Sproviero, E. M.; Gascon, J. A.; McEvoy, J. P.; Brudvig, G. W.; Batista, V. S. *J. Am. Chem. Soc.* 2008, *130*, 3428–3442.
14. Cramer, C. J.; Truhlar, D. G. *Phys. Chem. Chem. Phys.* 2009, *11*, 10757–10816.
15. Hu, H.; Yang, W. T. *Annu. Rev. Phys. Chem.* 2008, *59*, 573–601.
16. Kamerlin, S. C. L.; Haranczyk, M.; Warshel, A. *J. Phys. Chem. B* 2009, *113*, 1253–1272.
17. Rod, T. H.; Ryde, U. *J. Chem. Theory Comput.* 2005, *1*, 1240–1251.
18. Zhang, Y. *Theor. Chem. Acc.* 2006, *116*, 43–50.
19. Corminboeuf, C.; Hu, P.; Tuckerman, M. E.; Zhang, Y. K. *J. Am. Chem. Soc.* 2006, *128*, 4530–4531.
20. De Vivo, M.; Ensing, B.; Dal Peraro, M.; Gomez, G. A.; Christianson, D. W.; Klein, M. L. *J. Am. Chem. Soc.* 2007, *129*, 387–394.
21. Brunk, E. et al. *Chimia* 2011, *65*, 667–671.
22. Peters, M. B.; Yang, Y.; Wang, B.; Fusti-Molnar, L.; Weaver, M. N.; Merz Jr., K. M. *J. Chem. Theory Comput.* 2010, *6*, 2935–2947.
23. MacKerell Jr., A. D. et al. *J. Phys. Chem. B* 1998, *102*, 3586–3616.
24. Piquemal, J. P.; Williams-Hubbard, B.; Fey, N.; Deeth, R. J.; Gresh, N.; Giessner-Prettre, C. *J. Comput. Chem.* 2003, *24*, 1963–1970.
25. Gresh, N.; Cisneros, G. A.; Darden, T. A.; Piguemal, J.-P. *J. Chem. Theory Comput.* 2007, *3*, 1960–1986.
26. Chaudret, R.; Gresh, N.; Parisel, O.; Piguemal, J.-P. *J. Comput. Chem.* 2011, *32*, 2949–2957.
27. Sakharov, D. V.; Lim, C. *J. Comput. Chem.* 2008, *30*, 191–202.
28. Deeth, R. J. *Coord. Chem. Rev.* 2001, *212*, 11–34.
29. Deeth, R. J.; Anastasi, A.; Diedrich, C.; Randell, K. *Coord. Chem. Rev.* 2009, *253*, 795–816.
30. Tubert-Brohman, I.; Schmid, M.; Meuwly, M. *J. Chem. Theory Comput.* 2009, *5*, 530–539.

31. Root, D. M.; Landis, C. R.; Cleveland, T. *J. Am. Chem. Soc.* 1993, *115*, 4201–4209.
32. Cleveland, T.; Landis, C. R. *J. Am. Chem. Soc.* 1996, *118*, 6020–6030.
33. Landis, C. R.; Cleveland, T.; Firman, T. K. *J. Am. Chem. Soc.* 1998, *120*, 2641–2649.
34. Firman, T. K.; Landis, C. R. *J. Am. Chem. Soc.* 2001, *123*, 11728–11742.
35. Weinhold, F.; Landis, C. R. *Valency and Bonding*, Cambridge University Press: Cambridge, U.K., 2005.
36. Landis, C. R.; Cleveland, T.; Firman, T. K. *Science* 1996, *272*, 182.
37. Stote, R. H.; Karplus, M. *Proteins* 1995, *23*, 12–31.
38. Ponder, J. W.; Wu, C. J.; Ren, P. Y.; Pande, V. S.; Chodera, J. D.; Schnieders, M. J.; Haque, I. et al. *J. Phys. Chem. B* 2010, *114*, 2549–2564.
39. Wu, J. C.; Piquemal, J.-P.; Chaudret, R.; Reinhardt, P.; Ren, P. *J. Chem. Theory Comput.* 2010, *6*, 2059–2070.
40. Ryde, U. *Biophys. J.* 1999, *77*, 2777–2787.
41. Wu, R. B.; Lu, Z. Y.; Cao, Z. X.; Zhang, Y. K. *J. Chem. Theory Comput.* 2011, *7*, 433–443.
42. Wu, R. B.; Hu, P.; Wang, S. L.; Cao, Z. X.; Zhang, Y. K. *J. Chem. Theory Comput.* 2010, *6*, 337–343.
43. Dudev, T.; Lim, C. *Acc. Chem. Res.* 2007, *40*, 85–93.
44. Sousa, S. F.; Fernandes, P. A.; Ramos, M. J. *J. Am. Chem. Soc.* 2007, *129*, 1378–1385.
45. Wu, R. B.; Cao, Z. X.; Zhang, Y. K. *Prog. Chem.* 2012, *24*, 1175–1184.
46. Pang, X.; Han, K.; Cui, Q. *J. Comput. Chem.* 2013, *34*, 1620–1635.
47. Hu, L. H.; Ryde, U. *J. Chem. Theory Comput.* 2011, *7*, 2425–2463.
48. Babu, C. S.; Lim, C. *J. Phys. Chem. A* 2006, *110*, 691–699.
49. Yu, H. B.; Whitfield, T. W.; Harder, E.; Lamoureux, G.; Vorobyov, I.; Anisimov, V. M.; MacKerell Jr., A. D.; Roux, B. *J. Chem. Theory Comput.* 2010, *6*, 774–786.
50. Luo, Y.; Roux, B. *J. Phys. Chem. Lett.* 2010, *1*, 183–189.
51. Elstner, M.; Porezag, D.; Jungnickel, G.; Elsner, J.; Haugk, M.; Frauenheim, T.; Suhai, S.; Seifert, G. *Phys. Rev. B* 1998, *58*, 7260–7268.
52. Elstner, M.; Frauenheim, T.; Suhai, S. *THEOCHEM* 2003, *632*, 29–41.
53. Riccardi, D.; Schaefer, P.; Yang, Y.; Yu, H.; Ghosh, N.; Prat-Resina, X.; Konig, P. et al. *J. Phys. Chem. B* 2006, *110*, 6458–6469.
54. Xu, D.; Guo, H.; Cui, G. *J. Am. Chem. Soc.* 2007, *129*, 10814–10822.
55. Xu, D. G.; Guo, H. *J. Am. Chem. Soc.* 2009, *131*, 9780–9788.
56. Xu, D. G.; Xie, D. Q.; Guo, H. *J. Biol. Chem.* 2006, *281*, 8740–8747.
57. Xu, Q.; Guo, H. B.; Wlodawer, A.; Nakayama, T.; Guo, H. *Biochemistry* 2007, *46*, 3784–3792.
58. Chakravorty, D. K.; Wang, B.; Lee, C. W.; Giedroc, D. P.; Merz Jr., K. M. *J. Am. Chem. Soc.* 2012, *134*, 3367–3376.
59. Yang, Y.; Miao, Y. P.; Wang, B.; Cui, G. L.; Merz Jr., K. M. *Biochemistry* 2012, *51*, 2606–2618.
60. Lundberg, M.; Sasakura, Y.; Zheng, G. S.; Morokuma, K. *J. Chem. Theory Comput.* 2010, *6*, 1413–1427.
61. Gaus, M.; Cui, Q.; Elstner, M. *J. Chem. Theory Comput.* 2011, *7*, 931–948.
62. Elstner, M. *J. Phys. Chem. A* 2007, *111*, 5614–5621.
63. Yang, Y.; Yu, H.; York, D.; Cui, Q.; Elstner, M. *J. Phys. Chem. A* 2007, *111*, 10861–10873.
64. Gaus, M.; Goez, A.; Elstner, M. *J. Chem. Theory Comput.* 2012, *9*, 338–354.
65. Kohler, C.; Seifert, G.; Frauenheim, T. *Chem. Phys.* 2005, *309*, 23–31.
66. Köhler, C.; Seifert, G.; Gerstmann, U.; Elstner, M.; Overhof, H.; Frauenheim, T. *PCCP* 2001, *3*, 5109–5114.

67. Köhler, C.; Frauenheim, T.; Hourahine, B.; Seifert, G.; Sternberg, M. *J. Phys. Chem. A* 2007, *111*, 5622–5629.
68. Xu, X.; Truhlar, D. G. *J. Chem. Theory Comput.* 2012, *8*, 80–90.
69. Zhang, X.; Schwarz, H. *Theor. Chem. Acc.* 2011, *129*, 389–399.
70. Furche, F.; Perdew, J. P. *J. Chem. Phys.* 2006, *124*, 044103.
71. Jiang, W.; DeYonker, N. J.; Determan, J. J.; Wilson, A. K. *J. Phys. Chem. A* 2012, *116*, 870–885.
72. Jiang, W.; Laury, M. L.; Powell, M.; Wilson, A. K. *J. Chem. Theory Comput.* 2012, *8*, 4102–4111.
73. Bredow, T.; Jug, K. *Theor. Chem. Acc.* 2005, *113*, 1–14.
74. Stewart, J. J. P. *J. Mol. Model.* 2007, *13*, 1173–1213.
75. Retegan, M.; Martins-Costa, M.; Ruiz-Lopez, M. F. *J. Chem. Phys.* 2010, *133*, 064103.
76. Plotnikov, N. V.; Kamerlin, S. C. L.; Warshel, A. *J. Phys. Chem. B* 2011, *115*, 7950–7962.
77. Woods, C. J.; Manby, F. R.; Mulholland, A. J. *J. Chem. Phys.* 2008, *128*, 014109.
78. Heimdal, J.; Ryde, U. *Phys. Chem. Chem. Phys.* 2012, *14*, 12592–12604.
79. Gao, J. L.; Amara, P.; Alhambra, C.; Field, M. J. *J. Phys. Chem. A* 1998, *102*, 4714–4721.
80. Das, D.; Eurenius, K. P.; Billings, E. M.; Sherwood, P.; Chatfield, D. C.; Hodoscek, M.; Brooks, B. R. *J. Chem. Phys.* 2002, *117*, 10534–10547.
81. König, P. H.; Hoffmann, M.; Frauenheim, T.; Cui, Q. *J. Phys. Chem. B* 2005, *109*, 9082–9095.
82. Freindorf, M.; Gao, J. L. *J. Comput. Chem.* 1996, *17*, 386–395.
83. Riccardi, D.; Li, G.; Cui, Q. *J. Phys. Chem. B* 2004, *108*, 6467–6478.
84. Giese, T. J.; York, D. M. *J. Chem. Phys.* 2007, *127*, 194101.
85. Cui, Q.; Elstner, M.; Kaxiras, E.; Frauenheim, T.; Karplus, M. *J. Phys. Chem. B* 2001, *105*, 569–585.
86. Yang, Y.; Yu, H.; York, D.; Elstner, M.; Cui, Q. *J. Chem. Theory Comput.* 2008, *4*, 2067–2084.
87. Nielsen, S. O.; Bulo, R. E.; Moore, P. B.; Ensing, B. *Phys. Chem. Chem. Phys.* 2010, *12*, 12401–12414.
88. Park, K.; Götz, A. W.; Walker, R. C.; Paesani, F. *J. Chem. Theory Comput.* 2012, *8*, 2868–2877.
89. Hou, G.; Zhu, X.; Elstner, M.; Cui, Q. *J. Chem. Theory Comput.* 2012, *8*, 4293–4304.
90. Pople, J. A.; Beveridge, D. L. *Approximate Molecular Orbital Theory*, McGraw-Hill Companies: New York, 1970.
91. Riccardi, D.; Zhu, X.; Goyal, P.; Yang, S.; Hou, G.; Cui, Q. *Sci. China Chem.* 2012, *55*, 3–18.
92. Riccardi, D.; Schaefer, P.; Cui, Q. *J. Phys. Chem. B* 2005, *109*, 17715–17733.
93. Nam, K.; Gao, J. L.; York, D. M. *J. Chem. Theory Comput.* 2005, *1*, 2–13.
94. Walker, R. C.; Crowley, M. F.; Case, D. A. *J. Comput. Chem.* 2008, *29*, 1019–1031.
95. Wu, X.; Koslowski, A.; Thiel, W. *J. Chem. Theory Comput.* 2012, *8*, 2272–2281.
96. Schaefer, P.; Riccardi, D.; Cui, Q. *J. Chem. Phys.* 2005, *123*, Art. No. 014905.
97. Im, W.; Berneche, S.; Roux, B. *J. Chem. Phys.* 2001, *114*, 2924–2937.
98. Brooks III, C. L.; Karplus, M. *J. Chem. Phys.* 1983, *79*, 6312–6325.
99. Riccardi, D.; Cui, Q. *J. Phys. Chem. A* 2007, *111*, 5703–5711.
100. Benighaus, T.; Thiel, W. *J. Chem. Theory Comput.* 2008, *4*, 1600–1609.
101. Ghosh, N.; Cui, Q. *J. Phys. Chem. B* 2008, *112*, 8387–8397, PMC2562727.
102. Goyal, P.; Ghosh, N.; Phatak, P.; Clemens, M.; Gaus, M.; Elstner, M.; Cui, Q. *J. Am. Chem. Soc.* 2011, *133*, 14981–14997.

103. Ghosh, N.; Prat-Resina, X.; Cui, Q. *Biochemistry* 2009, *48*, 2468–2485.
104. Chimenti, M. S.; Khangulov, V. S.; Robinson, A. C.; Heroux, A.; Majumdar, A.; Schlessman, J. L.; Garcia-Moreno, B. *Structure* 2012, *20*, 1071–1085.
105. Lu, X.; Cui, Q. *J. Phys. Chem. B* 2013, *117*, 2005–2018.
106. Kastenholz, M. A.; Hünenberger, P. H. *J. Chem. Phys.* 2006, *124*, 124106-1–124106-27.
107. Kastenholz, M. A.; Hünenberger, P. H. *J. Chem. Phys.* 2006, *124*, 224501-1–224501-20.
108. Simonson, T. *J. Phys. Chem. B* 2000, *104*, 6509–6513.
109. Song, X. *J. Chem. Phys.* 2002, *116*, 9359–9363.
110. Warshel, A.; Sharma, P. K.; Kato, M.; Parson, W. W. *Biochim. Biophys. Acta* 2006, *1764*, 1647–1676.
111. Benighaus, T.; Thiel, W. *J. Chem. Theory Comput.* 2011, *7*, 238–249.
112. Zienau, J.; Cui, Q. *J. Phys. Chem. B* 2012, *116*, 12522–12534.
113. Xia, F.; Rudack, T.; Cui, Q.; Kötting, C.; Gerwert, K. *J. Am. Chem. Soc.* 2012, *134*, 20041–20044.
114. Wrewick, S. C.; Henshaw, T. F.; Hausinger, R. P.; Lindahl, T.; Sedgwick, B. *Nature* 2002, *419*, 174–178.
115. Falnes, P. O.; Johansen, R. F.; Seeberg, E. *Nature* 2002, *419*, 178–182.
116. Aas, P. A.; Otterlei, M.; Falnes, P. O.; Vagbo, C. B.; Skorpen, F.; Akbari, M.; Sundheim, O. et al. *Nature* 2003, *421*, 859–863.
117. Mishina, Y.; Duguid, E. M.; He, C. *Chem. Rev.* 2006, *106*, 215–232.
118. McDonough, M. A.; Loenarz, C.; Chowdhury, R.; Clifton, I. J.; Schofield, C. J. *Curr. Opin. Struct. Biol.* 2010, *20*, 659–672.
119. Rose, N. R.; McDonough, M. A.; King, O. N. F.; Kawamura, A.; Schofield, C. J. *Chem. Soc. Rev.* 2011, *40*, 4364–4397.
120. Yi, C. Q.; Yang, C. G.; He, C. *Acc. Chem. Res.* 2009, *42*, 519–529.
121. Konishi, N.; Nakamura, M.; Ishida, E.; Shimada, K.; Mitsui, E.; Yoshikawa, R.; Yamamoto, H.; Tsujikawa, K. *Clin. Cancer Res.* 2005, *11*, 5090–5097.
122. Shimada, K.; Nakamura, M.; Anai, S.; De Velasco, M.; Tanaka, M.; Tsujikawa, K.; Ouji, Y.; Konishi, N. *Cancer Res.* 2009, *69*, 3157–3164.
123. Scott, L. J. et al. *Science* 2007, *316*, 1341–1345.
124. Frayling, T. M. et al. *Science* 2007, *316*, 889–894.
125. Dina, C. et al. *Nat. Genet.* 2007, *39*, 724–726.
126. Thorleifsson, G. et al. *Nat. Genet.* 2009, *41*, 18–24.
127. Fischer, J.; Koch, L.; Emmerling, C.; Vierkotten, J.; Peters, T.; Bruning, J. C.; Ruther, U. *Nature* 2009, *458*, 894–898.
128. Church, C.; Moir, L.; McMurray, F.; Girard, C.; Banks, G. T.; Teboul, L.; Wells, S. et al. *Nat. Genet.* 2010, *42*, 1086–1092.
129. Gerken, T. et al. *Science* 2007, *318*, 1469–1472.
130. Jia, G. F.; Fu, Y.; Zhao, X.; Dai, Q.; Zheng, G. Q.; Yang, Y.; Yi, C. Q. et al. *Nat. Chem. Biol.* 2011, *7*, 885–887.
131. He, C. *Nat. Chem. Biol.* 2010, *6*, 863–865.
132. Sedgwick, B. *Nat. Rev. Mol. Cell Biol.* 2004, *5*, 148–157.
133. Duncan, T.; Trewick, S. C.; Koivisto, P.; Bates, P. A.; Lindahl, T.; Sedgwick, B. *Proc. Natl. Acad. Sci. USA* 2002, *99*, 16660–16665.
134. Ringvoll, J.; Nordstrand, L. M.; Vagbo, C. B.; Talstad, V.; Reite, K.; Aas, P. A.; Lauritzen, K. H. et al. *EMBO J.* 2006, *25*, 2189–2198.
135. Koivisto, P.; Robins, P.; Lindahl, T.; Sedgwick, B. *J. Biol. Chem.* 2004, *279*, 40470–40474.

136. Falnes, P. O. *Nucleic Acids Res.* 2004, *32*, 6260–6267.
137. Delaney, J. C.; Essigmann, J. M. *Proc. Natl. Acad. Sci. USA* 2004, *101*, 14051–14056.
138. Delaney, J. C.; Smeester, L.; Wong, C. Y.; Frick, L. E.; Taghizadeh, K.; Wishnok, J. S.; Drennan, C. L.; Samson, L. D.; Essigmann, J. M. *Nat. Struct. Mol. Biol.* 2005, *12*, 855–860.
139. Yu, B.; Edstrom, W. C.; Benach, J.; Hamuro, Y.; Weber, P. C.; Gibney, B. R.; Hunt, J. F. *Nature* 2006, *439*, 879–884.
140. Yang, C. G.; Yi, C. Q.; Duguid, E. M.; Sullivan, C. T.; Jian, X.; Rice, P. A.; He, C. *Nature* 2008, *452*, 961–964.
141. Han, Z.; Niu, T.; Chang, J.; Lei, X.; Zhao, M.; Wang, Q.; Cheng, W.; Wang, J.; Feng, Y.; Chai, J. *Nature* 2010, *464*, 1205–1209.
142. Yu, B.; Hunt, J. F. *Proc. Natl. Acad. Sci. USA* 2009, *106*, 14315–14320.
143. Yi, C.; Jia, G.; Hou, G.; Dai, Q.; Zheng, G.; Jian, X.; Yang, C. G.; Cui, Q.; He, C. *Nature* 2010, *468*, 330–333.
144. Liu, H.; Llano, J.; Gauld, J. W. *J. Phys. Chem. B* 2009, *113*, 4887–4898.
145. Cisneros, G. A. *Interdiscip. Sci. Comput. Life Sci.* 2010, *2*, 70–77.
146. Schneider, B.; Neidle, S.; Berman, H. M. *Biopolymers* 1997, *42*, 113–124.
147. Varnai, P.; Djuranovic, D.; Lavery, R.; Hartmann, B. *Nucleic Acids Res.* 2002, *30*, 5398–5406.
148. Lu, X. J.; Olson, W. K. *Nat. Protoc.* 2008, *3*, 1213–1227.
149. Fu, Y.; Jia, G.; Pang, X.; Wang, R.; Wang, X.; Li, C.; Dai, Q.; Han, K.; Cui, Q.; He, C. *Nat. Commun.* 2013, *4*, 1798.
150. Thiel, W. *Adv. Chem. Phys.* 1996, *93*, 703–757.
151. Zheng, G. S.; Witek, H. A.; Bobadova-Parvanova, P.; Irle, S.; Musaev, D. G.; Prabhakar, R.; Morokuma, K. *J. Chem. Theory Comput.* 2007, *3*, 1349–1367.
152. O'Brien, P. J.; Herschlag, D. *Chem. Biol.* 1999, *4*, R91–R105.
153. Copley, S. D. *Curr. Opin. Chem. Biol.* 2003, *7*, 265–272.
154. Schmidt, D. M. Z.; Mundorff, E. C.; Dojka, M.; Bermudez, E.; Ness, J. E.; Govindarajan, S.; Babbitt, P. C.; Minshull, J.; Gerlt, J. A. *Biochemistry* 2003, *42*, 8387–8393.
155. Zalatan, J. G.; Herschlag, D. *Nat. Chem. Biol.* 2009, *5*, 516–520.
156. Jonas, S.; Hollfelder, F. *Pure Appl. Chem.* 2009, *81*, 731–742.
157. Aharoni, A.; Gaidukov, L.; Khersonsky, O.; Gould, S.; Roodveldt, C.; Tawfik, D. S. *Nat. Genet.* 2005, *37*, 73–76.
158. Khersonsky, O.; Roodveldt, C.; Tawfik, D. S. *Curr. Opin. Chem. Biol.* 2006, *10*, 498–508.
159. Penning, T. M.; Jez, J. M. *Chem. Rev.* 2001, *101*, 3027–3046.
160. Kazlauskas, R. J. *Curr. Opin. Chem. Biol.* 2005, *9*, 195–201.
161. Glasner, M. E.; Gerlt, J. A.; Babbitt, P. C. *Curr. Opin. Chem. Biol.* 2006, *10*, 492–497.
162. Hult, K.; Berglund, P. *Trends Biotechnol.* 2007, *25*, 231–238.
163. Gerlt, J. A.; Babbitt, P. C. *Curr. Opin. Chem. Biol.* 2009, *13*, 10–18.
164. Nobeli, I.; Favia, A. D.; Thornton, J. M. *Nat. Biotechnol.* 2009, *27*, 157–167.
165. O'Brien, P. J.; Herschlag, D. *J. Am. Chem. Soc.* 1998, *120*, 12369–12370.
166. O'Brien, P. J.; Herschlag, D. *Biochemistry* 2001, *40*, 5691–5699.
167. Babtie, A. C.; Bandyopadhyay, S.; Olguin, L. F.; Hollfelder, F. *Angew. Chem. Int. Ed.* 2009, *48*, 3692–3694.
168. van Loo, B.; Jonas, S.; Babtie, A. C.; Benjdia, A.; Berteau, O.; Hyvönen, M.; Hollfelder, F. *Proc. Natl. Acad. Sci. USA* 2010, *107*, 2740–2745.
169. Hollfelder, F.; Herschlag, D. *Biochemistry* 1995, *38*, 12255–12264.

170. O'Brien, P. J.; Herschlag, D. *J. Am. Chem. Soc.* 1999, *121*, 11022–11023.
171. Nikolic-Hughes, I.; Rees, D. C.; Herschlag, D. *J. Am. Chem. Soc.* 2004, *126*, 11814–11819.
172. Zalatan, J. G.; Herschlag, D. *J. Am. Chem. Soc.* 2006, *128*, 1293–1303.
173. Zalatan, J. G.; Catrina, I.; Mitchell, R.; Grzyska, P. K.; O'Brien, P. J.; Herschlag, D.; Hengge, A. C. *J. Am. Chem. Soc.* 2007, *129*, 9789–9798.
174. Babtie, A.; Tokuriki, N.; Hollfelder, F. *Curr. Opin. Chem. Biol.* 2010, *14*, 200–207.
175. Tunon, I.; Lopez-Canut, V.; Ruiz-Pernia, J.; Ferrer, S.; Moliner, V. *J. Chem. Theory Comput.* 2009, *5*, 439–442.
176. Lopez-Canut, V.; Marti, S.; Bertran, J.; Moliner, V.; Tunon, I. *J. Phys. Chem. B* 2009, *113*, 7816–7824.
177. Lopez-Canut, V.; Roca, M.; Bertran, J.; Moliner, V.; Tunon, I. *J. Am. Chem. Soc.* 2010, *132*, 6955–6963.
178. López-Canut, V.; Roca, M.; Bertrán, J.; Moliner, V.; Tuñón, I. *J. Am. Chem. Soc.* 2011, *133*, 12050–12062.
179. Aqvist, J.; Kolmodin, K.; Florian, J.; Warshel, A. *Chem. Biol.* 1999, *6*, R71–R80.
180. Rosta, E.; Kamerlin, S. C. L.; Warshel, A. *Biochemistry* 2008, *47*, 3725–3735.
181. Wong, K. Y.; Gao, J. L. *Biochemistry* 2007, *46*, 13352–13369.
182. Garcia-Viloca, M.; Alhambra, C.; Truhlar, D. G.; Gao, J. L. *J. Am. Chem. Soc.* 2002, *124*, 7268–7269.
183. Kim, E. E.; Wyckoff, H. W. *J. Mol. Biol.* 1991, *218*, 449–464.
184. Strater, N.; Lipscomb, W. N.; Klabunde, T.; Krebs, B. *Angew. Chem. Int. Ed.* 1996, *35*, 2024–2055.
185. Steitz, T. A.; Steitz, J. A. *Proc. Natl. Acad. Sci. USA* 1993, *90*, 6498–6502.
186. Tesmer, J. J. G.; Sunahara, R. K.; Johnson, R. A.; Gosselin, G.; Gilman, A. G.; Sprang, S. R. *Science* 1999, *285*, 756–760.
187. Jedrzejas, M. J.; Setlow, P. *Chem. Rev.* 2001, *101*, 607–618.
188. Boehr, D. D.; Dyson, H. J.; Wright, P. E. *Chem. Rev.* 2006, *106*, 3055–3079.
189. Nagel, Z. D.; Klinman, J. P. *Nat. Chem. Biol.* 2009, *5*, 543–550.
190. Nashine, V. C.; Hammes-Schiffer, S.; Benkovic, S. J. *Curr. Opin. Chem. Biol.* 2010, *14*, 644–651.
191. Hou, G. H.; Cui, Q. *J. Am. Chem. Soc.* 2012, *134*, 229–246.
192. Bobyr, E.; Lassila, J. K.; Wiersma-Koch, H. I.; Fenn, T. D.; Lee, J. J.; Nikolic-Hughes, I.; Hodgson, K. O.; Rees, D. C.; Hedman, B.; Herschlag, D. *J. Mol. Biol.* 2011, *415*, 102–117.
193. Lassila, J. K.; Zalatan, J. G.; Herschlag, D. *Annu. Rev. Biochem.* 2011, *80*, 669–702.
194. O'Brien, P. J.; Herschlag, D. *Biochemistry* 2002, *41*, 3207–3225.
195. Hou, G. H.; Cui, Q. *J. Am. Chem. Soc.* 2013, *135*, 10457–10469.
196. Solomon, E. I.; Szilagyi, R. K.; George, S. D.; Basumallick, L. *Chem. Rev.* 2004, *104*, 419–458.
197. Solomon, E. I.; Hadt, R. G. *Coord. Chem. Rev.* 2011, *255*, 774–789.
198. Choi, M.; Davidson, V. L. *Metallomics* 2011, *3*, 140–151.
199. Hadt, R. G.; Sun, N.; Marshall, N. M.; Hodgson, K. O.; Hedman, B.; Lu, Y.; Solomon, E. *J. Am. Chem. Soc.* 2012, *134*, 16701–16716.
200. Marshall, N. M.; Garner, D. K.; Wilson, T. D.; Gao, Y. G.; Robinson, H.; Nilges, M. J.; Lu, Y. *Nature* 2009, *462*, 113–116.
201. Taylor, A. B.; Stoj, C. S.; Ziegler, L.; Kosman, D. J.; Hart, P. J. *Proc. Natl. Acad. Sci. USA* 2005, *102*, 15459–15464.

202. Bertini, I.; Bryant, D. A.; Ciurli, S.; Dikiy, A.; Fernandez, C. O.; Luchinat, C.; Safarov, N.; Vila, A. J.; Zhao, J. *J. Biol. Chem.* 2001, *276*, 47217–47226.

203. Jimenez, B.; Piccioli, M.; Moratal, J. M.; Donaire, A. *Biochemistry* 2003, *42*, 10396–10405.

204. Ma, L.; Hass, M. A. S.; Vierick, N.; Kristensen, S. M.; Ulstrup, J.; Led, J. J. *Biochemistry* 2003, *42*, 320–330.

205. Hass, M. A. S.; Vlasie, M. D.; Ubbink, M.; Led, J. J. *Biochemistry* 2009, *48*, 50–58.

206. Battistuzzi, G.; Borsari, M.; Canters, G. W.; de Waal, E.; Loschi, L.; Warmerdam, G.; Sola, M. *Biochemistry* 2001, *40*, 6707–6712.

207. Li, C.; Sato, K.; Monari, S.; Salard, I.; Sola, M.; Banfield, M. J.; Dennison, C. *Inorg. Chem.* 2011, *50*, 482–488.

208. Karp, D. A.; Gittis, A. G.; Stahley, M. R.; Fitch, C. A.; Stites, W. E.; Garcia-Moreno, E. B. *Biophys. J.* 2007, *92*, 2041–2053.

209. Isom, D. G.; Castaneda, C. A.; Velu, P. D.; Garcia-Moreno, E. B. *Proc. Natl. Acad. Sci. USA* 2010, *107*, 16096–16100.

210. Garcia-Moreno, E. B., Private communication.

211. Battistuzzi, G.; Borsari, M.; Canters, G. W.; de Waal, E.; Leonardi, A.; Ranieri, A.; Sola, M. *Biochemistry* 2002, *41*, 14293–14298.

212. Bortolotti, C. A.; Amadei, A.; Aschi, M.; Borsari, M.; Corni, S.; Sola, M.; Daidone, I. *J. Am. Chem. Soc.* 2012, *134*, 13670–13678.

213. Si, D.; Li, H. *J. Phys. Chem. A* 2009, *113*, 12979–12987.

214. Su, P.; Li, H. *Inorg. Chem.* 2010, *49*, 435–444.

215. Olsson, M. H. M.; Hong, G.; Warshel, A. *J. Am. Chem. Soc.* 2003, *125*, 5025–5039.

216. Cascella, M.; Magistrato, A.; Tavernelli, I.; Carloni, P.; Rothlisberger, U. *Proc. Natl. Acad. Sci. USA* 2006, *103*, 19641–19646.

217. van den Bosch, M.; Swart, M.; Snijdes, J. G.; Berendsen, H. J. C.; Mark, A. E.; Oostenbrink, C.; van Gunsteren, W. F.; Canters, G. W. *ChemBioChem* 2005, *6*, 738–746.

218. Li, H.; Webb, S. P.; Ivanic, J.; Jensen, J. H. *J. Am. Chem. Soc.* 2004, *126*, 8010–8019.

219. Ghosh, S.; Xie, X. J.; Dey, A.; Sun, Y.; Scholes, C. P.; Solomon, E. I. *Proc. Natl. Acad. Sci. USA* 2009, *106*, 4969–4974.

220. Steiner, D.; Oostenbrink, C.; van Gunsteren, W. F. *J. Comput. Chem.* 2012, *33*, 1467–1477.

221. Ullmann, R. T.; Ullmann, G. M. *J. Phys. Chem. B* 2011, *115*, 10346–10359.

222. Davidson, V. L. *Acc. Chem. Res.* 2008, *41*, 730–738.

223. Wallace, D. C. *Science* 1999, *283*, 1482–1488.

224. Beal, M. F. *Trends Neurosci.* 2000, *23*, 298–304.

225. Wallace, D. C. *Annu. Rev. Genet.* 2005, *39*, 359–407.

226. Li, G.; Zhang, X.; Cui, Q. *J. Phys. Chem. B* 2003, *107*, 8643–8653.

227. Blumberger, J. *Phys. Chem. Chem. Phys.* 2008, *10*, 5651–5667.

228. Patel, S.; Brooks III, C. L. *J. Comput. Chem.* 2004, *25*, 1–15.

229. Patel, S.; MacKerell Jr., A. D.; Brooks III, C. L. *J. Comput. Chem.* 2004, *25*, 1504–1514.

230. Lopes, P. E. M.; Roux, B.; MacKerell Jr., A. D. *Theor. Chem. Acc.* 2009, *124*, 11–28.

231. Finney, L. A.; O'Halloran, T. V. *Science* 2003, *300*, 931–936.

232. Waldron, K. J.; Rutherford, J. C.; Ford, D.; Robinson, N. J. *Nature* 2009, *460*, 823–830.

233. Waldron, K. J.; Robinson, N. J. *Nat. Rev. Microbiol.* 2009, *6*, 25–35.

234. Chen, P. R.; Wasinger, E. C.; Zhao, J.; van der Lelie, D.; Chen, L. X.; He, C. *J. Am. Chem. Soc.* 2007, *129*, 12350–12351.

235. Wegner, S. V.; Boyaci, H.; Chen, H.; Jensen, M. P.; He, C. *Angew. Chem. Int. Ed.* 2009, *48*, 2339–2341.
236. Valko, M.; Morris, H.; Cronin, M. T. D. *Curr. Med. Chem.* 2005, *12*, 1161–1208.
237. La Fontaine, S.; Mercer, J. F. *Arch. Biochem. Biophys.* 2007, *463*, 149–167.
238. Kambe, T.; Andrews, G. K. *Mol. Cell Biol.* 2009, *29*, 129–139.
239. Andrews, N. C. *Blood* 2008, *112*, 219–230.
240. Barnham, K. J.; Bush, A. I. *Curr. Opin. Chem. Biol.* 2008, *12*, 222–228.
241. Colvin, R. A.; Fontaine, C. P.; Laskowski, M.; Thomas, D. *Eur. J. Pharm.* 2003, *479*, 171–185.
242. Osman, D.; Cavet, J. S. *Adv. Microb. Physiol.* 2011, *58*, 175–232.
243. Ma, Z.; Jacobsen, F. E.; Giedroc, D. P. *Chem. Rev.* 2009, *109*, 4644–4681.
244. Rosenzweig, A. C. *Acc. Chem. Res.* 2001, *34*, 119–128.
245. Wimmer, R.; Herrmann, T.; Solioz, M.; Wuthrich, K. *J. Biol. Chem.* 1999, *274*, 22597–22603.
246. Banci, L.; Bertini, I.; Cantini, F.; Gonnelli, L.; Hadjiliadis, N.; Pierattelli, R.; Rosato, A.; Voulgaris, P. *Nat. Chem. Biol.* 2006, *2*, 367–368.
247. Changela, A.; Chen, K.; Xue, Y.; Holschen, J.; Outten, C. E.; O'Halloran, T. V.; Mondragon, A. *Science* 2003, *301*, 1383–1387.
248. Barondeau, D. P.; Getzoff, E. D. *Curr. Opin. Struct. Biol.* 2004, *14*, 765–774.
249. Dudev, T.; Lim, C. *Chem. Rev.* 2006, *103*, 773–787.
250. Dudev, T.; Lim, C. *Annu. Rev. Biochem.* 2008, *37*, 97–116.
251. Tai, H.; Lim, C. *J. Phys. Chem. A* 2006, *110*, 452–462.
252. Dudev, T.; Lim, C. *J. Phys. Chem. B* 2004, *108*, 4546–4557.
253. Dudev, T.; Chang, L.; Lim, C. *J. Am. Chem. Soc.* 2005, *127*, 4091–4103.
254. Dudev, T.; Lim, C. *J. Am. Chem. Soc.* 2010, *132*, 2321–2332.
255. Dudev, T.; Lim, C. *J. Am. Chem. Soc.* 2009, *131*, 8092–8101.
256. Roux, B.; Allen, T.; Berneche, S.; Im, W. *Q. Rev. Biophys.* 2004, *37*, 15–103.
257. Roux, B. *Biophys. J.* 2010, *98*, 2877–2885.
258. Varma, S.; Sabo, D.; Rempe, S. B. *J. Mol. Biol.* 2007, *376*, 13–22.
259. Bucher, D.; Guidoni, L.; Carloni, P.; Rothlisberger, U. *Biophys. J.* 2010, *98*, L47–L49.
260. Yu, H. B.; Roux, B. *Biophys. J.* 2009, *97*, L15–L17.
261. Roux, B.; Yu, H. *J. Chem. Phys.* 2010, *132*, 234101.
262. Chen, P.; He, C. *J. Am. Chem. Soc.* 2004, *126*, 728–729.
263. Nolan, E. M.; Lippard, S. J. *Chem. Rev.* 2008, *108*, 3443–3480.
264. Rao, L.; Cui, Q.; Xu, X. *J. Am. Chem. Soc.* 2010, *132*, 18092–18102.
265. Hobman, J. L. *Mol. Microbiol.* 2007, *63*, 1275–1278.
266. Hobman, J. L.; Wilkie, J.; Brown, N. L. *BioMetals* 2005, *18*, 429–436.
267. Tomasi, J.; Mennucci, B.; Cammi, R. *Chem. Rev.* 2005, *105*, 2999–3093.
268. Baker, N. A.; Sept, D.; Joseph, S.; Holst, M. J.; McCammon, J. A. *Proc. Natl. Acad. Sci. USA* 2001, *98*, 10037–10041.
269. Liptak, M.; Gross, K.; Seybold, P.; Feldgus, S.; Shields, G. *J. Am. Chem. Soc.* 2002, *124*, 6421–6427.
270. Parr, R. G.; Yang, W. T. *Density-Functional Theory of Atoms and Molecules*, Oxford University Press: New York, 1989.
271. Jian, X.; Wasinger, E. C.; Lockard, J. V.; Chen, L. X.; He, C. *J. Am. Chem. Soc.* 2009, *131*, 10869–10871.
272. Checa, S. K.; Espariz, M.; Audero, M. E. P.; Botta, P. E.; Spinelli, S. V.; Soncini, F. C. *Mol. Microbiol.* 2007, *63*, 1307–1318.

273. Jiang, L.; Althoff, E. A.; Clemente, F. R.; Doyle, L.; Rothlisberger, D.; Zanghellini, A.; Gallaher, J. L. et al. *Science* 2008, *319*, 1387–1391.

274. Rothlisberger, D.; Khersonsky, O.; Wollacott, A. M.; Jiang, L.; DeChancie, J.; Betker, J.; Gallaher, J. L. et al. *Nature* 2008, *453*, 190–194.

275. Siegel, J. B.; Zanghellini, A.; Lovick, H. M.; Kiss, G.; Lambert, A. R.; Clair, J. L. S.; Gallaher, J. L. et al. *Science* 2010, *329*, 309–313.

276. Köhler, V.; Wilson, Y. M.; Durrenberger, M.; Ghielier, D.; Churakova, E.; Quinto, T.; Knorr, L. et al. *Nat. Chem.* 2013, *5*, 93–99.

277. Sparta, M.; Shirvanyants, D.; Ding, F.; Dokholyan, N. V.; Alexandrova, A. N. *Biophys. J.* 2012, *103*, 767–776.

278. Ambroggio, X. I.; Kuhlman, B. *J. Am. Chem. Soc.* 2006, *128*, 1154–1161.

279. Kepp, K. P. *Chem. Rev.* 2012, *112*, 5193–5239.

280. Kozlowski, H.; Luczkowski, M.; Remelli, M.; Valensin, D. *Coord. Chem. Rev.* 2012, *256*, 2129–2141.

281. Hureau, C. *Coord. Chem. Rev.* 2012, *256*, 2164–2174.

282. Arena, G.; La Mendola, D.; Pappalardo, G.; Sovago, I.; Rizzarelli, E. *Coord. Chem. Rev.* 2012, *256*, 2202–2218.

283. Viles, J. H. *Coord. Chem. Rev.* 2012, *256*, 2271–2284.

284. Dahms, S. O.; Könnig, I.; Roeser, D.; Gührs, K.; Mayer, M. C.; Kaden, D.; Multhaup, G.; Than, M. E. *J. Mol. Biol.* 2012, *416*, 438–452.

285. Parthasarathy, S.; Long, F.; Miller, Y.; Xiao, Y.; McElheny, D.; Thurber, K.; Ma, B.; Nussinov, R.; Ishii, Y. *J. Am. Chem. Soc.* 2011, *133*, 3390–3400.

286. Miller, Y.; Ma, B.; Nussinov, R. *J. Am. Chem. Soc.* 2011, *133*, 2742–2748.

287. Miller, Y.; Ma, B.; Nussinov, R. *Coord. Chem. Rev.* 2012, *256*, 2245–2252.

288. Furlan, S.; La Penna, G. *Coord. Chem. Rev.* 2012, *256*, 2234–2244.

289. Quintanar, L.; Rivillas-Acevedo, L.; Grande-Aztatzi, R.; Gomez-Castro, C. Z.; Arcos-Lopez, T.; Vela, A. *Coord. Chem. Rev.* 2013, *257*, 429–444.

290. Kaminski, S.; Giese, T. J.; Gaus, M.; York, D. M.; Elstner, M. *J. Phys. Chem. A* 2012, *116*, 9131–9141.

291. Giese, T. J.; Chen, H.; Dissanayake, T.; Giambasu, G. M.; Heldenbrand, H.; Huang, M.; Kuechler, E. R. et al. *J. Chem. Theory Comput.* 2013, *9*, 1417–1427.

292. Cohen, A. J.; Mori-Sanchez, P.; Yang, W. T. *Chem. Rev.* 2012, *112*, 289–320.

293. Chan, G. K. L.; Sharma, S. *Annu. Rev. Phys. Chem.* 2011, *62*, 465–481.

294. Sherrill, C. D. *J. Chem. Phys.* 2010, *132*, 110902.

295. Barducci, A.; Bonomi, M.; Parrinello, M. *Wiley Interdiscip. Rev. Comput. Mol. Sci.* 2011, *1*, 826–843.

296. Ingledew, W. J.; Cobley, J. G. *Biochim. Biophys.* 1980, *590*, 141–158.

297. Takayama, S. J.; Irie, K.; Tai, H.; Kawahara, T.; Hirota, S.; Takabe, T.; Alcaraz, L. A.; Donaire, A.; Yamamoto, Y. *J. Biol. Inorg. Chem.* 2009, *14*, 821–828.

298. Giudici-Orticoni, M.; Guerlesquin, F.; Bruschi, M.; Nitschke, W. *J. Biol. Chem.* 1999, *274*, 30365–30369.

299. Olsson, M. H. M.; Ryde, U.; Roos, B. O. *Protein Sci.* 1998, *7*, 2659–2668.

300. Cambridge Structural Database, Version 5.33, Cambridge Crystallographic Data Centre: Cambridge, U.K., 2011.

301. Hou, G.; Zhu, X.; Cui, Q. *J. Chem. Theory Comput.* 2010, *6*, 2303–2314, PMC2918909.

Chapter 3

Development of AMOEBA Force Field with Advanced Electrostatics

Zhen Xia, Qiantao Wang, Xiaojia Mu, and Pengyu Ren

CONTENTS

3.1   INTRODUCTION

Electrostatic forces are crucial interatomic forces in molecular systems. Accurate representation of the electrostatic interactions remains a grand challenge in molecular modeling and simulations (Ren et al. 2011a). Fundamentally, the electrostatic interaction can be described by Coulomb's law, as employed in most classical force fields, including AMBER (Cornell et al. 1995), CHARMM

(MacKerell et al. 1998), GROMOS (Valdes et al. 2008), MM3 (Lii and Allinger 1989a,b), and OPLS (Jorgensen et al. 1996). In these force fields, the polarization effect is implicitly included in the parameters, by increasing or decreasing the magnitude of the partial charges in an average fashion. Because many force fields target water environment, the partial charges are typically overestimated compared to the gas-phase values obtained from high-level quantum mechanical calculations. For decades, the classic force fields have gone through extensive refinements, validations, and tests (Ponder and Case 2003, Rezac et al. 2008). This generation of force fields is now widely used in the studies of molecular structures, dynamics, and interactions. One challenge faced by such force fields, however, is that the electrostatics is unable to respond to environmental changes including dielectric constant, pH value, or nature of solvent.

It is possible to model the nonadditive nature of the polarization effect explicitly. Polarization refers to the redistribution of a particle's electron density due to an external electric field. The idea of explicit treatment of electrostatic polarization dates far back (Barker 1953). Only in the last decade or so has there been systematic development of polarizable force fields for biomolecular simulations. Different approaches have been introduced to incorporate the polarization effect, including induced dipole (Caldwell and Kollman 1995, Cieplak et al. 2009, Friesner 2006, Holt and Karlström 2008, Kaminski et al. 2002, Moghaddam et al. 2011, Molnar et al. 2009, Ren and Ponder 2002, 2003, Wang et al. 2006), Drude oscillator (Geerke et al. 2007, Lamoureux et al. 2003, 2006), and fluctuating charge models (Banks et al. 1999, Patel and Brooks 2004, Rappe and Goddard 1991).

The induced dipole model accounts for the polarization effect via atomic dipole induction. The atomic multipole optimized energetics for biomolecular applications (AMOEBA) force field, developed by Ponder et al. (2010) and Ren and Ponder (2002, 2003), is an example. In AMOEBA, the atomic multipole consists of charge, dipole, and quadrupole moments, which are derived from the ab initio quantum mechanical calculations using procedures such as Stone's distributed multipole analysis (DMA) (Stone 1985, 2005). The sum of interactions between fragments ab initio computed (SIBFA) is another elaborate model with emphasis on separability, anisotropy, nonadditivity, and transferability (Gresh et al. 1984, 2007, Piquemal et al. 2003, 2006). The polarization is treated with induced dipoles and distributed anisotropic polarizability tensors placed on the bond centers and on the heteroatom lone pairs (Garmer and Stevens 1989, Le Sueur and Stone 1993). Compared to AMOEBA, SIBFA is parameterized on the basis of quantum chemistry and calibrated on energy decomposition analysis. Karlström and coworkers have been developing the nonempirical molecular orbital (NEMO) polarizable model (Brdarski and Karlström 1998, Carignano et al. 1997, Holt and Karlström 2009), in which the atomic multipole moments are now obtained from ab initio calculation using the LoProp procedure

(Gagliardi et al. 2004, Holt and Karlström 2009). Yond and coworkers have been developing a new polarizable AMBER force field. The model parameters were optimized to the experimental static molecular polarizabilities obtained from the molecular refraction measurements (Wang et al. 2011a,b, 2012a,b).

The CHARMM polarizable force field developed by Roux et al. utilizes the classic Drude oscillator framework to model the electrostatic polarization (Anisimov et al. 2007, Drude et al. 1902, Lamoureux et al. 2003, 2006, Lopes et al. 2007, Vorobyov et al. 2005). In the Drude approach, each polarizable center includes a pair of point charges. For each heavy atom, a point partial charge is tethered via a harmonic spring. This point charge (the Drude oscillator) can react to the electrostatic environment and causes the displacement of the local electron density. The atomic polarizability depends on both charges on the Drude particle and the harmonic force constant.

The fluctuating charge approach is based on the charge equilibration (CHEQ) method (Rappe and Goddard 1991). Here, the chemical potential of atoms is equilibrated via the redistribution of charge density. In the fluctuating charge model, the magnitude of partial charge on each atom is allowed to adapt to different electrostatic and chemical environments. The variable partial charges are computed by minimizing the total electrostatic energy for a given molecular geometry. Friesner and Kaminski et al. also developed a polarizable protein force field (PFF) combining both induced dipole and fluctuating charge (Friesner 2006, Kaminski et al. 2002, 2003, 2004). Based on PFF, a polarizable simulations with second-order interaction model (POSSIM) was later proposed to reduce computational cost (Ponomarev and Kaminski 2011). A representative biological force field based on the fluctuating charge model is the CHARMM-FQ by Patel and Brooks et al. (Bauer and Patel 2009, Davis et al. 2009, Patel and Brooks 2004, Patel et al. 2009).

Sophisticate methods that follow quantum mechanics (QM) more rigorously to improve the electrostatic modeling also exist. For example, a Gaussian-based electrostatic model (GEM) has been developed (Cisneros et al. 2006, Gresh et al. 2007, Piquemal et al. 2003), aiming at replacing SIBFA's distributed multipoles with electron density. X-Pol, proposed by Gao and coworkers, combines the fragment-based electronic structure theory with molecular mechanical force field and incorporates all the valence interactions into QM (Song et al. 2009, Wang et al. 2012c, Xie and Gao 2007, Xie et al. 2009). In a quantum mechanics polarizable force field (QMPFF) (Donchev et al. 2005, 2006, 2008), developed by Donchev et al., a model was developed that consists of a nuclear charge and a negative electron cloud of exponential form located in the nuclear center.

Among these different approaches, fluctuating charge and Drude oscillator are easy to implement within existing fixed-charge force field framework. The induced dipole model requires more complicated algorithms; however, it fits nicely into the atomic multipole framework, which offers a more accurate description of electrostatic potential (ESP) than atomic charge models. Gaussian-based approach improves further the electrostatic representation with additional computational cost. An orthogonal issue is how to compute the polarization on the fly in simulations based on molecular dynamics. Either iterative induction or extended Lagrangian treatment is applicable in most of these approaches (Van Belle and Wodak 1995, Van Belle et al. 1992). The former can be accelerated with advanced linear algebra solver, and the latter requires a smaller time step and a separate low-temperature thermostat to be stable. The parameterization approaches of these force fields also vary in the degrees that rely on QM decomposition and empirical experimental data.

Overall development in modern force field based on improved electrostatic models is still limited. Only few polarizable models have reported parameters for biomolecules, including AMOEBA, SIBFA, PFF, and CHARMM-FQ. On one hand, the model and parameterization become much more complicated, and on the other, the expectation is higher, which also means there are more challenges to overcome. As comprehensive reviews that cover the basics of various modern polarizable force fields exist (Cieplak et al. 2009, Halgren 1992, Lopes et al. 2009, Ponder and Case 2003, Rick 2001), we choose to go in depth into the methodology and practical development of advanced electrostatic models using AMOEBA as an example. We will describe the fundamental concept of modeling electrostatic interactions using higher-order atomic multipole moments and treating polarization via dipole induction. The derivation of the electrostatic parameters in AMOEBA force field and recent applications of the polarizable force field will be presented. We will then conclude with remaining challenges and possible future directions for further improvement of electrostatic models for biomolecular simulations.

3.2 DEVELOPMENT OF ADVANCE ELECTROSTATIC MODEL

3.2.1 Introducing Multipole Moments

In classic force fields, the electrostatic energy is commonly modeled through Coulombic interactions among fixed charges located at the center of atoms or sometimes off-center sites:

$$U_{ij} = \frac{q_i q_j}{r_{ij}} \tag{3.1}$$

where

U_{ij} is the ESP energy between charges q_i and q_j
r_{ij} is the distance between two

The corresponding gradient or force can be derived with ease. However, the local atomic change distributions are spherically symmetrical, due to chemical bonding or the lone electron pairs. Therefore, introducing higher-order electrostatic moments allows those unsymmetrical features to be described. In AMOEBA force field, permanent atomic multipole moments, including the monopole, dipole, and quadrupole moments, are implemented at each atomic center:

$$M_i = [q_i, d_{ix}, d_{iy}, d_{iz}, Q_{ixx}, Q_{ixy}, Q_{ixz}, Q_{iyx}, Q_{iyy}, Q_{iyz}, Q_{izx}, Q_{izy}, Q_{izz}]^T \tag{3.2}$$

where
   M_i is the transposed permanent multipole vector at site i
   q_i is the point charge located at the atom i
   d is the dipole
   Q is the quadrupole moment

Adding point dipole and quadrupole moments could improve the accuracy of molecular electrostatics by orders of magnitudes in comparison with ab initio reference potential (Williams 1988). The use of additional point charge sites or Gaussian charge distributions would in principle lead to the same effect. However, it is difficult to uniquely determine the location of off-atom sites in general and the computational cost involving Gaussians is high.

The interaction energy between two multipole sites can be described as matrix formula, in the Cartesian form:

$$U_{ij} = M_i^T T_{ij} M_j \tag{3.3}$$

where T_{ij} is a matrix with the form of

$$T_{ij} = \begin{bmatrix} 1 & \dfrac{\partial}{\partial x_j} & \dfrac{\partial}{\partial y_j} & \dfrac{\partial}{\partial z_j} & \cdots \\[2mm] \dfrac{\partial}{\partial x_i} & \dfrac{\partial^2}{\partial x_i \partial x_j} & \dfrac{\partial^2}{\partial x_i \partial y_j} & \dfrac{\partial^2}{\partial x_i \partial z_j} & \cdots \\[2mm] \dfrac{\partial}{\partial y_i} & \dfrac{\partial^2}{\partial y_i \partial x_j} & \dfrac{\partial^2}{\partial y_i \partial y_j} & \dfrac{\partial^2}{\partial y_i \partial z_j} & \cdots \\[2mm] \dfrac{\partial}{\partial z_i} & \dfrac{\partial^2}{\partial z_i \partial x_j} & \dfrac{\partial^2}{\partial z_i \partial y_j} & \dfrac{\partial^2}{\partial z_i \partial z_j} & \cdots \\[2mm] \vdots & \vdots & \vdots & \vdots & \ddots \end{bmatrix} \left(\dfrac{1}{r_{ji}} \right). \tag{3.4}$$

For each atom type, the multipole moments need to be defined in a local coordinate frame set by covalently bonded neighboring atoms (Ren and Ponder 2003,

Ren et al. 2011b). Thus, the permanent atomic multipole moments remain constant with respect to their local frames as the molecules move and rotate in the simulation. The local multipole moments will be converted into the global frame by the rotation matrix prior to computation of the electrostatic interaction energies (Kong 1997).

Similar to the Coulomb energy, forces of the atomic multipole model can be derived from the derivative of the interaction matrix T, where $T^{n+1} = \nabla(T^n)$ and $T^0 = 1/r$ (Kong 1997). The extra torque caused by the derivatives of rotation matrix can be converted into forces at each atom and their frame-defining neighbors in order to make use of standard molecular dynamics or energy minimization algorithms (Ren and Ponder 2003).

3.2.2 Polarizable Model

The nature of polarization refers to the distortion of electron density in response to an electric field presented by the surrounding environment. Explicitly including polarization effect in molecular mechanic models can greatly increase the flexibility of the model to a wide range of electrostatic environments. Several different models have been developed to integrate the polarization effect, such as induced dipole, fluctuating charge, and Drude oscillator. Detailed comparison of those approaches can be found in the other reviews (Cieplak et al. 2009, Halgren 1992, Lopes et al. 2009, Ponder and Case 2003, Rick 2001). Here, we mainly focus on the methodology based on the distributed induced dipole model.

In a system, the electrostatic energy associated with induced dipole can be expressed as

$$U_{\text{ele}}^{\text{pol}} = -\frac{1}{2}\sum_i (\mu_i^{\text{pol}})^T E_i \tag{3.5}$$

where μ_i^{pol} is the induced dipole vector on any polarizable site i, which can be expressed as

$$\mu_i^{\text{pol}} = \alpha_i \left(\sum_{j\neq i \text{ if } i\in G}^{n} T_{ij}^1 M_j + \sum_{k\neq i}^{n} T_{ik}^{11}\mu_k \right) \tag{3.6}$$

where
  M is the permanent multipole
  T is the multipole interaction operator
  G is the *polarization group* (further explained later) that atom i belongs to

In Equation 3.6, the first term inside the parentheses on the right-hand side is the *direct* electric field, E, due to permanent multipoles from outside the group (index j). The second term corresponds to *mutual* induction by other induced dipoles (index k). Direction induction only occurs among groups of atoms while mutual polarization involves every atom pair. The induced dipole in Equation 3.6 can be solved iteratively to obtain the final induced dipoles. Comparing with Applequist's inducible dipole model (Applequist et al. 1972), the Thole model AMOEBA is based on involves damping at short range to avoid the *polarization catastrophe*. The damping is effectively achieved by smearing one of the atomic charge distributions when evaluating polarization energy, field, and gradient. In AMOEBA force field, the smearing function for charges has the functional form

$$\rho = \frac{3\beta}{4\pi}\exp(-\beta u^3) \qquad (3.7)$$

where $u = r_{ij}/(\alpha_i\alpha_j)^{1/6}$ is the effective distance as a function of linear separation r_{ij} and atomic polarizabilities of sites $i(\alpha_i)$ and $j(\alpha_j)$. The factor β is a dimensionless width parameter of the smeared charge distribution to effectively control the damping strength and therefore avoids polarization catastrophes at small separations.

The gradient of the polarization energy can be derived from Equation 3.5. We define a quantity $C = \alpha^{-1} - T^{11}$, which is symmetric (i.e., $C^t = C$). Equation 3.5 can be rewritten as

$$U = -\frac{1}{2}E^T C^{-1}E. \qquad (3.8)$$

Note the E here is the permanent field arising from other permanent atomic multipoles. Kong has described in details how to derive the analytical gradient of the polarization energy (Kong 1997).

There is a potential problem in applying this model to large molecules with intramolecular polarization. Typically, the intramolecular nonbonded interactions between atoms joined by three bonds (1–4 interactions) or less in between are ignored or scaled (i.e., masked) in molecular mechanics. The reasoning behind this is that such interactions are already included in the valence energy terms. We could have taken this approach and ignored the polarization due to permanent multipoles on atoms that are one, two, or three bonds away. However, to take advantage of the intramolecular polarization in modeling conformational energetics, a group-based polarization scheme was developed. As a result, the polarization energy is now

$$U = -\frac{1}{2}E_\mu^T C^{-1}E_p \qquad (3.9)$$

where the subscription p indicates the permanent field in group-based polarization process that produces the induced dipoles. The other E is the permanent field involved in the atomic-based interactions where 1–2 and 1–3 interactions are skipped and 1–4 and 1–5 interactions are scaled by 0.4 and 0.8, respectively.

The gradient then follows:

$$\frac{\partial U^{\text{pol}}}{\partial x_k} = -\frac{1}{2}\left(\frac{\partial E_\mu^T}{\partial x_k}C^{-1}E_p + E_\mu^T\frac{\partial C^{-1}}{\partial x_k}E_p + E_\mu^TC^{-1}\frac{\partial E_p}{\partial x_k}\right). \tag{3.10}$$

Note that $\mu = C^{-1}E_\mu$. A mathematical quantity $\nu = C^{-1}E_p$ is defined as the same fashion of μ. Note the induce–induce interaction is required to be present among atoms in the Thole type of mutual induction model, regardless of the chemical bonding. The mutual induction is crucial to achieve anisotropic molecular response with isotropic atomic polarizabilities. Otherwise the molecular polarizability is merely a sum of atomic polarizabilities and anisotropic tensor will be required. Models based on the latter approach exist (Dykstra 1993, Miller 1990). As a result, the matrix C is invariant regardless of how the permanent-induced interactions are treated.

Given that $C^{-1}C = I$ and $(\partial C^{-1}/\partial x_k)C + C^{-1}(\partial C/\partial x_k) = 0$, we have

$$\frac{\partial U^{\text{pol}}}{\partial x_k} = -\frac{1}{2}\left(\frac{\partial E_\mu^T}{\partial x_k}\nu - E_\mu^TC^{-1}\frac{\partial C}{\partial x_k}C^{-1}E_p + \mu^T\frac{\partial E_p}{\partial x_k}\right)$$

$$= -\frac{1}{2}\left[\frac{\partial T_\mu^1 M}{\partial x_k}\nu + \mu^T\frac{\partial T_p^1 M}{\partial x_k}\right] - \frac{1}{2}\mu^T\frac{\partial T^{11}}{\partial x_k}\nu. \tag{3.11}$$

When the masking rules for intramolecular direct induction and polarization energy are the same, it reduces to a simple expression

$$\frac{\partial U^{\text{pol}}}{\partial x_k} = -\mu^T\frac{\partial T^1 M}{\partial x_k} - \frac{1}{2}\mu^T\frac{\partial T^{11}}{\partial x_k}\mu \tag{3.12}$$

that is the exact gradient for a system such as water where there is no intramolecular polarization originating from internal permanent multipoles.

To compute the long-range electrostatic interactions, efficient particle-mesh Ewald (PME) method has been implemented (Darden et al. 1993, Sagui et al. 2004). One advantage with high-order moments is that a smaller real-space cutoff distance (e.g., 7 Å) can be used to achieve the same level of accuracy compared to fixed-charge model (~9 Å), which means that more work is shifted to the reciprocal space. This improves the computational for serial or limited parallel jobs. The detailed formula for real-space Ewald calculations of polarization energy, force, and torque can be found in a previous publication (Ren and

Ponder 2003). The additional algorithm introduced for large molecules with intramolecular group-based polarization was discussed in a subsequent paper (Ren et al. 2011b). As a principle, the corresponding terms in Equation 3.12 are replaced by those in Equation 3.12 in the real and reciprocal spaces. Similar changes shall be made to the polarization-related torque, where μ is replaced by the $\frac{1}{2}(\mu + \nu)$.

3.3 PARAMETERIZATION OF THE ELECTROSTATIC INTERACTION

3.3.1 Permanent Electrostatic Parameters

As discussed in previous sections, the electrostatic energy includes both permanent and polarized contributions. For each molecule, the atomic multipole moments can be derived from DMA (Stone 1981) of high-level QM calculations (current practical applications have generally been restricted at the second-order Møller–Plesset perturbation [MP2] theory due to the computational cost).

A local coordinate frame is defined for each type of atom. Three common definitions of the local framework, as illustrated in Figure 3.1, are available in AMOEBA to cover essentially all situations in organic chemistry. The first local coordinate frame is the so-called Z-then-X frame (Figure 3.1a). In this definition, the z-axis is defined first along the vector connecting the center atom and z-axis atom; the x-axis can then be defined as a line that is perpendicular to the z-axis and forms an acute angle with the vector pointing from the center atom to the x-axis defining atom. With the z- and x-axes defined, the y-axis is set with the right-hand coordinate system. Chiral centers can be treated with the y-components of multipoles. For nonchiral atoms, the y-component of dipole moment xy (yx) and yz (zy) components of quadrupole of the center atom should be zero. When dealing with molecules with twofold local symmetry or pseudosymmetry, the bisector frame (Figure 3.1b) is used. Taking water molecule as an example, the z-axis is defined to be along the bisector of the H–O–H angle, and the x-axis is defined perpendicular to the z-axis within the plane of the H–O–H angle. Since the two H atoms are identical, the x-axis can be either one of the two possible directions. This in turn requires the resulting multipoles for O to have zeroes for x-component of dipole, xy (yx) and xz (zx) components of quadrupole. Another good example for this frame is the carbon atom in aliphatic methylene. The third frame, which is called Z-bisector frame (Figure 3.1c), finds situations where symmetry or pseudosymmetry is not along the direction of the primary axis (i.e., the z-axis), such as the sulfur atom in dimethyl sulfoxide. This frame can also be used for threefold symmetry found in ammonia. The different local frames give rise to different rotation matrix for

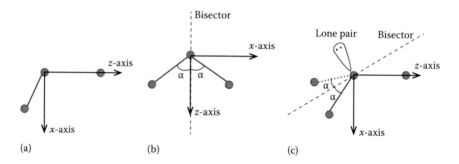

Figure 3.1   Illustrations of the three different local frame definitions for permanent atomic multipole moments in AMOEBA force field. In all definitions, the z-axis is always defined first and then the x-axis. The y-axis is chosen according to the right-handed coordinate system. In (a) Z-then-X and (b) bisector definition, all three atoms are in the same plane of the z- and x-axes. However, in the case of (c) Z-bisector, the z- and x-axes are in plane with the bisector.

translating multipoles from the local to global frames and also result in different treatment when converting torques into atomic forces.

The permanent multipole moments can be derived from the charge density of the molecule obtained from a single-point ab initio QM calculation. The starting structure can be either an experimental structure, for example, water, or a geometry-optimized structure from a QM calculation with similar levels of theory. Currently, MP2 theory is normally used in either geometry optimizations or single-point energy calculations. However, different basis set may be applied depending on the actual size of the molecule. Pople's 6-311G series basis set with additional polarization and diffusion functions is typically used with the DMA. Such initial atomic multipoles can be further optimized by fitting to the ESPs computed by using large basis sets such as Dunning's correlation consistent aug-cc-pVTZ. As it is often necessary to average the atomic multipoles over equivalent atom types (e.g., the H atoms in a CH_3 group), average multipoles over conformations, and move atomic multipoles when joining fragments (e.g., amino acids) together to ensure neutrality, the ESP fitting procedure offers a rigorous way to compensate such modifications (Ren et al. 2011b).

3.3.2   Polarization Parameters

3.3.2.1   Atomic Polarizability and the Damping Function

In a Thole-type model (Equation 3.5), molecular polarizability can be reproduced by using element-based isotropic atomic polarizability and a single universal value of the damping coefficient a for numerous small molecules (Stone 1981). Thole suggested a single value of 0.572 for β by fitting to

TABLE 3.1   ATOMIC POLARIZABILITY AND THE DAMPING FACTOR FOR VARIOUS ELEMENTS AVAILABLE IN AMOEBA FORCE FIELD

Element	Polarizability, α (Å^3)	Damping Factor, β
H	0.496	0.39
H (aromatic)	0.696	0.39
C	1.334	0.39
C (aromatic)	1.750	0.39
N	1.073	0.39
O	0.837	0.39
S	2.800	0.39
Na^+	0.120	0.39
K^+	0.780	0.39
Cl^-	4.000	0.39
Mg^{2+}	0.080	0.10
Ca^{2+}	0.550	0.16
Zn^{2+}	0.260	0.21

experiment molecular polarizability. However, the later studies of water by Ren and Ponder (Ponder and Case 2003) suggest that the polarization energy, which is not considered in Thole's work, is more sensitive to the damping factor. Thus, by fitting to interaction energies of a number of water clusters, a universal value of 0.39 for β was assigned to each atom type in AMOEBA. However, for divalent metal cations, stronger damping (i.e., $\beta < 0.39$) has been suggested to better model the electric field around (Jiao et al. 2006, Piquemal et al. 2006, Ponder et al. 2010). The variation in the damping strengths offers the ability to differentiate the electrostatic properties of these ions, which typically only differ in vdW parameters in the classical force fields (Table 3.1).

3.3.2.2 Intramolecular Polarization

As the size of a molecule getting larger, the number of conformations it can access also increases quickly. An obvious question associated with this is whether the same set of electrostatic parameters is able to correctly model the ESP of different conformers. It has been noted that the electrostatic parameters derived for alanine dipeptide, for example, vary significantly when different conformation of the dipeptide was used in parameterization (Price et al. 1991, Ren and Ponder 2002). Simply averaging the electrostatic parameters derived from different conformers does not give satisfactory ESPs for each of them.

In addition, it has also been suggested that when short-range polarization between bonded atoms is ignored, only marginal improvement was observed in comparison with nonpolarizable models even if intramolecular polarization was considered (Ren and Ponder 2002).

AMOEBA, as mentioned earlier, adopts a group-based scheme when calculating the induced dipole moment (Ren and Ponder 2002, Ren et al. 2011b). A *group*, typically, is defined as a rigid functional group with limited conformational degrees of freedom in a molecule. A good example of a group can be an amide group or a phenyl ring in an amino acid, for instance. For small molecules like water, each molecule is a group. When evaluating the induced dipole moment centered on atom i, the electric field due to all other atoms, except the field due to permanent multipoles of atoms within the same group, will be accounted, as shown in Equation 3.6. However, when evaluating the polarization energy and gradient, an atomic-based scaling rule is used as in typical classical mechanics. The permanent multipole and the induced dipole will be treated consistently, where 1–2 and 1–3 interactions are skipped, 1–4 interaction will be scaled by a factor of 0.4, and 1–5 interaction will be scaled by a factor of 0.8 for all bonded atoms. The intramolecular polarization, if any, needs to be subtracted from the QM multipoles to give the true *permanent* multipole moments (Ren and Ponder 2002). The devised intramolecular polarization, along with the ESP fitting discussed earlier, proves effective in producing conformation-independent permanent multipoles for flexible molecules and has been utilized in parameterizing electrostatic parameters for flexible molecules such as amino acids.

3.4   APPLICATIONS

Although AMOEBA force field is still under active development, it has been applied to various areas and problems in the past. AMOEBA's incorporation of permanent atomic multipoles and polarization effect allows one to explore electrostatic properties beyond the reach of fixed-charge models. Generally, AMOEBA performs well in both gas and solution phases for water, ions, small molecules, peptide, and proteins (Grossfield et al. 2003, Jiang et al. 2010, Jiao et al. 2006, 2008, Ren and Ponder 2002, Ren et al. 2011b, Shi et al. 2011, 2012, Wu et al. 2010, Zhang et al. 2012). Particularly, AMOEBA has also been demonstrated in protein–ligand binding calculations and x-ray crystallography high-resolution refinement where polarization and accurate electrostatics are critical (Boresch et al. 2003, Hamelberg and McCammon 2004, Jiao et al. 2008, 2009, Schnieders et al. 2011a, Yue et al. 2009). Several software packages have implemented the AMOEBA polarizable force field, including TINKER (Ponder 2012), Amber (Case et al. 2005, Weiner and Kollman 1981), OpenMM

(Eastman and Pande 2010), and Force Field X (Schnieders et al. 2011b). Among these, TINKER and FFX are capable of shared-memory parallelization, Amber PMEMD is MPI capable and OpenMM runs on GPUs.

POLTYPE is a program developed in an effort to automate the tedious parameterization procedure, including derivation of atomic multipoles from QM calculations, assigning atomic polarizability, vdW, and valence parameters, and fit the torsions around the rotatable bonds if requested by the users (Wu et al. 2012)

3.4.1   Hydration Free-Energy Calculations of Small Molecules

Small molecules serve multiple biological functions in the fields of pharmacology and biochemistry; they could be building blocks, substrates, and inhibitors in human bodies and interfere with the protein targets in drug design strategy (Arkin and Wells 2004, Leeson and Springthorpe 2007, Lipinski 2004, Veber et al. 2002, Zhang et al. 2009). Evaluating the hydration free energy (HFE) of small molecules is indispensable in predicting behaviors of biological molecules and part of the protein–ligand binding free-energy equation (Ponder et al. 2010). HFE is also useful in estimating the solubility for pharmaceutical and chemical engineering purposes. Thus, HFE is the natural test for a force field. During the past decades, there have been extensive simulation studies on HFE and increasing effort to reduce the mean unsigned error (MUE) of predicted HFE to about 1.0 kcal/mol of organic small molecules by improving the fixed-charge force fields (Mobley et al. 2007). Reducing this discrepancy is critical in protein–ligand binding calculations since large errors in HFE might draw the conclusions to wrong or an even opposite direction.

The major challenge in computing HFE with fixed-charge models is to derive atomic charges from gas-phase QM but corresponding to a liquid environment (Ponder et al. 2010). Even with perfect HFE, it is unclear how these charges would perform in a different chemical environment. With AMOEBA force field, HFE is computed via a free-energy alchemical transformation procedure using the Bennett acceptance ratio (BAR) method (Bennett 1976). Tests on a wide range of organic compounds show that it is possible to achieve an MUE less than 1.0 kcal/mol due to the incorporation of atomic multipoles and polarization, and such improvement lays the groundwork for the development of more reliable force field for protein and nucleic acid (Ponder et al. 2010, Ren et al. 2011b). Ponder et al. summarized the calculated HFE of 30 common biochemical small molecules using AMOEBA force fields; the rms error comparing to experimental values is 0.68 kcal/mol; results are shown in Figure 3.2 (Ponder et al. 2010). Shi et al. have explored and determined simulation protocols on HFE calculations (Shi et al. 2011). They found that with AMOEBA, incorporating diffuse basis functions (e.g., aug-cc-pVTZ as the basis set) is important in deriving the

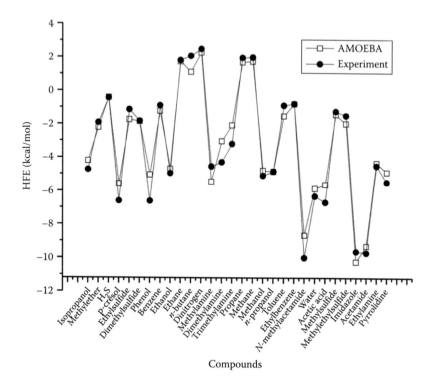

Figure 3.2   Hydration free energy of small molecules using AMOEBA. (From Ponder, J.W. et al., *J. Phys. Chem. B*, 114, 2549, 2010.)

permanent atomic multipoles. Note that these small molecules, while common in biological systems only represent a fraction of chemical space of synthetic ligands. Many other chemical groups such as phosphate and halogens are yet to be carefully examined.

3.4.2   Ion Modeling

Many of metal ions like sodium, potassium, magnesium, and calcium are essential to living organisms (Finney and O'Halloran 2003, Luoma 1983). There are approximately one-third of currently found protein structures in Protein Data Bank containing metal ions critical for biochemical functions (Chih-Hao et al. 2012). Due to the polyanionic nature, metal ions are fundamental components with the ability to serve both catalytic and structural roles for biomacromolecules, including metal-binding proteins (i.e., metalloproteins) and nucleic acids (Ben-David et al. 2013, Tainer et al. 1991, Wriggers et al. 1998).

To comprehensively understand the structure and function of biomacromolecules, a detailed illustration of the binding mechanism between biomolecules and metal ions is necessary. Because binding between biomolecules and metal ions requires specific structural coordination and condition, it is arduous to experimentally identify and characterize the metal ion–binding motifs. Ion–protein interactions usually involve conformational changes (Wriggers et al. 1998, Zhang et al. 2012) and electrostatic and coordinative interactions (Xiao et al. 2007); upon such convoluted process, the dynamics and energetics of protein or RNA folding will be influenced significantly.

It is in general difficult for fixed-charge models to capture structural and dynamic details of ions due to the deficiencies in representing the ionic size and high charge distribution. Therefore, with the incorporation of ion's polarization effect, polarizable force field has been applied and demonstrated by its success for a number of anions and cations, including Na^+, K^+, Mg^{2+}, Ca^{2+}, Zn^{2+}, Cl^-, Br^-, and I^- (Dang et al. 2012, Grossfield 2005, Grossfield et al. 2003, Jiao et al. 2006, Luo et al. 2013, Roux et al. 2007, Wu et al. 2010). With AMOEBA, ion-cluster solvation enthalpies and solvation free energies have been evaluated via MD simulations. In a recent report of Zinc-containing matrix metalloproteinases (MMPs), Zhang et al. examined Zn^{2+} coordination with organic compounds and protein side chains using AMOEBA (Zhang et al. 2012). They identified the importance of the polarization effect on both MMP complexes and zinc-finger proteins; especially, they found that polarization could actually determine the coordination geometry of Zn^{2+} that is fundamental in regulating metalloenzymes' catalytic activities. It was recently shown in a study of the Ca^{2+}–Cl^- system by Dang et al. that reproducing single-ion HFE does not necessarily guarantee correct modeling of interionic interactions (Dang et al. 2012). The balance between ion–ion and ion–water interactions is essential.

3.4.3 Protein–Ligand Binding

Almost every crucial biomolecular function within a cell, from enzyme catalysis to intracellular signaling, depends on the specific binding between proteins or protein and ligand; affinity is the characterization of the binding strength. Ren and coworkers have applied AMOEBA model to calculate the protein–ligand binding affinity for a few systems (Jiao et al. 2008, Shi et al. 2012, Yue et al. 2009, Zhang et al. 2012). As an initial attempt, trypsin–benzamidine system was selected for investigation (Jiao et al. 2008, Yue et al. 2009). For trypsin, benzamidine and its derivatives are well-characterized inhibitors (Figure 3.3). The net charges, relative size, and rigidity of benzamidine are ideal for the application of the polarizable potential. With alchemical transformation, the absolute binding free energy of benzamidinium and the relative binding free energies of a series of benzamidine analogs were computed, and the average error

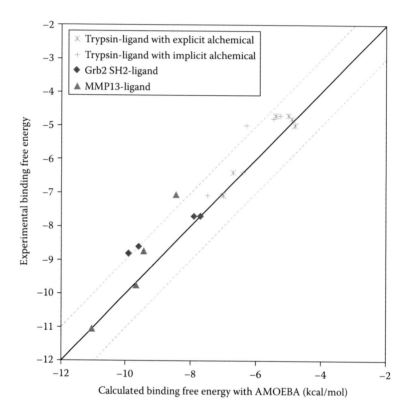

Figure 3.3 Comparison of AMOEBA-calculated binding free energy with experimental values. The trypsin binding to the same set of ligands was investigated using two different methods: alchemical perturbation in explicit solvent or implicit solvent (Jiao et al. 2009, Yue et al. 2009). The MMP13 (Zhang et al. 2012) and SH2 domain (Shi et al. 2012) binding simulations were all performed in explicit solvent.

was within 1.0 kcal/mol. The absolute binding affinities range from −6.7 to −7.3 kcal/mol, which were corroborated by experimental observations (Katz et al. 2001, Schwarzl et al. 2002, Talhout and Engberts 2001). Free-energy decomposition suggested that ignoring the polarization effect moving from water to protein environments will likely lead to an overestimation of ligand–protein binding free energy. It was also found that implicit solvent-based calculations could be reliable if alchemical transformations are applied to achieve the sampling as opposed to the typical end-state approaches using the complex structure (Jiao et al. 2009). Furthermore, the AMOEBA model was also used to examine the effect of conformational constraint on phosphorylated ligand binding to SH2 domain (Shi et al. 2012). In this later study, an *entropic paradox* during the ligand preorganization was observed, which refers to a trend that the binding of unconstrained peptides is entropically more favorable than the constrained counterparts (Shi et al. 2012).

The simulations not only reproduced the experimental trend but also explained this phenomenon by suggesting lower conformational entropy of the unconstrained peptide ligands in solution due to the specific intramolecular interactions (Figure 3.3). In the study of MMP binding with pyrimidine dicarboxamide inhibitors, the polarizable force field was able to capture the relative binding free energy upon subtle chemical changes in the ligands and demonstrated that the polarization effect due to Zn^{2+} is important (Figure 3.3).

3.4.4   X-Ray Structure Refinement and Prediction

Biomacromolecule's functions are largely dependent on their 3D conformation. X-ray crystallography is predominantly used to obtain a biomolecule's atomic structure (Deisenhofer et al. 1984, Mertens and Svergun 2010). However, using the collected diffraction data to resolve the 3D coordinates requires further calculation, refinement, and prior chemical knowledge. Polarizable atomic multipoles evaluated by PME (Darden et al. 1993) summation have been shown to improve x-ray crystallography refinement by a series of studies by Schnieders et al. (Fenn and Schnieders 2011, Fenn et al. 2010, 2011, Schnieders et al. 2009, 2012). In their work, symmetry operators were incorporated into PME electrostatics to improve the accuracy as well as efficiency. It is for the first time that the reliability and affordability of introducing PME to x-ray refinement are demonstrated of a range of system sizes encountered in macromolecular crystallography.

Recently, Schnieders et al. described a consistent procedure to predict structure, thermodynamic stability, and solubility of organic crystals (Fenn and Schnieders 2011, Schnieders et al. 2012). Solubility is a key parameter in determining the bioavailability and efficacy of pharmaceutical compounds. In addition, an organic compound may exist in multiple stable crystal forms (polymorphism). It is of critical importance to exploit a computational approach to reliably predict crystal structure and stabilities. By using the orthogonal space random walk (OSRW) sampling strategy (Zheng et al. 2008, 2009), structures and standard state solubility of a series of n-alkylamides were accurately calculated using the polarizable multipole force field.

3.5   CONCLUSIONS AND FUTURE DIRECTIONS

In the past decade, significant progress has been made in developing the next-generation polarizable force fields. Sophisticated models have been introduced to treat electrostatic interactions because of their long-range nature and at the same time strong dependence on the local chemical and physical environments. Moving beyond spherical atomic charge representation, atomic multipoles or Gaussian distributions will provide more accurate description of electrostatic interactions. Equally important, explicitly introducing the polarization effect could bring more flexibility and better transferability to force field development among a wide range of molecular systems in various environments. Polarizable force fields have demonstrated encouraging successes in a range of applications, from water to ions, from small organic molecules to proteins.

Despite these successful applications, challenges remain for the development of force fields with advanced electrostatics. A major hurdle is the increased computational expense due to complex algorithm. Many-body polarization requires computationally costly procedure including the self-consistent iteration to converge the induced dipoles. Better and more efficient configurational sampling algorithms are necessary in order to advance the application of polarizable force fields. Besides, although polarizable models provide more accurate descriptions of electrostatic interactions especially at the short range, the fixed-charge model remains useful in many situations after careful parameterization. The more advanced force fields can potentially be utilized in such parameterization or used in conjunction with fixed-charge force field in a multiscale fashion. Moreover, some other nontrivial electrostatic interactions, including short-range electrostatic penetration and charge-transfer effects, can be important in certain molecular systems and will likely become the focuses of future force field development.

ACKNOWLEDGMENTS

The authors are grateful to the support provided by Robert A. Welch Foundation (F-1691) and the National Institute of General Medical Sciences (R01 GM079686).

REFERENCES

Anisimov, V. M., I. V. Vorobyov, B. Roux, and A. D. MacKerell. 2007. Polarizable empirical force field for the primary and secondary alcohol series based on the classical drude model. *Journal of Chemical Theory and Computation* 3:1927–1946.

Applequist, J., J. R. Carl, and K.-K. Fung. 1972. An atom dipole interaction model for molecular polarizability. Application to polyatomic molecules and determination of atom polarizabilities. *Journal of the American Chemical Society* 94:2952–2960.

Arkin, M. R. and J. A. Wells. 2004. Small-molecule inhibitors of protein–protein interactions: Progressing towards the dream. *Nature Reviews Drug Discovery* 3:301–317.

Banks, J. L., G. A. Kaminski, R. H. Zhou et al. 1999. Parameterizing a polarizable force field from ab initio data. I. The fluctuating point charge model. *Journal of Chemical Physics* 110:741–754.

Barker, J. A. 1953. Statistical mechanics of interacting dipoles. *Proceedings of the Royal Society of London. Series A, Mathematical and Physical Sciences* 219:367–372.

Bauer, B. A. and S. Patel. 2009. Properties of water along the liquid–vapor coexistence curve via molecular dynamics simulations using the polarizable TIP4P-QDP-LJ water model. *Journal of Chemical Physics* 131:084709.

Ben-David, M., G. Wieczorek, M. Elias et al. 2013. Catalytic metal ion rearrangements underline promiscuity and evolvability of a metalloenzyme. *Journal of Molecular Biology* 425(6):1028–1038.

Bennett, C. H. 1976. Efficient estimation of free-energy differences from Monte-Carlo data. *Journal of Computational Physics* 22:245–268.

Boresch, S., F. Tettinger, M. Leitgeb, and M. Karplus. 2003. Absolute binding free energies: A quantitative approach for their calculation. *Journal of Physical Chemistry B* 107:9535–9551.

Brdarski, S. and G. Karlström. 1998. Modeling of the exchange repulsion energy. *Journal of Physical Chemistry A* 102:8182–8192.

Caldwell, J. W. and P. A. Kollman. 1995. Structure and properties of neat liquids using nonadditive molecular dynamics: Water, methanol, and *N*-methylacetamide. *Journal of Physical Chemistry* 99:6208–6219.

Carignano, M. A., G. Karlström, and P. Linse. 1997. Polarizable ions in polarizable water: A molecular dynamics study. *Journal of Physical Chemistry B* 101:1142–1147.

Case, D. A., T. E. Cheatham, T. Darden et al. 2005. The Amber biomolecular simulation programs. *Journal of Computational Chemistry* 26:1668–1688.

Chih-Hao, L., L. Yu-Feng, L. Jau-Ji, and Y. Chin-Sheng. 2012. Prediction of metal ion-binding sites in proteins using the fragment transformation method. *PLoS One* 7:e39252.

Cieplak, P., F. Y. Dupradeau, Y. Duan, and J. M. Wang. 2009. Polarization effects in molecular mechanical force fields. *Journal of Physics—Condensed Matter* 21:333102–333123.

Cisneros, G. A., J. P. Piquemal, and T. A. Darden. 2006. Generalization of the Gaussian electrostatic model: Extension to arbitrary angular momentum, distributed multipoles, and speedup with reciprocal space methods. *Journal of Chemical Physics* 125:184101.

Cornell, W. D., P. Cieplak, C. I. Bayly et al. 1995. A 2nd generation force-field for the simulation of proteins, nucleic-acids, and organic-molecules. *Journal of the American Chemical Society* 117:5179–5197.

Dang, L. X., T. B. Truong, and B. Ginovska-Pangovska. 2012. Note: Interionic potentials of mean force for Ca^{2+}–Cl^- in polarizable water. *Journal of Chemical Physics* 136:126101–126102.

Darden, T., D. York, and L. Pedersen. 1993. Particle Mesh Ewald—An N·Log(N) method for Ewald sums in large systems. *Journal of Chemical Physics* 98:10089–10092.

Davis, J. E., O. Raharnan, and S. Patel. 2009. Molecular dynamics simulations of a DMPC bilayer using nonadditive interaction models. *Biophysical Journal* 96:385–402.

Deisenhofer, J., O. Epp, K. Miki, R. Huber, and H. Michel. 1984. X-ray structure analysis of a membrane protein complex. Electron density map at 3 A resolution and a model of the chromophores of the photosynthetic reaction center from Rhodopseudomonas viridis. *Journal of Molecular Biology* 180:385–398.

Donchev, A. G., N. G. Galkin, A. A. Illarionov et al. 2006. Water properties from first prin-ciples: Simulations by a general-purpose quantum mechanical polarizable force field. *Proceedings of the National Academy of Sciences of the United States of America* 103:8613–8617.

Donchev, A. G., N. G. Galkin, A. A. Illarionov et al. 2008. Assessment of performance of the general purpose polarizable force field QMPFF3 in condensed phase. *Journal of Computational Chemistry* 29:1242–1249.

Donchev, A. G., V. D. Ozrin, M. V. Subbotin, O. V. Tarasov, and V. I. Tarasov. 2005. A quan-tum mechanical polarizable force field for biomolecular interactions. *Proceedings of the National Academy of Sciences of the United States of America* 102:7829–7834.

Drude, P., C. R. Mann, and R. A. Millikan. 1902. *The Theory of Optics*. Longmans, Green, and Co., New York.

Dykstra, C. E. 1993. Electrostatic interaction potentials in molecular force fields. *Chemical Reviews* 93:2339–2353.

Eastman, P. and V. S. Pande. 2010. OpenMM: A hardware-independent framework for molecular simulations. *Computing in Science & Engineering* 12:34–39.

Fenn, T. D. and M. J. Schnieders. 2011. Polarizable atomic multipole X-ray refinement: Weighting schemes for macromolecular diffraction. *Acta Crystallographica Section D* 67:957–965.

Fenn, T. D., M. J. Schnieders, A. T. Brunger, and V. S. Pande. 2010. Polarizable atomic mul-tipole X-ray refinement: Hydration geometry and application to macromolecules. *Biophysical Journal* 98:2984–2992.

Fenn, T. D., M. J. Schnieders, M. Mustyakimov et al. 2011. Reintroducing electrostatics into macromolecular crystallographic refinement: Application to neutron crystal-lography and DNA hydration. *Structure* 19:523–533.

Finney, L. A. and T. V. O'Halloran. 2003. Transition metal speciation in the cell: Insights from the chemistry of metal ion receptors. *Science* 300:931–936.

Friesner, R. A. 2006. Modeling polarization in proteins and protein–ligand complexes: Methods and preliminary results. *Advances in Protein Chemistry* 72:79–104.

Gagliardi, L., R. Lindh, and G. Karlström. 2004. Local properties of quantum chemical systems: The LoProp approach. *Journal of Chemical Physics* 121:4494–4500.

Garmer, D. R. and W. J. Stevens. 1989. Transferability of molecular distributed polarizabil-ities from a simple localized orbital based method. *Journal of Physical Chemistry* 93:8263–8270.

Geerke, D. P. and W. F. van Gunsteren. 2007. Calculation of the free energy of polarization: Quantifying the effect of explicitly treating electronic polarization on the transfer-ability of force-field parameters. *Journal of Physical Chemistry B* 111:6425–6436.

Gresh, N., G. A. Cisneros, T. A. Darden, and J. P. Piquemal. 2007. Anisotropic, polarizable molecular mechanics studies of inter- and intramolecular interactions and ligand–macromolecule complexes. A bottom-up strategy. *Journal of Chemical Theory and Computation* 3:1960–1986.

Gresh, N., P. Claverie, and A. Pullman. 1984. Theoretical studies of molecular conformation—Derivation of an additive procedure for the computation of intra-molecular interaction energies—Comparison with ab initio SCF computations. *Theoretica Chimica Acta* 66:1–20.

Grossfield, A. 2005. Dependence of ion hydration on the sign of the ion's charge. *Journal of Chemical Physics* 122:024506.

Grossfield, A., P. Ren, and J. W. Ponder. 2003. Ion solvation thermodynamics from simulation with a polarizable force field. *Journal of the American Chemical Society* 125:15671–15682.

Halgren, T. A. 1992. Representation of Vanderwaals (Vdw) interactions in molecular mechanics force-fields—Potential form, combination rules, and Vdw parameters. *Journal of the American Chemical Society* 114:7827–7843.

Hamelberg, D. and J. A. McCammon. 2004. Standard free energy of releasing a localized water molecule from the binding pockets of proteins: Double-decoupling method. *Journal of the American Chemical Society* 126:7683–7689.

Holt, A. and G. Karlström. 2008. Inclusion of the quadrupole moment when describing polarization. The effect of the dipole–quadrupole polarizability. *Journal of Computational Chemistry* 29:2033–2038.

Holt, A. and G. Karlström. 2009. Improvement of the NEMO potential by inclusion of intramolecular polarization. *International Journal of Quantum Chemistry* 109:1255–1266.

Jiang, J., Y. Wu, Z.-X. Wang, and C. Wu. 2010. Assessing the performance of popular quantum mechanics and molecular mechanics methods and revealing the sequence-dependent energetic features using 100 tetrapeptide models. *Journal of Chemical Theory and Computation* 6:1199–1209.

Jiao, D., P. A. Golubkov, T. A. Darden, and P. Ren. 2008. Calculation of protein–ligand binding free energy by using a polarizable potential. *Proceedings of the National Academy of Sciences of the United States of America* 105:6290–6295.

Jiao, D., C. King, A. Grossfield, T. A. Darden, and P. Ren. 2006. Simulation of Ca^{2+} and Mg^{2+} solvation using polarizable atomic multipole potential. *Journal of Physical Chemistry B* 110:18553–18559.

Jiao, D., J. Zhang, R. E. Duke et al. 2009. Trypsin-ligand binding free energies from explicit and implicit solvent simulations with polarizable potential. *Journal of Computational Chemistry* 30:1701–1711.

Jorgensen, W. L., D. S. Maxwell, and J. Tirado-Rives. 1996. Development and testing of the OPLS all-atom force field on conformational energetics and properties of organic liquids. *Journal of the American Chemical Society* 118:11225–11236.

Kaminski, G. A., R. A. Friesner, and R. H. Zhou. 2003. A computationally inexpensive modification of the point dipole electrostatic polarization model for molecular simulations. *Journal of Computational Chemistry* 24:267–276.

Kaminski, G. A., H. A. Stern, B. J. Berne et al. 2002. Development of a polarizable force field for proteins via ab initio quantum chemistry: First generation model and gas phase tests. *Journal of Computational Chemistry* 23:1515–1531.

Kaminski, G. A., H. A. Stern, B. J. Berne, and R. A. Friesner. 2004. Development of an accurate and robust polarizable molecular mechanics force field from ab initio quantum chemistry. *Journal of Physical Chemistry A* 108:621–627.

Katz, B. A., K. Elrod, C. Luong et al. 2001. A novel serine protease inhibition motif involving a multi-centered short hydrogen bonding network at the active site. *Journal of Molecular Biology* 307:1451–1486.

Kong, Y. 1997. Multipole electrostatic methods for protein modeling with reaction field treatment. Dissertation, Washington University, St. Louis, MO.

Lamoureux, G., E. Harder, I. V. Vorobyov, B. Roux, and A. D. MacKerell. 2006. A polarizable model of water for molecular dynamics simulations of biomolecules. *Chemical Physics Letters* 418:245–249.

Lamoureux, G., A. D. MacKerell, and B. Roux. 2003. A simple polarizable model of water based on classical Drude oscillators. *Journal of Chemical Physics* 119:5185–5197.

Le Sueur, C. R. and A. J. Stone. 1993. Practical schemes for distributed polarizabilities. *Molecular Physics* 78:1267–1291.

Leeson, P. D. and B. Springthorpe. 2007. The influence of drug-like concepts on decision-making in medicinal chemistry. *Nature Reviews Drug Discovery* 6:881–890.

Lii, J. H. and N. L. Allinger. 1989a. Molecular mechanics—The MM3 force-field for hydrocarbons. 2. Vibrational frequencies and thermodynamics. *Journal of the American Chemical Society* 111:8566–8575.

Lii, J. H. and N. L. Allinger. 1989b. Molecular mechanics—The MM3 force-field for hydrocarbons. 3. The Vanderwaals potentials and crystal data for aliphatic and aromatic-hydrocarbons. *Journal of the American Chemical Society* 111:8576–8582.

Lipinski, C. A. 2004. Lead- and drug-like compounds: The rule-of-five revolution. *Drug Discovery Today: Technologies* 1:337–341.

Lopes, P. E. M., G. Lamoureux, B. Roux, and A. D. MacKerell. 2007. Polarizable empirical force field for aromatic compounds based on the classical drude oscillator. *Journal of Physical Chemistry B* 111:2873–2885.

Lopes, P. E. M., B. Roux, and A. D. MacKerell. 2009. Molecular modeling and dynamics studies with explicit inclusion of electronic polarizability: Theory and applications. *Theoretical Chemistry Accounts* 124:11–28.

Luo, Y., W. Jiang, H. Yu, A. D. MacKerell, and B. Roux. 2013. Simulation study of ion pairing in concentrated aqueous salt solutions with a polarizable force field. *Faraday Discussions* 160:135–149.

Luoma, S. N. 1983. Bioavailability of trace metals to aquatic organisms—A review. *The Science of the Total Environment* 28:1–22.

MacKerell, A. D., D. Bashford, M. Bellott et al. 1998. All-atom empirical potential for molecular modeling and dynamics studies of proteins. *Journal of Physical Chemistry B* 102:3586–3616.

Mertens, H. D. T. and D. I. Svergun. 2010. Structural characterization of proteins and complexes using small-angle X-ray solution scattering. *Journal of Structural Biology* 172:128–141.

Miller, K. J. 1990. Calculation of the molecular polarizability tensor. *Journal of the American Chemical Society* 112:8543–8551.

Mobley, D. L., É. Dumont, J. D. Chodera, and K. A. Dill. 2007. Comparison of charge models for fixed-charge force fields: Small-molecule hydration free energies in explicit solvent. *Journal of Physical Chemistry B* 111:2242–2254.

Moghaddam, S., C. Yang, M. Rekharsky et al. 2011. New ultrahigh affinity host-guest complexes of cucurbit[7]uril with bicyclo[2.2.2]octane and adamantane guests: Thermodynamic analysis and evaluation of M2 affinity calculations. *Journal of the American Chemical Society* 133:3570–3581.

Molnar, L. F., X. He, B. Wang, and K. M. Merz, Jr. 2009. Further analysis and comparative study of intermolecular interactions using dimers from the S22 database. *Journal of Chemical Physics* 131:065102.

Patel, S. and C. L. Brooks III. 2004. CHARMM fluctuating charge force field for proteins: I parameterization and application to bulk organic liquid simulations. *Journal of Computational Chemistry* 25:1–15.

Patel, S., J. E. Davis, and B. A. Bauer. 2009. Exploring ion permeation energetics in gramicidin A using polarizable charge equilibration force fields. *Journal of the American Chemical Society* 131:13890–13891.

Piquemal, J.-P., L. Perera, G. A. Cisneros et al. 2006. Towards accurate solvation dynamics of divalent cations in water using the polarizable amoeba force field: From energetics to structure. *Journal of Chemical Physics* 125:054511–054517.

Piquemal, J. P., B. Williams-Hubbard, N. Fey et al. 2003. Inclusion of the ligand field contribution in a polarizable molecular mechanics: SIBFA-LF. *Journal of Computational Chemistry* 24:1963–1970.

Ponder, J. W. 2012. TINKER: Software tools for molecular design. Washington University, St. Louis, MO. http://dasher.wustl.edu/tinker/v6.0, accessed December 31, 2012.

Ponder, J. W. and D. A. Case. 2003. Force fields for protein simulations. In *Advances in Protein Chemistry*, D. Valerie (ed.). Academic Press, San Diego, CA, pp. 27–85.

Ponder, J. W., C. Wu, P. Ren et al. 2010. Current status of the AMOEBA polarizable force field. *Journal of Physical Chemistry B* 114:2549–2564.

Ponomarev, S. Y. and G. A. Kaminski. 2011. Polarizable Simulations with Second order Interaction Model (POSSIM) force field: Developing parameters for alanine peptides and protein backbone. *Journal of Chemical Theory and Computation* 7:1415–1427.

Price, S. L., C. H. Faerman, and C. W. Murray. 1991. Toward accurate transferable electrostatic models for polypeptides: A distributed multipole study of blocked amino acid residue charge distributions. *Journal of Computational Chemistry* 12:1187–1197.

Rappe, A. K. and W. A. Goddard. 1991. Charge equilibration for molecular-dynamics simulations. *Journal of Physical Chemistry* 95:3358–3363.

Ren, P., M. Marucho, J. Zhang, and N. A. Baker. 2011a. Biomolecular electrostatics and solvation: A computational perspective. *Quarterly Reviews of Biophysics* 45:427–491.

Ren, P. and J. W. Ponder. 2002. Consistent treatment of inter- and intramolecular polarization in molecular mechanics calculations. *Journal of Computational Chemistry* 23:1497–1506.

Ren, P. and J. W. Ponder. 2003. Polarizable atomic multipole water model for molecular mechanics simulation. *Journal of Physical Chemistry B* 107:5933–5947.

Ren, P., C. Wu, and J. W. Ponder. 2011b. Polarizable atomic multipole-based molecular mechanics for organic molecules. *Journal of Chemical Theory and Computation* 7:3143–3161.

Rezac, J., P. Jurecka, K. E. Riley et al. 2008. Quantum chemical benchmark energy and geometry database for molecular clusters and complex molecular systems (www. begdb.com): A users manual and examples. *Collection of Czechoslovak Chemical Communications* 73:1261–1270.

Rick, S. W. 2001. Simulations of ice and liquid water over a range of temperatures using the fluctuating charge model. *Journal of Chemical Physics* 114:2276–2283.

Roux, C., N. Gresh, L. E. Perera, J. P. Piquemal, and L. Salmon. 2007. Binding of 5-phospho-D-arabinonohydroxamate and 5-phospho-D-arabinonate inhibitors to zinc phosphomannose isomerase from *Candida albicans* studied by polarizable molecular mechanics and quantum mechanics. *Journal of Computational Chemistry* 28:938–957.

Sagui, C., L. G. Pedersen, and T. A. Darden. 2004. Towards an accurate representation of electrostatics in classical force fields: Efficient implementation of multipolar interactions in biomolecular simulations. *Journal of Chemical Physics* 120:73–87.

Schnieders, M. J., J. Baltrusaitis, Y. Shi et al. 2012. The structure, thermodynamics and solubility of organic crystals from simulation with a polarizable force field. *Journal of Chemical Theory and Computation* 8:1721–1736 (PMCID: PMC3348590).

Schnieders, M. J., T. D. Fenn, and V. S. Pande. 2011a. Polarizable atomic multipole X-ray refinement: Particle Mesh Ewald electrostatics for macromolecular crystals. *Journal of Chemical Theory and Computation* 7:1141–1156.

Schnieders, M. J., T. D. Fenn, V. S. Pande, and A. T. Brunger. 2009. Polarizable atomic multipole X-ray refinement: Application to peptide crystals. *Acta Crystallographica Section D* 65:952–965.

Schnieders, M. J., T. D. Fenn, J. Wu, W. Yang, and P. Ren. 2011b. Force field X open source, platform independent modules for molecular biophysics simulations. http://ffx.kenai.com, accessed December 31, 2012.

Schnieders, M. J., T. S. Kaoud, C. Yan, K. N. Dalby, and P. Ren. 2012. Computational insights for the discovery of non-ATP competitive inhibitors of MAP kinases. *Current Pharmaceutical Design* 18:1173–1185.

Schwarzl, S. M., T. B. Tschopp, J. C. Smith, and S. Fischer. 2002. Can the calculation of ligand binding free energies be improved with continuum solvent electrostatics and an ideal-gas entropy correction? *Journal of Computational Chemistry* 23:1143–1149.

Shi, Y., C. Wu, J. W. Ponder, and P. Ren. 2011. Multipole electrostatics in hydration free energy calculations. *Journal of Computational Chemistry* 32:967–977.

Shi, Y., C. Z. Zhu, S. F. Martin, and P. Ren. 2012. Probing the effect of conformational constraint on phosphorylated ligand binding to an SH2 domain using polarizable force field simulations. *Journal of Physical Chemistry B* 116:1716–1727.

Song, L., J. Han, Y. L. Lin, W. Xie, and J. Gao. 2009. Explicit polarization (X-Pol) potential using ab initio molecular orbital theory and density functional theory. *Journal of Physical Chemistry A* 113:11656–11664.

Stone, A. J. 1981. Distributed multipole analysis, or how to describe a molecular charge distribution. *Chemical Physics Letters* 83:233–239.

Stone, A. J. 1985. Distributed multipole analysis: Methods and applications. *Molecular Physics* 56:1047–1064.

Stone, A. J. 2005. Distributed multipole analysis: Stability for large basis sets. *Journal of Chemical Theory and Computation* 1:1128–1132.

Tainer, J. A., V. A. Roberts, and E. D. Getzoff. 1991. Metal-binding sites in proteins. *Current Opinion in Biotechnology* 2:582–591.

Talhout, R. and J. B. F. N. Engberts. 2001. Thermodynamic analysis of binding of p-substituted benzamidines to trypsin. *European Journal of Biochemistry* 268:1554–1560.

Valdes, H., K. Pluhackova, M. Pitonak, J. Rezac, and P. Hobza. 2008. Benchmark database on isolated small peptides containing an aromatic side chain: Comparison between wave function and density functional theory methods and empirical force field. *Physical Chemistry Chemical Physics* 10:2747–2757.

Van Belle, D., M. Froeyen, G. Lippens, and S. J. Wodak. 1992. Molecular-dynamics simulation of polarizable water by an extended Lagrangian method. *Molecular Physics* 77:239–255.

Van Belle, D. and S. J. Wodak. 1995. Extended Lagrangian formalism applied to temperature control and electronic polarization effects in molecular dynamics simulations. *Computer Physics Communications* 91:253–262.

Veber, D. F., S. R. Johnson, H.-Y. Cheng et al. 2002. Molecular properties that influence the oral bioavailability of drug candidates. *Journal of Medicinal Chemistry* 45:2615–2623.

Vorobyov, I. V., V. M. Anisimov, and A. D. MacKerell. 2005. Polarizable empirical force field for alkanes based on the classical drude oscillator model. *Journal of Physical Chemistry B* 109:18988–18999.

Wang, J., P. Cieplak, Q. Cai et al. 2012a. Development of polarizable models for molecular mechanical calculations. 3. Polarizable water models conforming to thole polarization screening schemes. *Journal of Physical Chemistry B* 116:7999–8008.

Wang, J. M., P. Cieplak, J. Li et al. 2011a. Development of polarizable models for molecular mechanical calculations I: Parameterization of atomic polarizability. *Journal of Physical Chemistry B* 115:3091–3099.

Wang, J. M., P. Cieplak, J. Li et al. 2011b. Development of polarizable models for molecular mechanical calculations II: Induced dipole models significantly improve accuracy of intermolecular interaction energies. *Journal of Physical Chemistry B* 115:3100–3111.

Wang, J. M., P. Cieplak, J. Li et al. 2012b. Development of polarizable models for molecular mechanical calculations. 4. van der Waals parametrization. *Journal of Physical Chemistry B* 116:7088–7101.

Wang, Y. J., C. P. Sosa, A. Cembran, D. G. Truhlar, and J. L. Gao. 2012c. Multilevel X-Pol: A fragment-based method with mixed quantum mechanical representations of different fragments. *Journal of Physical Chemistry B* 116:6781–6788.

Wang, Z. X., W. Zhang, C. Wu et al. 2006. Strike a balance: Optimization of backbone torsion parameters of AMBER polarizable force field for simulations of proteins and peptides (vol 27, pg 781, 2006). *Journal of Computational Chemistry* 27:994–994.

Weiner, P. K. and P. A. Kollman. 1981. AMBER: Assisted model building with energy refinement. A general program for modeling molecules and their interactions. *Journal of Computational Chemistry* 2:287–303.

Williams, D. E. 1988. Representation of the molecular electrostatic potential by atomic multipole and bond dipole models. *Journal of Computational Chemistry* 9:745–763.

Wriggers, W., E. Mehler, F. Pitici, H. Weinstein, and K. Schulten. 1998. Structure and dynamics of calmodulin in solution. *Biophysical Journal* 74:1622–1639.

Wu, J. C., G. Chattre, and P. Ren. 2012. Automation of AMOEBA polarizable force field parameterization for small molecules. *Theoretical Chemistry Accounts* 131:1138–1148 (PMCID: PMC3322661).

Wu, J. C., J.-P. Piquemal, R. Chaudret, P. Reinhardt, and P. Ren. 2010. Polarizable molecular dynamics simulation of Zn(II) in water using the AMOEBA force field. *Journal of Chemical Theory and Computation* 6:2059–2070.

Xiao, B., R. Heath, P. Saiu et al. 2007. Structural basis for AMP binding to mammalian AMP-activated protein kinase. *Nature* 449:496–500.

Xie, W. and J. Gao. 2007. The design of a next generation force field: The X-POL potential. *Journal of Chemical Theory and Computation* 3:1890–1900.

Xie, W., M. Orozco, D. G. Truhlar, and J. Gao. 2009. X-Pol potential: An electronic structure-based force field for molecular dynamics simulation of a solvated protein in water. *Journal of Chemical Theory and Computation* 5:459–467.

Yue, S., J. Dian, M. J. Schnieders, and R. Pengyu. 2009. Trypsin-ligand binding free energy calculation with AMOEBA. In *Engineering in Medicine and Biology Society, 2009, EMBC 2009. Annual International Conference of the IEEE*, Minneapolis, MN, pp. 2328–2331.

Zhang, J., P. L. Yang, and N. S. Gray. 2009. Targeting cancer with small molecule kinase inhibitors. *Nature Reviews Cancer* 9:28–39.

Zhang, J., W. Yang, J.-P. Piquemal, and P. Ren. 2012. Modeling structural coordination and ligand binding in zinc proteins with a polarizable potential. *Journal of Chemical Theory and Computation* 8:1314–1324.

Zheng, L. Q., M. G. Chen, and W. Yang. 2008. Random walk in orthogonal space to achieve efficient free-energy simulation of complex systems. *Proceedings of the National Academy of Sciences of the United States of America* 105:20227–20232.

Zheng, L. Q., M. G. Chen, and W. Yang. 2009. Simultaneous escaping of explicit and hidden free energy barriers: Application of the orthogonal space random walk strategy in generalized ensemble based conformational sampling. *Journal of Chemical Physics* 130:234105.

Self-Assembly of Biomolecules

Chapter 4

Molecular Simulations of Protein Folding Dynamics and Thermodynamics

Deepak R. Canchi, Charles English,
Camilo A. Jimenez-Cruz, and Angel E. Garcia

CONTENTS

4.1 INTRODUCTION

Proteins are biological macromolecules and are present in and vital to all living organisms.

They carry out a wide variety of functions—catalyze a range of reactions in the cell, provide structural rigidity to the cell, control transport across the cell membrane, act as sensors and switches in cell signaling, cause motion, bind to foreign bodies in immune response, and control gene function. Structurally, proteins are heteropolymers of the 20 naturally occurring α-amino acids. The amino acids can have varied chemical characters—apolar, polar, acidic, or basic—depending on the identity of the side chain group attached to the α-carbon. The amino acids are connected to each other through the peptide bond and the sequence in which the amino acids are connected is known as the primary structure of the protein. Proteins adopt a well-defined 3D structure, also called the tertiary or native structure, in solution. This usually consists of secondary structures—locally organized structural elements such as α-helices, β-sheets, turns—held together by noncovalent interactions and, in some cases, by disulfide bonds. Proteins in solution are not static entities—rather they undergo conformational fluctuations and transitions. It has been realized that the diverse array of functions carried out by proteins is intimately connected to its structure and dynamics.

While studying ribonuclease A, Anfinsen [1] postulated that under appropriate conditions (temperature, pressure, solvent composition, etc.), the native structure of a protein is unique, stable, and kinetically accessible. Anfinsen postulated that the native state of the protein lies at the minimum of the free energy of the system, that is, protein in solution at normal physiological conditions, and that the native structure is unique, stable, and kinetically accessible. The process by which an extended configuration reaches the native structure is called protein folding. The number of possible conformations available to a heteropolymer grows exponentially with the number of monomers in the chain. For a protein of typical size, the time necessary to sample all the different possible conformations in order to reach the native state is astronomically large under reasonable assumptions. However, protein folding occurs spontaneously in μs or ms timescales. This contradiction is known as Levinthal's paradox and has been explained in terms of a funneled energy landscape [2–4].

Thus, the *protein folding problem* can be stated as threefold:

- To prove the validity of Anfinsen's dogma, that is, finding the 3D native structure of a protein given its sequence.
- To calculate thermodynamic properties. This amounts to characterizing metastable states and relative populations.
- To describe the kinetics. This is done by calculating the rates of transition between the identified metastable states.

Proteins in solution are in conformational equilibrium with their unfolded forms, with the folded state being favored at ambient conditions. The process of increasing the population of the unfolded states is known as denaturation. This can be achieved by various means—by increasing (and decreasing, in some cases) the temperature, applying pressure, changing the pH, or adding cosolvents to the solution. Denaturation is an important process in biochemical studies, as thermodynamic functions such as changes in Gibbs free energy, ΔG; enthalpy changes, ΔH; and entropy changes, ΔS, that provide information about the stability of the folded state can be obtained only by perturbing the equilibrium to populate the unfolded (denatured) states. Quantities such as spectral properties, heat capacity, and enzyme activity can provide information about the population of different states from which the equilibrium is constant and hence other thermodynamics quantities can be derived. Many proteins undergo a sharp, cooperative transition from native to denatured states upon increasing the temperature, leading to the characteristic sigmoidal curve for the experimental observable. Such transitions can be treated in a two-state model, and the two-state assumption is considered valid if the van't Hoff enthalpy and the calorimetrically measured enthalpy are equal [5,6]. Proteins are also, counterintuitively, destabilized by increasing the pressure and unfold with a decrease in partial molar volume [7,8]. This has been explained by noting that the packing of the folded state is not optimal and the volume of the *water swollen* unfolded state is smaller than the folded state due to hydration of interior hydrophobic residues [9–11]. Phase diagrams for the pressure–temperature behavior of the folding/unfolding transition of proteins have been obtained, from experiment as well as computation [12–17]. A free-energy surface of the form given in the equation as follows is fit to experimental data, and it is observed that most proteins studied show an elliptic phase diagram [18]:

$$\Delta G_{u}(P,T) = \Delta G_0 - \Delta S_0(T - T_0) + \Delta V_0(P - P_0) + \Delta \alpha_0(T - T_0)(P - P_0)$$

$$- \Delta C_{p}\left[T\left(\ln\left(\frac{T}{T_0}\right) - 1 \right) + T_0 \right] + \frac{\Delta \beta_0}{2}(P - P_0)^2. \tag{4.1}$$

The equation can be considered to be a Taylor expansion of ΔG to second order around a reference point T_0 and P_0. In the previous equation, $\Delta \alpha$, $\Delta \beta$, and ΔC_p are expansivity, compressibility, and heat capacity changes, respectively, and other terms have their usual meaning. The boundary of the native state is given by the curve $\Delta G = 0$, and it can be shown that the curve is elliptical if the condition $T_0\Delta\alpha^2 > \Delta C_p\Delta\beta$ is satisfied.

Further investigation of the effects of pressure and cosolvents from the standpoint of kinetics and mechanism by all-atom simulations is largely an unexplored area of research. Change in solvent conditions could be reflected in the folding/unfolding rates (e.g., the Chevron plot) as well as the mechanism of the folding process. Certain folding pathways may be preferred over others when solvent conditions are changed, or new pathways may be made accessible. New intermediates may appear upon changing solvent conditions and rate-limiting steps for the folding process may be different. One may also ask if the reaction coordinate(s) for the folding process that is calculated in ambient, aqueous solution provides a sufficient description even when solution conditions are altered. Protein folding is a rare event on the timescales accessible by all-atom molecular dynamics (MD) simulations, making it difficult to observe multiple folding events required for the determination of kinetics from a direct MD simulation. However, this limitation can be circumvented by exploiting the advances made in simulation methodologies and parallel/distributed computing capabilities. A complete, accurate, and detailed description of the folding mechanism has been elusive [19]. All-atom MD simulation–based methods in explicit solvent are the most promising method for obtaining such knowledge. Progress in energy functions, statistical methods to enhance the sampling of rare events, and increases in computational power offer an optimistic outlook to simulations as a powerful tool to explore dynamics at microscopic scales. We provide a brief overview of some of the simulation methods in the following and comment on their suitability for problems of our interest.

4.2   MOLECULAR DYNAMICS

An MD simulation consists of the iterative numerical integration of the classical equations of motion for a set of particles with defined interactions between them. Macroscopic properties can then be obtained from microscopic states (positions and velocities of each particle) by using the framework of statistical mechanics. In an MD simulation, each particle can represent one or more real objects, and the model describing the interactions between these particles is known as the force field. The interactions between the particles are divided into two classes. The bonded interactions consist of vibrations, angle vibrations, and torsion potentials. The nonbonded are usually van der Waals' interactions and electrostatic interactions. Because of their long-range nature, nonbonded interactions are the most computationally expensive part of the simulation. In principle, they require N^2 distance calculations for N atoms, but under specific conditions, specialized algorithms can reduce these $N \log N$ calculations [20]. In an all-atom MD simulation of a biomolecule, the solvent can be modeled either by explicitly including all the water molecules or by the use of an effective term that mimics the solvent (mostly water) effects (so-called implicit solvent). This is done because typically, in order to simulate experimental conditions, the number of water atoms can account for up to 90% of the total number of atoms in the simulation. Thus, the use of implicit solvent results in a great speedup at the cost of the loss of solvent effect details.

Here, we will concentrate mostly on explicit solvent simulation results and methods. In many instances, the methods are very similar, but the limitations of the methods may be different. For a review of implicit solvent models, we refer the reader to the review article by Chen et al. [21]. The explicit form and parameters of the force fields and water models are calculated from ab initio quantum mechanical calculations and are refined to reproduce experimental data on small solutes. Commonly used force fields are Amber [22–24], Gromos [25], OPLS [26,27], and CHARMM [28,29]. There are multiple versions and modifications of these force fields. Recent developments in force fields [30–34] have led to important advances in the computational biology field and suggest that equilibrium properties of proteins can be accurately characterized by current state-of-the-art methods [34–42]. However, some other more subtle details like temperature dependence of conformational propensities and the kinetics of folding are still being evaluated and require further development and comparison to experimental data.

Even if current models are able to represent the fundamental features of a biomolecule correctly, sampling of timescales of interest for large biomolecules is yet to be feasible. Relevant processes in biomolecules span a wide range of timescales. Hydrogen vibrations occur in femtoseconds, while global rearrangements like protein folding can take from microseconds to seconds. These differences in timescales highlight the complexity of the highly dimensional energy function with many local minima. Assuming that the chosen force field correctly describes the system under study, a single event of protein folding from a random, extended configuration under natural conditions may take 10^8–10^{14} integration steps. This can take weeks, months, or years of CPU time depending on the system size, even in a highly parallelized computing setup. Moreover, one needs several of these events in order to claim adequate equilibrium sampling of the configurational space. The development of a specialized massively parallel computer, Anton, has enabled the execution of MD single trajectories in the millisecond timescale [43,44]. Using this computer power, Shaw et al. have shown that existing force fields and its modifications were sufficiently accurate to fold 12 proteins with different folds, with reasonable (0.5–4.8 A RMSD) structural resolution [45]. Other, more approaches include the development of software that uses GPUs [46].

4.3 PROTEIN KINETICS

Even when MD simulations can produce unbiased folding trajectories of proteins, due to the stochastic nature of the dynamics of proteins, the mechanism (or pathways) of protein folding cannot be trivially extracted from these simulations. We provide a brief overview of the algorithms that have been proposed to investigate the kinetics of protein folding.

4.3.1   Transition Path Sampling

Transition path sampling (TPS) is a method that provides insight into reaction mechanism by harvesting dynamical trajectories that connect two stable states [47–51]. This is achieved by constructing an importance sampling scheme in trajectory space, biased such that reactive trajectories are sampled. The method requires only stable states to be defined, and no assumption of a reaction coordinate and the transition state is required. Rather, these are deduced from the analysis of harvested reactive trajectories, the transition path ensemble. In the original formulation of TPS [51], deterministic shooting and shifting moves are described to sample the transition paths. The calculation of the rates is carried out by determining the reversible work to constrain an ensemble of trajectories to end in product region. The identification of the transition state ensemble and the reaction coordinate(s) are carried out by calculating the committor and the committor distributions. Significant improvements have been made over the original algorithms of TPS. A stochastic shooting algorithm has been proposed to overcome the low efficiency of deterministic shooting moves for diffusive transitions over rough energy landscapes [52]. A more efficient method, called transition interface sampling (TIS), has been put forward for calculation of rate constants [53]. Finally, methods have been proposed to identify reaction coordinates without the expensive computation of committor distributions [54–56]. Starting from applications to relatively simple systems [57–60], TPS has been applied successfully to more complex biological systems. Examples include folding of beta-hairpin [61] and Trp-cage miniprotein [62,63], DNA base pairing [64], closing transition of DNA polymerase [65], and more recently, studying the reaction catalyzed by lactate dehydrogenase [66,67] and the conformational changes in photoactive yellow protein [68], mapping the conformational changes of a riboswitch [69], and studying ligand-induced structural transition in adeylate kinase [70], among others. A recent review on the use of TPS on proteins has been published by Bolhuis et al. [71].

4.3.2   Chain-of-Images Methods

It is often of interest to obtain transition paths that bridge conformations of biomolecules. For example, these conformations could be the crystal structures of enzymes in active/inactive states or intermediates/kinetic traps along folding pathways of proteins. In such cases, it is desirable to find pathways that connect them in configurational space and estimate the barriers separating these states. These pathways are parameterized by some order parameter(s) describing the transition, with no explicit reference to time as in TPS. The basic idea of path-finding algorithms such as nudged elastic band (NEB) method [72,73] or string methods [74,75] is to construct a chain of states as an initial guess to the transition pathway and allow the chain to evolve according to the underlying energy (free-energy) landscape and diffusion dynamics in the order parameters. Different update rules have been proposed and lead to slightly different paths such as minimal energy path [74], minimum free-energy path [76], or most probable transition path [77].

One of the most powerful chain-of-states methods is the string method, proposed by Vanden-Eijnden and coworkers, and its variations. In the zero-temperature string (ZTS) method [74], a smooth curve with an intrinsic parameterization called the string is evolved according to a differential equation that converges to the MEP. In practice, the string is represented by a discrete number of images, each of which evolves according to the component of potential force normal to the string. After a certain number of iterations, the string is reparameterized, that is, the evolved images are moved along the string to impose the parameterization. A popular choice is to maintain equally spaced images along the string. This ensures that the images do not pool up in either of the metastable basins. For complex landscapes, the finite-temperature string (FTS) method [78,79] has been proposed wherein the potential force of ZTS has been replaced by an average thermodynamic force, which can be obtained by constrained sampling of hyperplanes normal to the string. In the finite-temperature case, the MEP is generalized to *transition tubes*, which are regions in configurational space to which reactive paths stay confined with high probability. Using results from stochastic process theory, it was shown that the converged hyperplanes of FTS are the local approximation of the isocommittor surfaces of the transition, implying that trajectories launched from these surfaces have the same probability of reaching first one of the metastable region than the other [75]. A framework for statistical analysis of transition paths, called transition path theory, has been developed in which a detailed characterization of the reaction mechanism can be made [80,81]. Algorithmic improvements of string methods have been recently proposed [82,83]. The string method has also been extended to incorporate collective variables [76,84]. A variant of string method, proposed by Roux and coworkers, uses multiple, short, unbiased simulations to iterate the string instead of computing the mean force through constrained simulations [77]. This method has also been called the dynamic string method and has been shown to be less sensitive to the particular choice of collective variables than other variants of the string method, generating pathways that follow the reactive flux [85]. The dynamic string method evaluates the local drift vector at each image point by averaging the displacement in collective variables over an ensemble of unbiased, short trajectories that start from the desired point. The notable applications of string methods to biomolecular systems include hydrophobic collapse of a hydrated chain [86], conformational activation of Src kinase [87], and more recently, diffusion pathways of CO in myoglobin [88].

The framework of TPT was also applied in a recent Markov model study of Pin WW folding [89] and the conformation change between the prepower stroke and rigor structures of the converter domain of myosin VI [90].

4.3.3 Milestoning

The strategy of milestoning, proposed by Elber and coworkers [91–93], can be used in conjunction with the string method to compute kinetics and timescales of transitions. Milestoning involves computing local kinetic information between *milestones* placed along a presumed reaction coordinate by running short MD simulations and gluing them together to obtain a global kinetic picture [93]. It has been proposed that the optimal milestones are the isocommittor surfaces of the transition, which can be computed by the string methods described earlier. Refinements over the original algorithm have been proposed [94]. Recent applications of milestoning to biomolecular systems are allosteric transition of hemoglobin [95], kinetics of helix unfolding [96], folding of protein A [97], and an atomic description of the recovery stroke of myosin [98].

4.3.4 Markov Models

Protein folding is studied in the liquid phase, and the random forces exerted by the solvent on the protein make the process of protein folding inherently stochastic [99]. To draw statistically significant conclusions about the folding process, one needs to produce a multitude of trajectories. This has been made possible by increase in parallel computing power, advent of distributed computing environments [100], and improved software to exploit parallelism. The large amount of data generated has to be analyzed to capture dynamical processes and provide a physical interpretation, which can be done by creating Markov state models to describe the data. A Markov model is constructed by lumping the configurations generated in the multiple simulations together into discrete states and constructing a transition matrix $T(\tau)$ to describe the transitions between the states [99]. The elements of the transition matrix, T_{ij}, indicate the probability of the system to undergo a transition from state i to j, in a time interval of τ. Known as the lag time, τ, is the time interval required for a trajectory to lose memory of how the system entered a state. In other words, it represents the time interval over which transitions between the states of the system are history independent, for the Markovian assumption to hold. The lag time must be longer than the longest equilibration time among the states used to describe the process. The eigenvalues of the transition matrix provide information on the timescales of the process, and the eigenvectors describe the aggregate transitions associated with the timescale. The separation of timescales, fast intrastate motions, and slow interstate transitions is a crucial feature that allows the modeling of conformational dynamics as stochastic transitions between discrete states. The partitioning of the conformational space must reflect this property, and decomposition based on intuitive notions of order parameter for which the potential of mean force (PMF) is known or geometrical clustering approaches may not be sufficient [101]. Approaches based on kinetic clustering have been proposed, where conformations are grouped together to maximize the metastability of the state. These include the Perron cluster analysis methods developed by Schutte et al. [102] and the iterative split and lump procedure of Chodera et al. [103]. These methods have been successfully applied to polyalanines [104], the FS helical peptide and the trpzip hairpin [103], the villin headpiece (HP-35 NleNle) [105], NTL9 [106], lambda 6-85 [107], and the FiP35 WW domain [108]. From further analysis of the Markov models, reactive fluxes, folding pathways, and their relative probabilities can be calculated to provide a more comprehensive mechanistic description of the folding process [89,106,107,109].

4.4 REPLICA EXCHANGE MOLECULAR DYNAMICS FOR CALCULATING PROTEIN THERMODYNAMICS

In this section, we discuss the simulation of protein thermodynamics using replica exchange molecular dynamics (REMD) in some detail and highlight physical insights gained from these studies. REMD, based on the parallel tempering Monte Carlo algorithm [110,111], is an elegant and efficient way to sample the configurational space of complex systems [112,113]. REMD is best suited for calculating thermodynamics of protein folding, since it produces equilibrium properties over a broad range of temperatures and other thermodynamic variables. In REMD, several copies (replicas) of identical systems are simulated in parallel at different temperatures. State exchange moves between replicas are attempted periodically and accepted with probability

$$P_{\text{acc}} = \min(1, \exp[-\Delta]), \tag{4.2}$$

where

$$\Delta = \exp[(\beta_i - \beta_j)(U(\vec{r}_j^N) - U(\vec{r}_i^N))], \tag{4.3}$$

where $U(\vec{r}_j^N)$ is the potential energy of the system in the state j and $\beta_j = 1/k_B T_j$. Here, k_B is the Boltzmann constant and T_j is the temperature of the jth replica. If the exchange is successful, the momenta of all atoms are scaled by the factor $(T_i/T_j)^{1/2}$, such that the kinetic terms in the Boltzmann factor cancel out [112]. This protocol generates a correctly weighted ensemble of configurations at every temperature. As a result of an REMD simulation, equilibrium constants as a function of temperature can be computed and thermodynamic quantities can be calculated. A pictorial description of the method can be seen in Figure 4.1. The REMD method has been successfully used to study the free-energy landscape (FEL) and folding/unfolding equilibrium of peptides [16,32,114–124], proteins [17,125–130], and RNA [131,132], as well as pressure [15,126,131,133], side chain charge states [134], denaturants [135,136], other cosolvents [137–139], and confinement [140] effects on protein stability. Practically for the exchanges to occur, the energy distributions of the different temperatures need to overlap sufficiently [141]. This limits the use of REMD to small systems since the overlap decreases proportionally to $1/m^{1/2}$ with m as the mass of the system requiring many more replicas to span the same

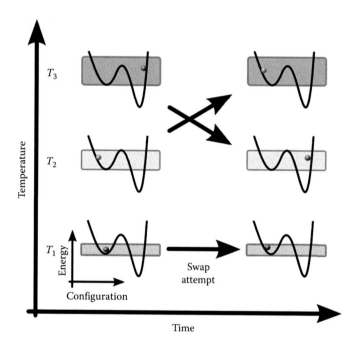

Figure 4.1   Graphical description of the REMD method. Several copies of the system are simulated at different temperatures. An exchange is attempted periodically and accepted with a probability that ensures correct Boltzmann weight. Exchanges allow individual replicas to perform a random walk in temperature space, thus enhancing the configurational sampling of lower temperatures. When a replica is trapped in a local minimum at low temperatures, it can be swapped to higher temperatures where the height of the barrier is comparable to the thermal energy $k_B T$, allowing the replica to escape the minima.

temperature range. A method of replica exchange with dynamical scaling of the temperatures, proposed by Rick [142,143], enables to simulate explicitly fewer replicas by varying the temperature (dynamically) of virtual replicas between large temperature ranges. This method can reduce the number of replicas required to simulate a given temperature range by a factor of 2–4. Zhang and Ma have suggested a single-copy tempering method that uses a runtime estimate of the thermal average energy to guide a continuous temperature-space random walk [144]. Rhee and Pande have developed a variation of REMD that uses multiplexed replicas with independent MD runs at each temperature [129]. Exchanges of temperature between replicas are tried using stored configurations of energies for different temperatures. These energy histograms can be updated as the simulations progress. This method is applicable to large-scale distributed computing, like folding@home.

REMD can be combined with other generalized ensemble methods [144–149]. REMD has been extended to allow swaps on generalized ensembles where the thermodynamic quantity exchanged between replicas can be taken to be an arbitrary parameter such as charge states of amino acids with pH as a coupling parameter [150], mixing implicit and explicit solvent models [151], or a Hamiltonian [152]. The REMD with solute tempering method, REST, includes exchanges between solute degrees of freedom while maintaining the same temperature on the solvent [153]. This method is not always more efficient than REMD [154]. Combinations of other enhanced sampling methods, like metadynamics [155] and well-tempered metadynamics [156], have also been combined with REMD to improve sampling of the FEL of a complex system along predetermined order parameters [123,157].

The efficiency and errors in implementing REMD simulations have been discussed in the literature. The four main issues are as follows:

1. The use of an appropriate heat bath coupling algorithm that samples a canonical ensemble [158] and to ensure proper seeding of pseudorandom numbers in stochastic thermostats [159]
2. The distribution of temperatures between replicas such that there are sufficient exchange rates and individual trajectories span and cycle through the whole temperature range [160]
3. The exchange attempt frequency [161–163]
4. The range of temperatures spanned by the replicas such that the sampling is efficient [164–166]

Issue (1)—Thermostat can be easily solved by using stochastic dynamics, the Nose–Hoover [167,168] or the Andersen [169] thermostats that have been proven to sample from a canonical ensemble [170]. Bennet's method of energy histogram overlap can be used to verify that a canonical ensemble is being sampled (i.e., to verify that the energies sampled obey a Boltzmann distribution) [117]. In Bennet's method, given the histograms of energies sampled, $H(E|T)$ at a temperature T, the histograms must satisfy

$$\log \frac{H(E|T_1)}{H(E|T_2)} = (\beta_2 - \beta_1)E + \text{const.} \tag{4.4}$$

Issue (2)—The distributions of temperatures. Temperatures among replicas can be distributed exponentially, which will provide a uniform acceptance ratio in the limit of constant heat capacity, but may fail in regions where there is a sharply peaked heat capacity. This will be particularly troublesome when using implicit solvent models or coarse-grained models. Rathore et al. developed a method of calculating replica exchange acceptance rates by approximating the sample energy histograms by Gaussians and calculating the area of overlap of energy histograms [160]. The Gaussian distribution approximation is very good in most cases, except when there are multimodal distributions (than can give rise to highly peaked heat capacities). For protein systems in explicit solvent, this approximation is very good. The exchange acceptance rate, R_{acc}, is given by

$$R_{\text{acc}}(\beta_1, \beta_2) = \iint\limits_{E_1, E_2} P_{\beta_1} P_{\beta_2} \min[1, \exp[\Delta\beta\Delta E] dE_1 \, dE_2, \tag{4.5}$$

where $P_\beta(E)$ is the probability density of obtaining energy E at temperature β. If $P_\beta(E)$ is assumed to be Gaussian, with average $\langle E \rangle$ and width σ, the area of overlap of the distributions is given by

$$A_{\text{overlap}} = \text{erfc}\left[\frac{\langle E_2 \rangle - \langle E_1 \rangle}{\sqrt{8}\,\sigma}\right] = \text{erfc}\left[\frac{\Delta E}{\sqrt{8}\,\sigma}\right], \tag{4.6}$$

where $\text{erfc}(x)$ is the complementary error function. Rathore et al. showed that A_{overlap} is related to P_{acc} and can be used to distribute temperatures between neighboring replicas. Within the Gaussian approximation, P_{acc} can be integrated analytically [171],

$$R_{\text{acc}}(\beta_1, \beta_2) = \frac{1}{2}\left[1 + \text{erf}\left[\frac{\langle E_2 \rangle - \langle E_1 \rangle}{\sqrt{2(\sigma_1^2 + \sigma_1^2)}}\right]\right] + \frac{1}{2}\exp\left[\Delta\beta(\langle E_2 \rangle - \langle E_1 \rangle) + \left(\frac{\Delta\beta}{2}\right)^2 (\sigma_1^2 + \sigma_2^2)\right]$$

$$\times \text{erfc}\left[\frac{\Delta\beta(\sigma_1^2 + \sigma_2^2) + \langle E_2 \rangle - \langle E_1 \rangle}{\sqrt{2(\sigma_1^2 + \sigma_1^2)}}\right]. \tag{4.7}$$

Using this formula and given the first and second moments of the energy distributions as a function temperature, we can iteratively solve for the neighboring T_{i+1} given that we know T_i. We have shown that Equations 4.6 and 4.7 produce very similar temperature distributions, while the geometric distribution underestimates the temperatures at higher temperatures, thus yielding a higher exchange rate than the desired rate and requiring a larger number of replicas to span a desired temperature range. When the distributions are not well represented by Gaussians, there is a robust, iterative method described by Trebts et al. that should always work [172]. Trebts et al. have described ways to optimize the distribution of temperatures by monitoring the diffusivity in the simulated random walk in temperature space and monitoring the bottlenecks of the simulation.

Issue (3)—Exchange attempt frequency. Periole and Mark [162] and Abraham and Gready [161] have argued that frequency exchange rates should not be done more frequently than the potential energy correlation time at the relevant temperatures (~5 ps for protein systems at ~300 K). Sindhikara et al. argue that as long as the move is thermodynamically correct and computationally inexpensive, it is worth doing, and therefore, exchange attempts should be done often to improve the Monte Carlo sampling [163,173]. Exchange attempts at periods larger than the potential energy correlation time may be relevant if one uses Equation 4.5 or 4.6 to estimate exchange rates, since the modeling of the sampled energies as a Gaussian distribution presumes uncorrelated times [161]. Sindhikara et al. and others have shown that frequent exchanges do not affect the thermodynamics. In some instances, it may be worth sacrificing efficiency of the sampling to gain information that can be used to model other relevant quantities, like folding rates, from the REMD sampling. For example, Yang et al. [115] and Buchete and Hummer [174] have proposed methods for extracting kinetics information from REMD by using the information in constant temperature trajectory segments between exchanges to model local diffusion coefficients and drifts along order parameters. In this case, exchanges should be done at the same rate (or multiples) as the rate at which configurations are saved.

Issue (4)—Efficiency. Rosta and Hummer [165,166] and Zhang et al. have shown that the efficiency of REMD in sampling is diminished at higher temperatures where the system may show non-Arrhenius behavior [175]. Rosta and Hummer define the efficiency of sampling as the rate with which the error in an estimated equilibrium property, as measured by the variance of the estimator over repeated simulations, decreases with simulation time. The relative efficiency of REMD and MD simulations is given by the ratio of the number of transitions between the two states averaged over all replicas at the different temperatures and the number of transitions at the single temperature of the MD run. High efficiency of REMD is achieved by including replica temperatures in which the frequency of transitions is higher than that at the temperatures of interest [165,166]. In some instances, we may be interested in calculating the thermodynamic stability of a protein over a broad range of temperatures covering the folding and unfolding states and may need to compromise REMD efficiency.

4.5 SIMULATIONS OF PROTEIN THERMODYNAMICS

The marginal stability of proteins can be altered by change in solvent conditions, such as application of pressure or addition of cosolvents [7,12,14,36,176]. These effects have been studied computationally by means of REMD simulations and provided insights into these processes [16,17,131,135]. REMD simulations describe the equilibrium ensemble over a broad range of thermodynamic states like temperatures and pressures. These ensembles can be analyzed together to provide a description of the stability of the system in terms of the thermodynamic variables like temperature and pressure. For the case of pressure and temperature, the change in the Gibbs free energy in the folding/unfolding process can be described in terms of six measurable thermodynamic quantities. Typically, REMD calculations are simulated at constant volume, so the calculated thermodynamic potential is the Helmholtz free energy. However, the Helmholtz free energy is related to the Gibbs free energy by the Legendre transform, $G = H + pV$. To obtain a pressure–temperature stability diagram, we can fit the calculated fraction folded, the average pressure, and average potential energy and the Gibbs free-energy differences as a function of temperature. The free-energy difference between the folded and unfolded states as a function of P and T is given in Equation 4.1. The system free-energy difference can be fitted by minimizing an χ^2 function as a function of the expansion coefficients in Equation 4.1 [17,18]. Multiple REMD simulations at various states (P, T, urea concentration [135], etc.) can also be combined in the fitting. Ensemble averages of the difference in free energy, energy, and volume at each of the states are used in the fitting. A typical fitting contains few hundred *observables* to fit the P–T diagram [17,126,135]. When available experimentally, these parameters compare reasonably well with experimentally measured quantities for the system [126,177]. The ability to fold reversibly proteins (e.g., protein A [125,178] and Trp-cage [126]) under various solvent, temperature, and pressure conditions enables us to further explore the mechanism of interaction of proteins with denaturants like urea [135,136] and with protecting osmolytes [137] and also explore the effect of different protonation states of the protein on its stability [134]. These calculations can also be used to validate and help the development of more accurate force fields [32,116]. Though we currently sample the folding equilibrium using REMD simulations [111,112], other enhanced sampling methods may also be used if they prove to be more efficient [144,147,149].

4.6   TRP-CAGE MINIPROTEIN AS A MODEL FOR FOLDING SIMULATIONS

Obtaining the folding/unfolding equilibrium of a biomolecule by molecular simulation presents a major computational challenge, even for small systems. Trp-cage miniprotein is a model protein designed to understand protein folding pathways and stability [179]. It is a 20-residue protein with a nontrivial fold and structurally consists of an α-helix, a 3_{10} helix, and a polyproline segment. The sequence is Ac-NLYIQWLKDGGPSSGRPPPS-Nme. The protein fold is stabilized by a hydrophobic core consisting of a tryptophan (W) residue and an ion pair. Furthermore, the α-helix is stabilized by tertiary contacts. The melting behavior of Trp-cage has been characterized by various methods in literature [177,180–182]. It has been demonstrated that the folding of Trp-cage can be described well by a two-state model [177]. Its small size and fast folding kinetics (folding time of 4 μs) make it a computationally attractive target, and many studies using both implicit and explicit solvent models have been reported [17,39,127,128,130,183–185]. Trp-cage shows the thermodynamic features observed for globular proteins, such as the temperature dependence of unfolding free energy and enthalpy [17], and thus provides a good justification for the choice as a model system.

4.6.1   Pressure Unfolding

Day et al. used REMD to study the thermodynamics of Trp-cage at two different volumes [126]. The fit of the stability using the Hawley equation (Equation 4.1) shows that the protein exhibits an elliptical stability diagram, shown in Figure 4.2. This diagram is similar to the measured diagram for various globular proteins. Table 4.1 shows the parameters obtained by fitting the REMD simulated states to Equation 4.1. These parameters show that the volume change, ΔV, is very small (~1–2 mL/mol), consistent with the high pressure (500 MPa) required to unfold the protein at room temperature. The folding transition temperature (317 K) is close to the experimental value (321 K), and the heat capacity differences are also very similar to the experimental measurements [177]. However, the calculated entropy and enthalpy differences are

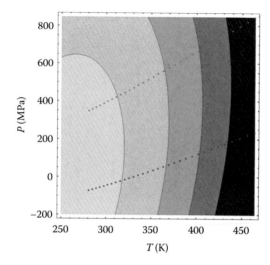

Figure 4.2 Pressure–temperature stability diagram for the Trp-cage protein. The contours for $\Delta G(P, T)$ are elliptical. The free-energy profile shows small pressure dependence and the protein will require high pressures to unfold at low temperatures, suggesting a small volume change upon folding. The dotted lines represent the states sampled during the REMD simulations at low and high densities.

TABLE 4.1 THERMODYNAMIC PARAMETERS DESCRIBING THE SIMULATED $\Delta G(T, p)$	
Calculated (at 298 K, 1 atm)	**Measured [177]**
$\Delta\alpha = 0.038 \pm 0.026$ kJ/mol	
$\Delta\beta = -0.07 \pm 0.18$ kJ/mol MPa^2	
$\Delta C_\pi = 0.33 \pm 0.03$ kJ/mol K	$\Delta C_\pi = 0.30 \pm 0.1$ kJ/mol K
$\Delta S = 0.062 \pm 0.004$ kJ/mol K	$\Delta S = 0.16$ kJ/mol K
$\Delta V = 1.2 \pm 10.3$ mL/mol	
$\Delta\Gamma_0 = 1.48 \pm 0.59$ kJ/mol	$\Delta\Gamma_0 = 3.2$ kJ/mol
$T_\phi = 317$ K	$T_\varphi = 321$ K

approximately a factor of two smaller than the measured values, but they compensate and give $\Delta G \sim 0$ at 317 K. These differences in ΔS and ΔH explain the different temperature profiles described by the calculations and experiments. In the calculations, the fraction of folded protein changes slowly with temperature, while the experimental curves show rapid, cooperative transitions in a narrow temperature range. The differences in temperature behavior are indicative of deficiencies in the force fields.

4.6.2   Urea Denaturation

REMD simulations were used by Canchi et al. to study the unbiased equilibrium folding/unfolding of the Trp-cage miniprotein at different concentrations of urea, using Amber ff94 [22] and ff99SB force fields [23], TIP3P model for water [186], and the Kirkwood–Buff model for urea [187]. They simulated systems at three different concentrations of urea (i.e., ~2 M, 4 M, 6 M, and 0 M [water]). To obtain thermodynamic information about the denaturation process, they modeled the unfolding free energy as a function of temperature, pressure, and concentration, using an equation similar to Equation 4.1, with additional terms added to include the effect of urea on free energy (m_1), entropy (m_2), volume (m_3), and heat capacity (m_4) changes:

$$\Delta G_u(P,T,[C]) = \Delta G_0 + m_1[C] - (\Delta S_0 + m_2[C])(T - T_0) + (\Delta V_0 + m_3[C])(P - P_0)$$

$$+ \Delta \alpha_0 (T - T_0)(P - P_0) - (\Delta C_p + m_4[C]) \left[T \left(\ln \left(\frac{T}{T_0} \right) - 1 \right) + T_0 \right]$$

$$+ \frac{\Delta \beta_0}{2} (P - P_0)^2. \tag{4.8}$$

Figure 4.3a shows the free energy at the reference state, $T = T_0 = 300$ K and $P = P_0 = 0.1$ MPa. The free energy is linear with urea concentration, which shows that the all-atom model employed is able to capture the experimental trend. From the thermodynamic model, the free energy can be decomposed into enthalpic and entropic contributions. In Figure 4.3a, we see that the sign of ΔG_u as well as its behavior with urea concentration is dictated by the enthalpy term. The entropic term favors unfolding as expected and is very weakly concentration dependent. It has been argued that preferential binding of urea to the protein is entropically favorable, as it involves displacement of water by larger urea molecules [188–190]. In view of this argument, the weak concentration dependence of entropy is intriguing. The quantity $-\partial \Delta G_u / \partial C$ can be thought of as a generalized m-value, dependent on temperature and pressure, which reduces to the traditional m-value at the reference state, $T = T_0$ and $P = P_0$. Figure 4.3b shows the effective m-value, $m(P, T) = -\partial \Delta G_u / \partial C$, as a function of temperature at two different pressures and along an isochore. The increase in this quantity with temperature and pressure indicates that denaturation by urea is aided by both of these factors, with the effect of pressure being markedly stronger. The m-value calculated from the model at 300 K and 0.1 MPa is 0.41 ± 0.03 kJ/mol M. At 300 K and 200 MPa, the m-value is calculated to be 0.79 kJ/mol M, almost double the value at ambient conditions. The strong pressure dependence arises from the term $m_3(P - P_0)$, whose contribution to volume changes (i.e., $\partial \Delta G_u / \partial P$) is negligible. The temperature dependence is weaker compared to pressure, giving an m-value of 0.66 kJ/mol M

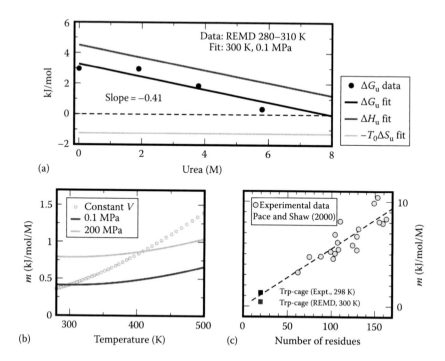

Figure 4.3 Thermodynamic features of urea denaturation. (a) Free energy of unfolding as function of urea concentration, along with enthalpic and entropic contributions. (b) $m = -\partial \Delta G_u / \partial C$ as a function of temperature along two isobars (0.1 and 200 MPa) and an isochore. (c) Comparison of m-values from simulation and experiment. Light and dark squares are m-values for Trp-cage from simulation and experiments [192], respectively. The dashed line is an extrapolation of experimental m-values for larger proteins taken from Pace and Shaw [191]. (Reprinted from Canchi, D. et al., *J. Am. Chem. Soc.*, 137(7), 2338, 2010.)

at 500 K and 0.1 MPa. The contribution of the m_2 term to the temperature dependence of m-value is negligible. Figure 4.3c shows the experimental m-values for larger proteins obtained from Pace and Shaw [191] as a function of number of residues and the linear extrapolation of these data as a dashed line. The m-values obtained from Amber ff94 and ff99SB simulations are 0.41 kJ/mol M [135] and 1.81 kJ/mol M [136], respectively, which are of the same order of magnitude as the experimentally measured m-value of 1.3 kJ/mol M [192].

In addition to describing the thermodynamics of urea denaturation, analysis of the interactions of urea with the protein has revealed that urea interacts with all groups in the protein and not exclusively (or preferentially) with the protein. A thermodynamic analysis of the interaction of urea with small solutes and peptides by Guinn et al. has verified this observation [193].

4.7 CONCLUSIONS

We have provided an overview of the recently developed methods for studying protein folding dynamics, kinetics, and thermodynamics. We have also shown examples where calculations of the thermodynamics of a small miniprotein compare well with experimental results on the same system. Discrepancies in the results point to possible deficiencies in the classical force fields. Some of these deficiencies may be resolved with further parameterization of the force fields, but others may require the inclusion of other effects not commonly present in classical force fields, like charge polarizability and directional hydrogen bonding (Kortemme, 2003 #49). The development of powerful computers and sampling methods should enable the description of protein folding kinetics models that can be directly compared to experimental observables.

ACKNOWLEDGMENT

The authors want to acknowledge funding by the National Science Foundation, MCB-1050966.

REFERENCES

1. Anfinsen, C., Principles that govern folding of protein chains. *Science* 1973;**181**: 223–230.
2. Bryngelson, J. et al., Funnels, pathways, and the free energy landscape of protein folding: A synthesis. *Proteins—Structure Function and Genetics* 1995;**21**: 167–195.
3. Onuchic, J., Z. Luthey-Schulten, and P. Wolynes, Theory of protein folding: The energy landscape perspective. *Annual Review of Physical Chemistry* 1997;**48**: 545–600.
4. Onuchic, J. and P.G. Wolynes, Theory of protein folding. *Current Opinion in Structural Biology* 2004;**14**: 70–75.
5. Makhatadze, G.I. and P. Privalov, Energetics of protein structure. *Advances in Protein Chemistry* 1995;**47**: 307–425.
6. Chan, H., S. Shimizu, and H. Kaya, Cooperativity principles in protein folding. *Energetics of Biological Macromolecules, Method in Enzymology* 2004;**380**: 350–379.
7. Heremans, K., High-pressure effects on proteins and other biomolecules. *Annual Review of Biophysics and Bioengineering* 1982;**11**: 1–21.
8. Silva, J., D. Foguel, and C. Royer, Pressure provides new insights into protein folding, dynamics and structure. *Trends in Biochemical Sciences* 2001;**26**: 612–618.
9. Hummer, G. et al., The pressure dependence of hydrophobic interactions is consistent with the observed pressure denaturation of proteins. *Proceedings of the National Academy of Sciences of the United States of America* 1998;**95**(4): 1552–1555.
10. Sarupria, S. et al., Studying pressure denaturation of a protein by molecular dynamics simulations. *Proteins* 2010;**78**: 1641–1651.

11. Roche, J. et al., Cavities determine the pressure unfolding of proteins. *Proceedings of the National Academy of Sciences of the United States of America* 2012;**109**: 6945–6950.

12. Brandts, J.F., R.J. Oliveira, and C. Westort, Thermodynamics of protein denaturation effect of pressure on the denaturation of RNase A. *Biochemistry* 1970;**9**(4): 1038–1047.

13. Hwaley, S., Reversible pressure–temperature denaturation of chymotrypsinogen. *Biochemistry* 1971;**10**: 2436–2442.

14. Panick, G. et al., Exploring the temperature–pressure phase diagram of staphylococcal nuclease. *Biochemistry* 1999;**38**(13): 4157–4164.

15. Paschek, D. and A. Garcia, Reversible temperature and pressure denaturation of a protein fragment: A replica exchange molecular dynamics study. *Physical Review Letters* 2004;**93**: 238105.

16. Paschek, D., S. Gnanakaran, and A.E. Garcia, Simulations of the pressure and temperature unfolding of an alpha-helical peptide. *Proceedings of the National Academy of Sciences of the United States of America* 2005;**102**(19): 6765–6770.

17. Paschek, D., S. Hempel, and A.E. Garcia, Computing the stability diagram Trp-cage miniprotein. *Proceedings of the National Academy of Sciences of the United States of America* 2008;**105**(46): 17754–17759.

18. Smeller, L., Pressure–temperature phase diagrams of biomolecules. *Biochimica et Biophysica Acta—Protein Structure and Molecular Enzyme* 2002;**1595**(1–2): 11–29.

19. Dill, K.A. and J. MacCallum, The protein-folding problem, 50 years on. *Science* 2012;**338**: 1042–1046.

20. Essmann, U. et al., A smooth particle mesh Ewald method. *Journal of Chemical Physics* 1995;**103**(19): 8577–8593.

21. Chen, J., C. Brooks, and J. Khandogin, Recent advances in implicit solvent-based methods for biomolecular simulations. *Current Opinion in Structural Biology* 2008;**18**: 140–148.

22. Cornell, W.D. et al., A second generation force field for the simulation of proteins, nucleic acids, and organic molecules. *Journal of the American Chemical Society* 1995;**117**: 5179–5197.

23. Hornak, V. et al., Comparison of multiple amber force fields and development of improved protein backbone parameters. *Proteins—Structure Function and Bioinformatics* 2006;**65**(3): 712–725.

24. Duan, Y. et al., A point-charge force field for molecular mechanics simulations of proteins based on condensed-phase quantum mechanical calculations. *Journal of Computational Chemistry* 2003;**24**: 1999–2012.

25. Oostenbrink, C. et al., A biomolecular force field based on the free enthalpy of hydration and solvation: The GROMOS force-field parameter sets 53A5 and 53A6. *Journal of Computational Chemistry* 2004;**25**: 1656–1676.

26. Jorgensen, W., D. Maxwell, and J. Tirado Rivas, Development and testing of the OPLS all-atom force field on conformational energetics and properties of organic liquids. *Journal of the American Chemical Society* 1996;**118**: 11225–11236.

27. Kaminski, G. et al., Evaluation and reparametrization of the OPLS-AA force field for proteins via comparison with accurate quantum chemical calculations on peptides. *Journal of Physical Chemistry B* 2001;**105**: 6474–6487.

28. MacKerell, A. et al., All-atom empirical potential for molecular modeling and dynamics studies of proteins. *Journal of Physical Chemistry B* 1998;**102**: 3586–3616.

29. MacKerell, A.D., M. Feig, and C. Books, Extending the treatment of backbone energetics in protein force fields: Limitations of gas-phase quantum mechanics in reproducing protein conformational distributions in molecular dynamics simulations. *Journal of Computational Chemistry* 2004;**25**: 1400–1415.

30. Lindorff-Larsen, K. et al., Improved side-chain torsion potentials for the Amber ff99SB protein force field. *Proteins—Structure Function and Bioinformatics* 2010;**78**(8): 1950–1958.

31. Piana, S., K. Lindorff-Larsen, and D.E. Shaw, How robust are protein folding simulations with respect to force field parameterization? *Biophysical Journal* 2011;**100**: l47–l49.

32. Best, R.B. and G. Hummer, Optimized molecular dynamics force fields applied to the helix-coil transition of polypeptides. *Journal of Physical Chemistry B* 2009;**113**(26): 9004–9015.

33. Wickstrom, L., A. Okur, and C. Simmerling, Evaluating the performance of the ff99SB force field based on NMR scalar coupling data. *Biophysical Journal* 2009;**97**(3): 853–856.

34. Best, R.B., N.V. Buchete, and G. Hummer, Are current molecular dynamics force fields too helical? *Biophysical Journal* 2008;**95**(1): L7–L9.

35. Lange, O., D. van der Spoel, and J. de Groot, Scrutinizing molecular mechanics force fields on the submicrosecond timescale with NMR data. *Biophysical Journal* 2010;**99**: 647–655.

36. Lindorff-Larsen, K. et al., Systematic validation of protein force fields against experimental data. *OPLOS One* 2012;**7**: e32131.

37. Beauchamp, K. et al., Are protein force fields getting better? A systematic benchmark on 524 diverse NMR measurements. *Journal of Chemical Theory and Computation* 2012;**8**: 1409–1414.

38. Piana, S., K. Lindorff-Larsen, and D.E. Shaw, Protein folding kinetics and thermodynamics from atomistic simulation. *Proceedings of the National Academy of Sciences of the United States of America* 2012;**109**: 17845–17850.

39. Simmerling, C., B. Strockbine, and A.E. Roitberg, All-atom structure prediction and folding simulations of a stable protein. *Journal of the American Chemical Society* 2002;**124**(38): 11258–11259.

40. Freddolino, P. et al., Ten-microsecond molecular dynamics simulation of a fast-folding WW domain. *Biophysical Journal* 2008;**94**: L75–L77.

41. Freddolino, P. et al., Force field bias in protein folding simulations. *Biophysical Journal* 2009;**96**: 3772–3780.

42. Best, R.B. and J. Mittal, Balance between a and beta structures in ab initio protein folding. *Journal of Physical Chemistry B* 2010;**114**: 8790–8798.

43. Shaw, D. et al., Anton, a special-purpose machine for molecular dynamics simulation. *Communications of the ACM* 2008;**51**: 91–97.

44. Shaw, D. et al., Atomic-level characterization of the structural dynamics of proteins. *Science* 2010;**330**: 341–346.

45. Lindorff-Larsen, K. et al., How fast-folding proteins fold. *Science* 2011;**334**: 517–520.

46. Stone, J. et al., Accelerating molecular modeling applications with graphics processors. *Journal of Computational Chemistry* 2007;**28**: 2618–2640.

47. Pratt, L., A statistical method for identifying transition states in high dimensional problems. *Journal of Chemical Physics* 1986;**85**: 6780–6786.
48. Bolhuis, P.G., C. Dellago, and D. Chandler, Sampling ensembles of deterministic transition pathways. *Faraday Discussions* 1998;**110**: 421–436.
49. Dellago, C. et al., Transition path sampling and the calculation of rate constants. *Journal of Chemical Physics* 1998;**108**(5): 1964–1977.
50. Bolhuis, P.G. et al., Transition path sampling: Throwing ropes over rough mountain passes, in the dark. *Annual Review of Physical Chemistry* 2002;**53**: 291–318.
51. Dellago, C., P.G. Bolhuis, and P.L. Geissler, Transition path sampling. *Advances in Chemical Physics* 2002;**123**: 1–78.
52. Bolhuis, P.G., Transition path sampling on diffusive barriers. *Journal of Physics— Condensed Matter* 2003;**15**(1): S113–S120.
53. van Erp, T.S., D. Moroni, and P.G. Bolhuis, A novel path sampling method for the calculation of rate constants. *Journal of Chemical Physics* 2003;**118**(17): 7762–7774.
54. Best, R.B. and G. Hummer, Reaction coordinates and rates from transition paths. *Proceedings of the National Academy of Sciences of the United States of America* 2005;**102**(19): 6732–6737.
55. Peters, B. and B.L. Trout, Obtaining reaction coordinates by likelihood maximization. *Journal of Chemical Physics* 2006;**125**(5): 054108.
56. Ma, A. and A.R. Dinner, Automatic method for identifying reaction coordinates in complex systems. *Journal of Physical Chemistry B* 2005;**109**(14): 6769–6779.
57. Dellago, C., P.G. Bolhuis, and D. Chandler, Efficient transition path sampling: Application to Lennard–Jones cluster rearrangements. *Journal of Chemical Physics* 1998;**108**(22): 9236–9245.
58. Geissler, P.L., C. Dellago, and D. Chandler, Kinetic pathways of ion pair dissociation in water. *Journal of Physical Chemistry B* 1999;**103**(18): 3706–3710.
59. Bolhuis, P.G., C. Dellago, and D. Chandler, Reaction coordinates of biomolecular isomerization. *Proceedings of the National Academy of Sciences of the United States of America* 2000;**97**(11): 5877–5882.
60. Bolhuis, P.G. and D. Chandler, Transition path sampling of cavitation between molecular scale solvophobic surfaces. *Journal of Chemical Physics* 2000;**113**(18): 8154–8160.
61. Bolhuis, P.G., Transition-path sampling of beta-hairpin folding. *Proceedings of the National Academy of Sciences of the United States of America* 2003;**100**(21): 12129–12134.
62. Juraszek, J. and P.G. Bolhuis, Sampling the multiple folding mechanisms of Trp-cage in explicit solvent. *Proceedings of the National Academy of Sciences of the United States of America* 2006;**103**(43): 15859–15864.
63. Juraszek, J. and P.G. Bolhuis, Rate constant and reaction coordinate of Trp-Cage folding in explicit water. *Biophysical Journal* 2008;**95**(9): 4246–4257.
64. Hagan, M.F. et al., Atomistic understanding of kinetic pathways for single base-pair binding and unbinding in DNA. *Proceedings of the National Academy of Sciences of the United States of America* 2003;**100**(24): 13922–13927.
65. Radhakrishnan, R. and T. Schlick, Orchestration of cooperative events in DNA synthesis and repair mechanism unraveled by transition path sampling of DNA polymerase beta's closing. *Proceedings of the National Academy of Sciences of the United States of America* 2004;**101**(16): 5970–5975.

66. Quaytman, S. and S. Schwartz, Reaction coordinate of an enzymatic reaction revealed by transition path sampling. *Proceedings of the National Academy of Sciences of the United States of America* 2007;**104**: 12253–12258.

67. Antoniou, D. and S. Schwartz, Toward identification of the reaction coordinate directly from the transition state ensemble using the kernel PCA method. *Journal of Physical Chemistry B* 2011;**115**: 2465–2469.

68. Vreede, J., J. Juraszek, and P.G. Bolhuis, Predicting the reaction coordinates of millisecond light-induced conformational changes in photoactive yellow protein. *Proceedings of the National Academy of Sciences of the United States of America* 2010;**107**(6): 2397–2402.

69. Giambasu, G. et al., Mapping L1 ligase ribozyme conformational switch. *Journal of Molecular Biology* 2012;**423**: 106–122.

70. Matsunaga, Y. et al., Minimum free energy path of ligand-induced transition in adenylate kinase. *PLOS Computational Biology* 2012;**8**: e1002555.

71. Juraszek, J., J. Vreede, and P.G. Bolhuis, Transition path sampling of protein conformational changes. *Chemical Physics* 2012;**396**: 30–44.

72. Henkelman, G., B.P. Uberuaga, and H. Jonsson, A climbing image nudged elastic band method for finding saddle points and minimum energy paths. *Journal of Chemical Physics* 2000;**113**(22): 9901–9904.

73. Sheppard, D., R. Terrell, and G. Henkelman, Optimization methods for finding minimum energy paths. *Journal of Chemical Physics* 2008;**128**(13): 134106.

74. Weinan, E., W.Q. Ren, and E. Vanden-Eijnden, String method for the study of rare events. *Physical Review B* 2002;**66**(5): 052301.

75. Weinan, E., W.Q. Ren, and E. Vanden-Eijnden, Transition pathways in complex systems: Reaction coordinates, isocommittor surfaces, and transition tubes. *Chemical Physics Letters* 2005;**413**(1–3): 242–247.

76. Maragliano, L. et al., String method in collective variables: Minimum free energy paths and isocommittor surfaces. *Journal of Chemical Physics* 2006;**125**(2): 24106.

77. Pan, A.C., D. Sezer, and B. Roux, Finding transition pathways using the string method with swarms of trajectories. *Journal of Physical Chemistry B* 2008;**112**(11): 3432–3440.

78. Weinan, E., W.Q. Ren, and E. Vanden-Eijnden, Finite temperature string method for the study of rare events. *Journal of Physical Chemistry B* 2005;**109**(14): 6688–6693.

79. Ren, W. et al., Transition pathways in complex systems: Application of the finite-temperature string method to the alanine dipeptide. *Journal of Chemical Physics* 2005;**123**(13): 134109.

80. Weinan, E. and E. Vanden-Eijnden, Towards a theory of transition paths. *Journal of Statistical Physics* 2006;**123**(3): 503–523.

81. Weinan, E. and E. Vanden-Eijnden, Transition-path theory and path-finding algorithms for the study of rare events. *Annual Review of Physical Chemistry* 2010;**61**: 391–420.

82. Weinan, E., W.Q. Ren, and E. Vanden-Eijnden, Simplified and improved string method for computing the minimum energy paths in barrier-crossing events. *Journal of Chemical Physics* 2007;**126**(16): 164103.

83. Vanden-Eijnden, E. and M. Venturoli, Revisiting the finite temperature string method for the calculation of reaction tubes and free energies. *Journal of Chemical Physics* 2009;**130**(19): 194103.

84. Maragliano, L. and E. Vanden-Eijnden, On-the-fly string method for minimum free energy paths calculation. *Chemical Physics Letters* 2007;**446**(1–3): 182–190.

85. Johnson, M. and G. Hummer, Characterization of a dynamic string method for the construction of transition pathways in molecular reactions. *Journal of Physical Chemistry B* 2012;**116**: 8573–8583.

86. Miller, T.F., E. Vanden-Eijnden, and D. Chandler, Solvent coarse-graining and the string method applied to the hydrophobic collapse of a hydrated chain. *Proceedings of the National Academy of Sciences of the United States of America* 2007;**104**(37): 14559–14564.

87. Gan, W.X., S.C. Yang, and B. Roux, Atomistic view of the conformational activation of Src kinase using the string method with swarms-of-trajectories. *Biophysical Journal* 2009;**97**(4): L8–L10.

88. Maragliano, L. et al., Mapping the network of pathways of CO diffusion in myoglobin. *Journal of the American Chemical Society* 2010;**132**(3): 1010–1017.

89. Noe, F. et al., Constructing the equilibrium ensemble of folding pathways from short off-equilibrium simulations. *Proceedings of the National Academy of Sciences of the United States of America* 2009;**106**(45): 19011–19016.

90. Ovchinnikov, V., M. Karplus, and E. Vanden-Eijnden, Free energy of conformational transition paths in biomolecules: The string method and its application to myosin VI. *Journal of Chemical Physics* 2011;**134**: 085103.

91. Elber, R. et al., Bridging the gap between long time trajectories and reaction pathways, *Advances in Chemical Physics* 2003;**126**: 93–129.

92. Faradjian, A.K. and R. Elber, Computing time scales from reaction coordinates by milestoning. *Journal of Chemical Physics* 2004;**120**(23): 10880–10889.

93. West, A.M.A., R. Elber, and D. Shalloway, Extending molecular dynamics time scales with milestoning: Example of complex kinetics in a solvated peptide. *Journal of Chemical Physics* 2007;**126**(14): 145104.

94. Majek, P. and R. Elber, Milestoning without a reaction coordinate. *Journal of Chemical Theory and Computation* 2010;**6**(6): 1805–1817.

95. Elber, R., A milestoning study of the kinetics of an allosteric transition: Atomically detailed simulations of deoxy Scapharca hemoglobin. *Biophysical Journal* 2007;**92**(9): L85–L87.

96. Kuczera, K., G.S. Jas, and R. Elber, Kinetics of helix unfolding: Molecular dynamics simulations with milestoning. *Journal of Physical Chemistry A* 2009;**113**(26): 7461–7473.

97. Ghosh, A., R. Elber, and H.A. Scheraga, An atomically detailed study of the folding pathways of protein A with the stochastic difference equation. *Proceedings of the National Academy of Sciences of the United States of America* 2002;**99**(16): 10394–10398.

98. Elber, R. and A. West, Atomically detailed simulation of the recovery stroke in myosin by Milestoning. *Proceedings of the National Academy of Sciences of the United States of America* 2010;**107**(11): 5001–5005.

99. Swope, W.C., J.W. Pitera, and F. Suits, Describing protein folding kinetics by molecular dynamics simulations. 1. Theory. *Journal of Physical Chemistry B* 2004;**108**(21): 6571–6581.

100. Pande, V.S. et al., Atomistic protein folding simulations on the submillisecond time scale using worldwide distributed computing. *Biopolymers* 2003;**68**(1): 91–109.

101. Swope, W.C. et al., Describing protein folding kinetics by molecular dynamics simulations. 2. Example applications to alanine dipeptide and beta-hairpin peptide. *Journal of Physical Chemistry B* 2004;**108**(21): 6582–6594.

102. Schutte, C. et al., A direct approach to conformational dynamics based on hybrid Monte Carlo. *Journal of Computational Physics* 1999;**151**(1): 146–168.

103. Chodera, J.D. et al., Automatic discovery of metastable states for the construction of Markov models of macromolecular conformational dynamics. *Journal of Chemical Physics* 2007;**126**(15): 155101.

104. Noe, F. et al., Hierarchical analysis of conformational dynamics in biomolecules: Transition networks of metastable states. *Journal of Chemical Physics* 2007;**126**(15): 155102.

105. Bowman, G.R. et al., Progress and challenges in the automated construction of Markov state models for full protein systems. *Journal of Chemical Physics* 2009;**131**(12): 124101.

106. Voelz, V.A. et al., Molecular simulation of ab initio protein folding for a millisecond folder NTL9(1–39). *Journal of the American Chemical Society* 2010;**132**(5): 1526–1528.

107. Bowman, G., V. Voelz, and V. Pande, Atomistic folding simulations of the five-helix bundle protein lambda(6–85). *Journal of the American Chemical Society* 2011;**133**: 664–667.

108. Lane, T. et al., Markov state model reveals folding and functional dynamics in ultra-long MD trajectories. *Journal of the American Chemical Society* 2011;**133**: 18413–18419.

109. Berezhkovskii, A., G. Hummer, and A. Szabo, Reactive flux and folding pathways in network models of coarse-grained protein dynamics. *Journal of Chemical Physics* 2009;**130**(20): 205102.

110. Hukushima, K. and K. Nemoto, Exchange Monte Carlo method and application to spin glass simulations. *Journal of the Physical Society of Japan* 1996;**65**(6): 1604–1608.

111. Hansmann, U.H.E., Parallel tempering algorithm for conformational studies of biological molecules. *Chemical Physics Letters* 1997;**281**(1–3): 140–150.

112. Sugita, Y. and Y. Okamoto, Replica-exchange molecular dynamics method for protein folding. *Chemical Physics Letters* 1999;**314**(1–2): 141–151.

113. Hansmann, U. and Y. Okamoto, New Monte Carlo algorithms for protein folding. *Current Opinion in Structural Biology* 1999;**9**: 177–183.

114. Garcia, A.E. and K.Y. Sanbonmatsu, Exploring the energy landscape of a beta hairpin in explicit solvent. *Proteins* 2001;**42**(3): 345–354.

115. Yang, S.C. et al., Folding time predictions from all-atom replica exchange simulations. *Journal of Molecular Biology* 2007;**372**(3): 756–763.

116. Garcia, A.E. and K.Y. Sanbonmatsu, Alpha-helical stabilization by side chain shielding of backbone hydrogen bonds. *Proceedings of the National Academy of Sciences of the United States of America* 2002;**99**(5): 2782–2787.

117. Sanbonmatsu, K. and A. Garcia, Structure of Met-enkephalin in explicit aqueous solution using replica exchange molecular dynamics. *Proteins—Structure Function and Genetics* 2002;**46**: 225–234.

118. Yang, W. et al., Heterogeneous folding of the trpzip hairpin: Full atom simulation and experiment. *Journal of Molecular Biology* 2004;**336**: 241–251.

119. Zhou, R., B.J. Berne, and R.S. Germain, The free energy landscape for beta hairpin folding in explicit water. *Proceedings of the National Academy of Sciences of the United States of America* 2001;**98**: 14931–14936.
120. Baumketner, A. and J. Shea, The structure of the Alzheimer amyloid beta 10–35 peptide probed through replica-exchange molecular dynamics simulations in explicit solvent. *Journal of Molecular Biology* 2007;**366**: 275–285.
121. Nymeyer, H., Energy landscape of the Trpzip2 peptide. *Journal of Physical Chemistry B* 2009;**113**: 8288–8295.
122. De Simone, A. and P. Derreumaux, Low molecular weight oligomers of amyloid peptides display beta-barrel conformations: A replica exchange molecular dynamics study in explicit solvent. *Journal of Chemical Physics* 2010;**132**: 165103.
123. Bussi, G. et al., Free-energy landscape for beta hairpin folding from combined parallel tempering and metadynamics. *Journal of the American Chemical Society* 2006;**128**: 13435–13441.
124. Best, R.B., D. de Sancho, and J. Mittal, Residue-specific alpha-helix propensities from molecular simulation. *Biophysical Journal* 2012;**102**: 1462–1467.
125. Garcia, A.E. and J.N. Onuchic, Folding a protein in a computer: An atomic description of the folding/unfolding of protein A. *Proceedings of the National Academy of Sciences of the United States of America* 2003;**100**(24): 13898–13903.
126. Day, R., D. Paschek, and A.E. Garcia, Microsecond simulations of the folding/unfolding thermodynamics of the Trp-cage miniprotein. *Proteins—Structure Function and Bioinformatics* 2010;**78**(8): 1889–1899.
127. Paschek, D., H. Nymeyer, and A.E. Garcia, Replica exchange simulation of reversible folding/unfolding of the Trp-cage miniprotein in explicit solvent: On the structure and possible role of internal water. *Journal of Structural Biology* 2007;**157**(3): 524–533.
128. Pitera, J.W. and W. Swope, Understanding folding and design: Replica-exchange simulations of "Trp-cage" fly miniproteins. *Proceedings of the National Academy of Sciences of the United States of America* 2003;**100**(13): 7587–7592.
129. Rhee, Y.M. and V.S. Pande, Multiplexed-replica exchange molecular dynamics method for protein folding simulation. *Biophysical Journal* 2003;**84**: 775–786.
130. Zhou, R., Trp-cage: Folding free energy landscape in explicit water. *Proceedings of the National Academy of Sciences of the United States of America* 2003;**100**: 13280–13285.
131. Garcia, A.E. and D. Paschek, Simulation of the pressure and temperature folding/unfolding equilibrium of a small RNA hairpin. *Journal of the American Chemical Society* 2008;**130**(3): 815–817.
132. Bowman, G.R. et al., Structural insight into RNA hairpin folding intermediates. *Journal of the American Chemical Society* 2008;**130**: 9676–9678.
133. Okumura, H. and Y. Okamoto, Temperature and pressure dependence of alanine dipeptide studied by multibaric-multithermal molecular dynamics simulations. *Journal of Physical Chemistry B* 2008;**112**: 12038–12049.
134. Jimenez-Cruz, C., G.I. Makhatadze, and A. Garcia, Protonation/deprotonation effects on the stability of the Trp-cage miniprotein. *Physical Chemistry Chemical Physics* 2011;**13**: 17056–17063.
135. Canchi, D., D. Paschek, and A.E. Garcia, Equilibrium study of protein denaturation by urea. *Journal of the American Chemical Society* 2010;**137**(7): 2338–2344.

136. Canchi, D.R. and A. Garcia, Backbone and side-chain contributions in protein denaturation by urea. *Biophysical Journal* 2011;**100**: 1526–1533.

137. Canchi, D. et al., Molecular mechanism for the preferential exclusion of TMAO from protein surfaces. *Journal of Physical Chemistry B* 2012;**116**: 12095–12104.

138. Yoshida, K., T. Yamaguchi, and Y. Okamoto, Replica-exchange molecular dynamics simulation of small peptide in water and in ethanol. *Chemical Physics Letters* 2005;**412**: 280–284.

139. Kamiya, N. et al., Folding of the 25 residue A beta(12–36) peptide in TFE/water: Temperature-dependent transition from a funneled free-energy landscape to a rugged one. *Journal of Physical Chemistry B* 2007;**111**: 5351–5356.

140. Bhattacharya, A., R. Best, and J. Mittal, Smoothing of the GB1 hairpin folding landscape by interfacial confinement. *Biophysical Journal* 2012;**103**: 596–600.

141. Garcia, A.E., H.D. Herce, and D. Paschek, Simulations of temperature and pressure unfolding of peptides and proteins with replica exchange molecular dynamics. *Annual Reports in Computational Chemistry* 2006;**2**: 83–95.

142. Barsegov, V., G. Morrison, and D. Thirumalai, Role of internal chain dynamics on the rupture kinetic of adhesive contacts. *Physical Review Letters* 2008;**100**(24): 248102.

143. Rick, S., Replica exchange with dynamical scaling. *Journal of Chemical Physics* 2007;**126**: 054102.

144. Zhang, C. and J. Ma, Enhanced sampling and applications in protein folding in explicit solvent. *Journal of Chemical Physics B* 2010;**132**: 244101.

145. Sugita, Y. and Y. Okamoto, Replica-exchange multicanonical algorithm and multicanonical replica-exchange method for simulating systems with rough energy landscape. *Chemical Physics Letters* 2000;**329**: 261–270.

146. Mitsutake, A., Y. Sugita, and Y. Okamoto, Generalized-ensemble algorithms for molecular simulations of biopolymers. *Biopolymers* 2001;**60**: 96–123.

147. Kim, J., T. Keyes, and J.E. Straub, Generalized replica exchange method. *Journal of Chemical Physics* 2010;**132**(22): 224107.

148. Li, H. et al., Finite reservoir replica exchange to enhance canonical sampling in rugged energy surfaces. *Journal of Chemical Physics* 2006;**125**: 144902.

149. Rauscher, S., C. Neale, and R. Pomes, Simulated tempering distributed replica sampling, virtual replica exchange, and other generalized-ensemble methods for conformational sampling. *Journal of Chemical Theory and Computation* 2009;**5**(10): 2640–2662.

150. Shi, C., J. Wallace, and J. Shen, Thermodynamic coupling of protonation and conformational equilibria in proteins: Theory and simulation. *Biophysical Journal* 2012;**102**: 1590–1597.

151. Okur, A. et al., Improved efficiency of replica exchange simulations through use of a hybrid explicit/implicit solvation model. *Journal of Chemical Theory and Computation* 2006;**2**: 420–433.

152. Fukunishi, H., O. Watanabe, and S. Takada, On the Hamiltonian replica exchange method for efficient sampling of biomolecular systems: Application to protein structure prediction. *Journal of Chemical Physics* 2002;**116**: 9058–9067.

153. Liu, P. et al., Replica exchange with solute tempering: A method for sampling biological systems in explicit water. *Proceedings of the National Academy of Sciences of the United States of America* 2005;**102**: 13749–13754.

154. Huang, X. et al., Replica exchange with solute tempering: Efficiency in large scale systems. *Journal of Physical Chemistry B* 2007;**111**: 5405–5410.
155. Laio, A. and M. Parrinello, Escaping free-energy minima. *Proceedings of the National Academy of Sciences of the United States of America* 2002;**99**: 1252–12566.
156. Barducci, A., G. Bussi, and M. Parrinello, Well-tempered metadynamics: A smoothly converging and tunable free-energy method. *Physical Review Letters* 2008;**100**: 020603.
157. Bonomi, M. and M. Parrinello, Enhanced sampling in the well-tempered ensemble. *Physical Review Letters* 2010;**104**: 190601.
158. Rosta, E., N.V. Buchete, and G. Hummer, Thermostat artifacts in replica exchange molecular dynamics simulations. *Journal of Chemical Theory and Computation* 2009;**5**: 1393–1399.
159. Sindhikara, D. et al., Bad seeds sprout perilous dynamics: Stochastic thermostat induced trajectory synchronization in biomolecules. *Journal of Chemical Theory and Computation* 2009;**5**: 1624–1631.
160. Rathore, N., M. Chopra, and J. de Pablo, Optimal allocation of replicas in parallel tempering simulations. *Journal of Chemical Physics* 2005;**122**: 024111.
161. Abraham, M. and J. Gready, Ensuring mixing efficiency of replica-exchange molecular dynamics simulations. *Journal of Chemical Physics* 2008;**4**: 1119–1128.
162. Periole, X. and A. Mark, Convergence and sampling efficiency in replica exchange simulations of peptide folding in explicit solvent. *Journal of Chemical Physics* 2007;**126**: 014903.
163. Sindhikara, D., Y. Meng, and A. Roitberg, Exchange frequency in replica exchange molecular dynamics. *Journal of Chemical Physics* 2008;**128**: 024103.
164. Andrec, M. et al., Protein folding pathways from replica exchange simulations and a kinetic network model. *Proceedings of the National Academy of Sciences of the United States of America* 2005;**102**: 6801–6806.
165. Rosta, E. and G. Hummer, Error and efficiency of replica exchange molecular dynamics simulations. *Journal of Chemical Physics* 2009;**131**: 165102.
166. Rosta, E. and G. Hummer, Error and efficiency of simulated tempering simulations. *Journal of Chemical Physics* 2010;**132**: 034012.
167. Hoover, W., Canonical dynamics—Equilibrium phase space distributions. *Physical Review A* 1985;**31**: 1695–1697.
168. Nose, S., A molecular dynamics method for simulations in the canonical ensemble. *Molecular Physics* 1984;**52**: 255–268.
169. Andersen, H., Molecular dynamics simulations at constant pressure or temperature. *Journal of Chemical Physics* 1980;**72**: 2384–2393.
170. Martyna, G., M. Klein, and M. Tuckerman, Nose-Hoover chains—The canonical ensemble via continuous dynamics. *Journal of Chemical Physics* 1992;**97**: 2635–2643.
171. Garcia, A.E., Molecular dynamics simulations of protein folding. In *Protein Structure Prediction*, M. Zaki and C. Bystroff (eds.). Humana Press: Totowa, NJ, 2007, pp. 315–330.
172. Trebst, S., M. Troyer, and U. Hansmann, Optimized parallel tempering simulations of proteins. *Journal of Chemical Physics* 2006;**124**: 174903.
173. Sindhikara, D., D. Emerson, and A. Roitberg, Exchange often and properly in replica exchange molecular dynamics. *Journal of Chemical Theory and Computation* 2010;**6**: 2804–2808.

174. Buchete, N.V. and G. Hummer, Peptide folding kinetics from replica exchange molecular dynamics. *Physical Review E* 2008;**77**: 030902.

175. Nymeyer, H., How efficient is replica exchange molecular dynamics? An analytic approach. *Journal of Chemical Theory and Computation* 2008;**4**: 626–636.

176. Timasheff, S.N., The control of protein stability and association by weak interactions with water—How do solvents affect these processes? *Annual Review of Biophysics and Biomolecular Structure* 1993;**22**: 67–97.

177. Streicher, W.W. and G.I. Makhatadze, Unfolding thermodynamics of Trp-cage, a 20 residue miniprotein, studied by differential scanning calorimetry and circular dichroism spectroscopy. *Biochemistry* 2007;**46**(10): 2876–2880.

178. Noel, J. et al., Mirror images as naturally competing conformations in protein folding. *Journal of Physical Chemistry B* 2012;**116**: 6880–6888.

179. Barua, B. et al., The Trp-cage: Optimizing the stability of a globular miniprotein. *Protein Engineering Design & Selection* 2008;**21**(3): 171–185.

180. Qiu, L.L. et al., Smaller and faster: The 20-residue trp-cage protein folds in 4 micros. *Journal of the American Chemical Society* 2002;**124**: 12952–12953.

181. Ahmed, Z. et al., UV-resonance Raman thermal unfolding study of Trp-cage shows that it is not a simple two-state miniprotein. *Journal of the American Chemical Society* 2005;**127**(31): 10943–10950.

182. Neuweiler, H., S. Doose, and M. Sauer, A microscopic view of miniprotein folding: Enhanced folding efficiency through formation of an intermediate. *Proceedings of the National Academy of Sciences of the United States of America* 2005;**102**(46): 16650–16655.

183. Snow, C.D., B. Zagrovic, and V.S. Pande, The trp cage: Folding kinetics and unfolded state topology via molecular dynamics simulations. *Journal of the American Chemical Society* 2002;**124**: 14548–14549.

184. Chowdhury, S. et al., Ab initio folding simulation of the Trp-cage mini-protein approaches NMR resolution. *Journal of Molecular Biology* 2003;**327**(3): 711–717.

185. Ding, F. et al., Ab initio RNA folding by discrete molecular dynamics: From structure prediction to folding mechanisms. *RNA—A Publication of the RNA Society* 2008;**14**(6): 1164–1173.

186. Jorgensen, W. et al., Comparison of simple potential functions for simulating liquid water. *Journal of Chemical Physics* 1983;**79**(2): 926–935.

187. Weerasinghe, S. and P.E. Smith, A Kirkwood-Buff derived force field for mixtures of urea and water. *Journal of Physical Chemistry B* 2003;**107**(16): 3891–3898.

188. Stumpe, M. and H. Grubmuller, Interaction of urea with amino acids: Implications for urea-induced protein denaturation. *Journal of the American Chemical Society* 2007;**129**: 16126–16131.

189. Rossky, P., Protein denaturation by urea: Slash and bond. *Proceedings of the National Academy of Sciences of the United States of America* 2008;**105**: 16825–16826.

190. Sagle, L. et al., Investigating the hydrogen-bonding model of urea denaturation. *Journal of the American Chemical Society* 2009;**131**: 9304–9310.

191. Pace, C.N. and K.L. Shaw, Linear extrapolation method of analyzing solvent denaturation curves. *Proteins—Structure Function and Genetics* 2000;**4**: 1–7.

192. Wafer, L.N.R., W.W. Streicher, and G.I. Makhatadze, Thermodynamics of the Trp-cage miniprotein unfolding in urea. *Proteins—Structure Function and Bioinformatics* 2010;**78**(6): 1376–1381.

193. Guinn, E.J. et al., Quantifying why urea is a protein denaturant, whereas glycine betaine is a protein stabilizer. *Proceedings of the National Academy of Sciences of the United States of America* 2011;**108**: 16932–16937.
194. Kortemme, T., A.V. Morozov, and D. Baker, An orientation-dependent hydrogen bonding potential improves prediction of specificity and structure for proteins and protein–protein complexes. *Journal of Molecular Biology* 2003;**326**(4): 1239–1259.

Chapter 5

Minimal Models for the Structure and Dynamics of Nucleic Acids

Changbong Hyeon and Devarajan (Dave) Thirumalai

CONTENTS

5.1   INTRODUCTION

The description of reality using models, the detail of which depends on the phenomenon of interest, requires an appropriate level of abstraction, which depends on the question of interest. For example, near a critical point, exponents that describe the vanishing of order parameter or divergence of correlation length are universal, depending only on the dimensionality (d), and are impervious to atomic details. These findings, which are rooted in the concepts of universality and renormalization group [1], are also applicable to the properties of polymers [2]. For example, the size of a long homopolymer and the distribution of end-to-end distance depend only on the solvent quality, the degree of polymerization, and d, but not on the details of monomer structure [2]. However, to describe the dynamics occurring on length scales that are on the order of a few nm, one has to contend with chemical properties of the monomer.

Without rigorous theoretical underpinnings, intuitive arguments and phenomenology are often used in modeling complex biological processes. Here, also the level of description depends on length scales. In nucleic acids, at short length scales ($l < \sim 5$ Å), detailed chemical environment determines the basic forces (hydrogen bonds and dispersion forces) between two nucleotides. On the scale $l \sim 1\text{–}3$ nm interactions between two bases, base stacks and grooves of the nucleic acids become relevant. Understanding how RNA folds ($l \sim 1\text{–}3$ nm) requires energy functions that provide at least a CG description of nucleotides and interactions between them in the native state and excitations around the folded structure. On the persistence-length scale $l_p \approx 150$ bp ≈ 50 nm [3] and beyond, it suffices to treat double-stranded DNA (dsDNA) as a stiff elastic filament without explicitly capturing the base pairs. If $l \sim O(1)$ μm, dsDNA behaves like a self-avoiding polymer [4]. On the scale of chromosomes ($l \sim$ mm), a much coarser description suffices. Thus, models for nucleic acids vary because the scale of structural organization changes from nearly mm in chromosome to several nm in the folded states of RNA.

5.2 POLYMER MODELS FOR dsDNA AND CHROMOSOME STRUCTURE

The length, L, of dsDNA exceeds a few μm with persistence length, $l_p \approx 50$ nm. On these scales, global properties of dsDNA, such as the end-to-end distance and the dependence of l_p on salt concentration, are not greatly affected by fluctuations of individual base pairs. Consequently, dsDNA can be treated as a fluctuating elastic material, for which the wormlike chain (WLC) is a suitable polymer model. On much longer scales ($L \sim 1$ mm, $L/l_p \gg 1$), which is relevant to chromosome, the genomic material can be described as a flexible polymer. Using these scale-dependent models, a number of predictions for DNA organization and dynamics can be made.

5.2.1 Looping Dynamics

Loop formation in biopolymers is an elementary process in the self-assembly of DNA, RNA, and proteins. However, understanding cyclization kinetics is complicated because multiple-length scales and internal chain modes are intertwined in bringing distant parts of DNA into proximity. For a short chain, the cyclization time, τ_c, scales as $L^{3/2}$, while $\tau_c \sim L^2$ when L increases [5,6]. The problem of cyclization becomes more challenging in the looping dynamics of dsDNA, an elementary process that is relevant in gene regulation and DNA condensation. In the CG model, a single pitch of a double helix, formed by 10.5 bp, could represent one interaction center (Figure 5.1a). Thus, l_p encompasses (14–15) CG interaction centers ($l_p \approx 150$ bp). The parameters for bond and bending potentials along the chain, consisting of multiple CG centers, are selected to reproduce the persistence length of dsDNA [7,8], allowing us to study various dynamical aspects of dsDNA, stretching, looping, or supercoiling from the perspective of polymer physics. The ease of loop formation and the associated kinetics is characterized by L/l_p. For $L/l_p \sim O(1)$, energetic cost required to bend dsDNA makes the cyclization difficult for short chains. In contrast, when $L/l_p \gg 1$, the cyclization between two ends gets harder because of entropic cost. Theory and simulations using CG model showed that τ_c is minimized when $L/l_p \approx 2$–3 [9–11] (Figure 5.1a). Of note, in looping of dsDNA responsible for gene regulation in prokaryotes, $L \approx 100$ bp ($L/l_p \approx 0.7$). For such a dsDNA with $L \approx 100$ bp, sequence effects are also relevant [12–15].

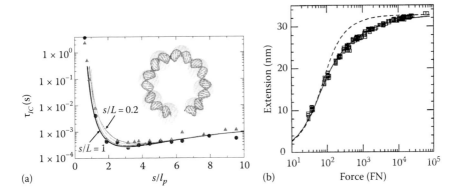

Figure 5.1 Applications to DNA. (a) Loop formation times between two regions in dsDNA separated by s along the contour from simulations using CG model that represents a single pitch of DNA helix as a monomer unit. Lines are theoretical results. (Adapted from Hyeon, C. and Thirumalai, D., *J. Chem. Phys.*, 124, 104905, 2006.) (b) Extension as a function of mechanical force for 97 kb λ-DNA. Symbols are experimental results and the dashed line is the fit using WLC model. (Adapted from Bustamante, C. et al., *Science*, 265, 1599, 1994.)

(*continued*)

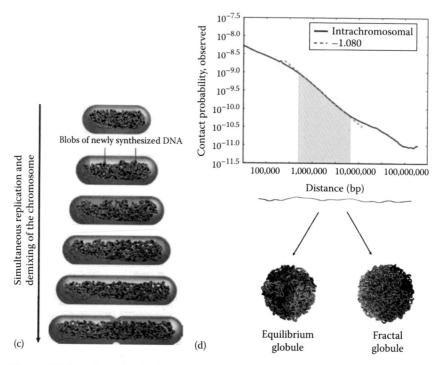

Figure 5.1 (continued) Applications to DNA. (c) Model of bacterial chromosomal separation from simulations of tightly confined polymer chain. The newly synthesized DNA is extruded to the periphery of the unreplicated nucleoid and the two strings of blobs drift apart and segregate due to the excluded-volume interactions and conformational entropy. (Adapted from Jun, S. and Wright, A., *Nat. Rev. Microbiol.*, 8, 600, 2010.) (d) Top figure shows scaling law $P(s) \sim s^{-1.08}$ where $P(s)$ is the contact probability for a given genomic distance s, measured by Hi-C, a method that probes the 3D architecture of whole genomes by coupling proximity-based ligation with massively parallel sequencing. The exponent in the power law decay is distinct from $s^{-1.5}$ for an equilibrated globule (bottom left), whereas $s^{-1.08}$ scaling (dashed lines from CG simulations in the top figure) is explained using a fractal globule (bottom right), a knot-free, polymer conformation, which enables reversible folding and unfolding at any genomic locus. (Adapted from Lieberman-Aiden, E. et al., *Science*, 326, 289, 2009.)

5.2.2 Stretching dsDNA

Fluctuations of dsDNA on scales comparable to l_p can be described using WLC model, which parameterizes dsDNA as a polymer that resists bending on scale $\sim l_p$. Smith et al. measured the response of a 97 kbp dsDNA (contour length $L \approx 33.0$ μm) from λ-phage to a stretching force, f [3,16] (Figure 5.1b). In the absence of f, λ-DNA conformations are determined by thermal fluctuations, whereas loss in chain entropy must be overcome to stretch dsDNA with $f \neq 0$. The free energy of stretching a semiflexible chain under a constant tension f is equivalent to a quantum mechanical problem of a dipolar rotor with moment of inertia l_p in an electric field f. An extrapolation formula is obtained by numerically solving the quantum mechanical problem that accurately describes the measured force as a function of extension (Figure 5.1b). Fits to experimental data yield L of λ-DNA 32.80 ± 0.10 μm and solving the quantum mechanical problem that accurately describes the measured force as a function of extension (Figure 5.1b). Fits to experimental data yield L of λ-DNA 32.80 ± 0.10 μm and $l_p \approx 53.4 \pm 2.3$ nm, thus confirming that dsDNA is a semiflexible chain.

5.2.3 Confined Polymers and Bacterial Chromosome Segregation

Replication and passage of genetic information to daughter cells are major events in cell reproduction. These complex events are remarkably accurate even in the simplest organisms. Although chromosome segregation is likely to be complex and well orchestrated, it has recently been proposed that confinement-induced entropic forces due to restrictions in cellular space are sufficient to drive chromosome segregation in bacteria [17,18] (Figure 5.1c). This proposal was formulated using molecular simulations of tightly confined self-avoiding polymer chains in cylindrical space, which show that the chains segregate and become spatially organized reminiscent of that observed in bacteria. In such highly confined spaces, polymer conformations are determined by ξ, the size of a renormalized structural unit, the Flory radius R_F in the absence of confinement, and the length (P) and diameter D of the cylinder. In *Escherichia coli*, the values are $\xi = 87$ nm, $R_F = 3.3$ μm, $D = 0.24$ μm, and $P = 1.3$ μm. Armed with the results for confined polymers, a concentric shell model for bacterial chromosome was proposed [17,18] in which the nucleoid was modeled as an inner and outer cylinder. The unreplicated *mother* strand, a self-avoiding chain, is restricted to the inner compartment, whereas the *daughter* chain (obtained in simulations by adding monomers at a set time in the Monte Carlo simulations) is free to explore the entire nucleoid volume. The results of the simulations show that the newly added (or replicated) chain segregates to the periphery of the nucleoid, driven by gain in entropy, and becomes spatially organized as they are synthesized (Figure 5.1c). CG modeling combined with polymer theory leads to the discovery that entropic forces alone are sufficient to drive chromosome segregation in bacteria, with proteins perhaps playing a secondary role in poising the state of the chromosome for enabling the entropy-driven mechanism.

5.2.4   Chromosome Folding

In eukaryotic cells, chromosomes fold into globules that spatially occupy well-defined volumes known as chromosome territories [19]. In this process, widely separated gene-rich regions are brought into close proximity. Knowledge of the spatial arrangement of chromosomes is important in describing gene activity and the state of the cell. Polymer physics concepts have been used to describe the structures of folded chromosome using constraints derived from experiments. These calculations have provided considerable insights into their compartmentalization in the nucleus [20]. A number of models, such as the random walk model and models that connect mega-based size domains by chromatin loops, have been used to describe higher structures of chromatin. The experimental resolution is roughly 1 Mb ($\approx$ 340 µm), and consequently, coarse graining in this context must be on length scales on the order of a µm. Recently, folding principles for human genome were proposed using data for long-range contacts between distinct loci as constraints [21]. Experiments showed that contact probability, $P(s)$, between loci in a chromosome, which is separated by genomic distance s (measured in units of bp), exhibits a power law decay in the range ~500 kb to ~7 Mb. The observed dependence $P(s) \sim s^{-1}$ can be rationalized using polymer models (Figure 5.1d) introduced a number of years ago in describing collapse of homopolymers [22]. If chromosome folds up into an equilibrium globule (polymer in a poor solvent), then $P(s)$ should scale as $P(s) \sim s^{-1.5}$, which cannot account for the experimental observations. An alternate model suggests that interface DNA can organize itself into a fractal globule, which is compact and not entangled as an equilibrium globule would be. Monte Carlo simulations of a polymer with 4000 beads (1 bead = 1200 bp ~ 0.4 µm) were used to generate conformations of fractal and equilibrium globules. The power law decay of $P(s)$, with exponent $\approx$–1, is consistent with measurements (Figure 5.1d). More importantly, the unknotted fractal globule loci that are close in genomic sequence are also in proximity in 3D spatial arrangement, which clearly is relevant for gene activity.

5.3   RNA FOLDING

The discovery that RNA can serve as enzymes has provided a great impetus to describe their folding in quantitative terms. RNA folding landscape is rugged because of interplay of several competing factors. First, phosphate groups are negatively charged, which implies that polyelectrolyte effects oppose folding. Valence, size, and shape of counterions, necessary to induce compaction and folding [23], can dramatically alter the thermodynamics and kinetics of RNA folding. Second, the nucleotides purine and pyrimidine bases have different sizes but are chemically similar. Third, folded RNA molecules have a complex architectural organization [24]. Many, not all, of the nucleotides engage the Watson–Crick base pairing [25], while other regions form bulges, loops, etc. The associations between non-base-paired regions are critical in bringing the pieces of secondary structures into a number of distinct folds whose stability can be dramatically altered by counterions [26]. Fourth, the lack of chemical diversity in the bases results in RNA easily adopting alternate misfolded con-formations, which means that the stability gap between the folded and mis-folded structures is not too large. Thus, the homopolymer nature of the RNA monomers, the critical role of counterions in shaping the folding landscape, and the presence of low-energy excitations around the folded state make RNA folding a challenging problem [27].

At first glance, it might appear that it is difficult to develop coarse-grained models for RNA, which are polyelectrolytes, that fold into compact struc-tures as the electrostatic interactions are attenuated by adding counterions. Moreover, recent studies have shown valence, size, and shape of counterions profoundly influence RNA folding [26,28–31]. Despite the complexity, it is possible to devise physics-based models that capture the essential aspects of RNA folding and dynamics. In order to provide a framework for understand-ing and anticipating the outcomes of increasingly sophisticated experiments involving RNA, we have developed two classes of models. These models are particularly useful in probing the effect of mechanical force in modulating the folding landscape of simple hairpins to ribozymes. In the following sec-tions, we discuss two coarse-graining strategies for representing RNA mol-ecules (Figure 5.2) and assess their usefulness in reproducing experimental observations.

5.3.1 Three-Interaction-Site Model [32]

From the general architecture of RNA molecules, it is immediately clear that they are composed of a series of nucleotides that are connected together via chemically identical ribose sugars and charged phosphates that make up its backbone. Protruding from the backbone are four possible aromatic bases that may form hydrogen-bonding interactions with other bases, typically following the well-known Watson–Crick pairing rules. Local base-stacking interactions may also play an important role in stabilizing the folded struc-ture. Taking into account the aforementioned cursory observations, we con-structed a coarse-grained model of RNA by representing each nucleotide by three beads with interaction sites corresponding to the ribose sugar group,

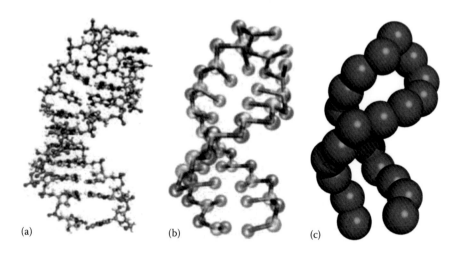

(a)   (b)   (c)

Figure 5.2   A schematic illustration of the various levels of coarse graining for models of RNA. The detailed all-atom representation (a) can be reduced to include three beads for each nucleotide corresponding to the base, sugar, and phosphate moieties as in the TIS model (b). Further coarse-graining results in each bead being represented by a single nucleotide (c) and is referred to as the self-organized polymer (SOP) model. The details of energy functions in the TIS and SOP models are discussed in the main text.

the phosphate group, and the base. In the three-interaction-site (TIS) model, the bases are covalently linked to the ribose center, and the sugar and phosphates compose the backbone. Thus, an RNA molecule with N nucleotides is composed of $3N$ interaction centers. The potential energy of a conformation is given by

$$V_T = V_{SR} + V_{LR}$$
$$V_{SR} = V_{Bonds} + V_{Angles} + V_{Dihedrals} \tag{5.1}$$
$$V_{LR} = V_{NC} + V_{Elec} + V_{Stack}.$$

The short-range interactions (V_{SR}) include the bond, angle, and dihedral terms (V_{Bonds}, V_{Angles}, and $V_{Dihedrals}$, respectively), which account for the chain connectivity and the angular degrees of freedom as is commonly used in coarse-grained models of this type [33]. The long-range interactions (V_{LR}) are composed of the native interaction term, V_{NC}; pairwise additive electrostatic term between the phosphates, V_{Elec}; and base-stacking interaction term that stabilize, the hairpin, V_{Stack}. We now describe the long-range interaction terms in detail.

The native interaction term between the bases mimics the hydrophobicity of the purine/pyrimidine group, and a Lennard–Jones interaction between the nonbonded interaction centers is as follows:

$$V_{NC} = \sum_{i=1}^{N-1} \sum_{j=i+1}^{N} V_{B_i B_j}(r) + \sum_{i=1}^{N} \sum_{m=1}^{2N-1} {'}V_{B_i (SP)_m}(r) + \sum_{m=1}^{2N-4} \sum_{n=m+3}^{2N-1} V_{(SP)_m (SP)_n}(r). \quad (5.2)$$

A native contact is defined for two noncovalently bonded beads whose separation in the native structure is within a cutoff distance r_c (= 7.0 Å). Two beads that are beyond r_c in the native structure are considered to be *nonnative*. Pairs of beads that are considered native have the following potential:

$$V_{\alpha,\beta}(r) = C_h \left[\left(\frac{r_{ij}^o}{r} \right)^{12} - 2 \left(\frac{r_{ij}^o}{r} \right)^6 \right] \quad (5.3)$$

For beads that are nonnative, the interactions are described by

$$V_{\alpha,\beta}(r) = C_R \left[\left(\frac{a}{r} \right)^{12} + \left(\frac{a}{r} \right)^6 \right] \quad (5.4)$$

where a = 3.4 Å and C_R = 1 kcal/mol. The electrostatic potential between the phosphate groups is assumed to be pairwise additive:

$$V_{Elec} = \sum_{i=1}^{N-1} \sum_{j=i+1}^{N} V_{P_i P_j}(r) \quad (5.5)$$

where we assume a Debye–Hückel interaction, which accounts for the effect of screening due to counterions and it is given by

$$V_{P_i P_j} = \frac{z_{P_i} z_{P_j} e^2}{4\pi\varepsilon_0 \varepsilon_r r} e^{-r/l_D} = z_{P_i} z_{P_j} \frac{k_B T l_B}{r} \left(\frac{80}{\varepsilon_r} \right) e^{-r/l_D} \quad (5.6)$$

where
  $z_{P_i} = -1$ is the charge on the phosphate ion
  $l_B \approx 7$ Å is the Bjerrum length at $T = T_r$
  l_D is the Debye length, $l_D^{-1} = \sqrt{2e^2 I / \varepsilon_r \varepsilon_0 k_B T} = \sqrt{8\pi l_B I (80/\varepsilon_r)}$

For ε_r, we used ε_r = 10. To calculate the ionic strength, the concentration of the ions, c, is used. Since the Debye screening length $\sim l_B^{-1/2} \sim \sqrt{T}$, the strength of the electrostatic interaction between the phosphate groups is temperature dependent, even when we ignore the variations of ε with T. At room temperature ($T \sim 300$ K), the electrostatic repulsion $V_{P_i P_j} \sim 0.5$ kcal/mol between the phosphate groups at $r \sim 5.8$ Å, which is the closest distance between them. It follows that the V_{Elec} between phosphate groups across the base pairing ($r = 16$–18 Å) is almost negligible.

Finally, simple RNA secondary structures are stabilized largely by stacking interactions whose context-dependent values are known. The orientationally dependent stacking interaction term is taken to be

$$V_i(\{\phi\}, \{\psi\}, \{r\}; T) = \Delta G_i(T) \times e^{-\alpha_{st} \{\sin^2(\phi_{1i} - \phi_{1i}^o) + \sin^2(\phi_{2i} - \phi_{2i}^o) + \sin^2(\phi_{3i} - \phi_{3i}^o) + \sin^2(\phi_{4i} - \phi_{4i}^o)\}}$$

$$\times e^{-\beta_{st} \{(r_{ij} - r_{ii}^o)^2 + (r_{i+1 j-1} - r_{2i}^o)^2\}} \tag{5.7}$$

$$\times e^{-\gamma_{st} \{\sin^2(\psi_{1i} - \psi_{1i}^o) + \sin^2(\psi_{2i} - \psi_{2i}^o)\}}$$

where $\Delta G(T) = \Delta H - T\Delta S$. The bond angles $\{\phi\}$ are $\phi_{1i} \equiv \angle S_i B_i B_j$, $\phi_{2i} \equiv \angle B_i B_j S_j$, $\phi_{3i} \equiv \angle S_{i+1} B_{i+1} B_{j-1}$, and $\phi_{4i} \equiv \angle B_{i+1} B_{j-1} S_{j-1}$, where S_i and B_i denote the position of ith ribose and base group, respectively. The distance between two paired bases $r_{ij} = |B_i - B_j|$, $r_{i+1 j-1} = |B_{i+1} - B_{j-1}|$ and ψ_{1i} and ψ_{2i} are the dihedral angles formed by the four beads $B_i S_i S_{i+1} B_{i+1}$ and $B_{j-1} S_{j-1} S_j B_j$, respectively. The superscript o refers to angles and distances in the protein data bank (PDB) structure. The values of α_{st}, β_{st}, and γ_{st} are 1.0, 0.3, and 1.0 Å^{-2}, respectively. The values for ΔH and ΔS were taken from Turner's thermodynamic data set [34,35].

Once the appropriate model has been formulated, simulations can be performed to follow the dynamics of the RNA molecule of interest for comparison to experiments. A combination of forced unfolding and force-quench refolding of a number of RNA molecules has been used to map the energy landscape of RNA. These experiments identify kinetic barriers and the nature of intermediates by using mechanical unfolding or refolding trajectories that monitor end-to-end distance $R(t)$ of the molecule in real time (t) or from force-extension curves (FECs). The power of simulations is that they can be used to deduce structural details of the intermediates that cannot be unambiguously inferred using $R(t)$ or FECs. As such, forced-unfolding simulations are performed by applying a constant force to the bead at one end of the molecule under conditions that mimic the experiments.

5.3.2 Forced Unfolding of P5GA Using the TIS Model

To date, laser optical tweezer experiments have used f to unfold or refold by force quench by keeping T fixed [36]. A fuller understanding of RNA folding landscape can be achieved by varying T and f. Calculations using the TIS model for even a simple hairpin show that the phase diagram is rich when both T and f are varied. Using the fraction of native contacts, $\langle Q \rangle$, as an order parameter, the diagram of states in the (f, T) plane shows that P5GA hairpin behaves approximately as a *two-state* folder. In the absence of force $f = 0$ pN, the folding unfolding transition midpoint is at $T_m = 341$ K. As force increases, T_F decreases monotonically such that the transition midpoints (T_m, f_m) form a phase boundary separating the folded $(\langle Q \rangle > 0.5$ and $\langle R \rangle < 3$ nm) and unfolded states. The phase boundary is sharp at low T_m and large f_m, but it is broad at low force. The locus of points separating the unfolded and folded states is given by (see Figure 5.3a)

$$f_c \sim f_o \left(1 - \left(\frac{T}{T_m} \right)^\alpha \right)$$

$$(5.8)$$

where
    f_o is the critical force at low temperatures
    $\alpha \ (= 6.4)$ is a sequence-dependent exponent

The large value of α suggests a weak first-order transition.

The thermodynamic relation $\log K_{eq}(f) = \Delta F_{UF}/k_B T + f \cdot \Delta x_{UF}/k_B T$ and the dependence of $\log K_{eq}$ (K_{eq} is computed as time averages of the traces in Figure 5.3b) on f are used to estimate ΔF_{UF} and Δx_{UF}, which is the equilibrium distance separating the native basin of attraction (NBA) and the basin corresponding to the ensemble of unfolded states (UBA). The transition midpoint $K(f_m) = 1$ gives $f_m \approx 6$ pN (Figure 5.3c), which is in excellent agreement with the value obtained from the equilibrium phase diagram (Figure 5.3a). From the slope, $\partial \log K_{eq}(f)/\partial f = 1.79$ pN^{-1}, $\Delta x_{UF} \approx 7.5$ nm, we found, by extrapolation to $f = 0$, that $\Delta F_{UF} \approx 6.2$ kcal/mol under the assumption that Δx_{UF} is constant and independent of f.

In the single-molecule force experiments on RNA [37], the time interval between the hopping transitions from folded to unfolded states at the midpoint of force was measured at a single temperature. We calculated the dynamics along the phase boundary (T_m, f_m) to evaluate the variations in the free-energy profiles and the dynamics of transition from the NBA to UBA. Along the boundary (T_m, f_m), there are substantial changes in the free-energy landscape. The free-energy barrier $\Delta F^\ddagger$ increases dramatically at low T and high f. The weakly first-order phase transition at $T \approx T_m$ and low f becomes increasingly stronger as we move along the (T_m, f_m) boundary to low T and high f.

The two basins of attraction (NBA and UBA) are separated by a free-energy barrier whose height increases as force increases (or temperature decreases) along (T_m, f_m) (Figure 5.3e). The hopping time τ_h along (T_m, f_m) is

$$\tau_h = \tau_o \exp \left(\frac{\Delta F^\ddagger}{k_B T} \right)$$

$$(5.9)$$

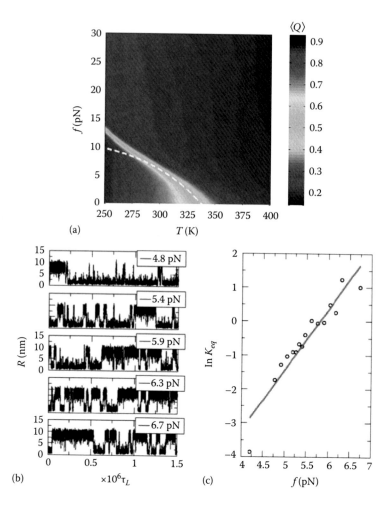

Figure 5.3 (a) Phase diagram for the P5GA hairpin in terms of f and T. This panel shows the diagram of states obtained using the fraction of native contacts as the order parameter. The values of the thermal average of the fraction of native contacts, $\langle Q \rangle$, in gray scale are shown on the right. The dashed line is a fit to the locus of points in the (f, T) plane that separates the folded hairpin from the unfolded states (Equation 5.8). (b) Time traces of R at various values of constant force at $T = 305$ K. At $f = 4.8$ pN, $f < f_m \approx 6$ pN $\langle R \rangle$ fluctuates around at low values, which shows that the NBA is preferentially populated (first panel). As $f \sim f_m$ (third panel), the hairpin hops between the folded state (low R value) and unfolded states ($R \approx 10$ nm). The transitions occur over a short time interval. These time traces are similar to that seen in Figure 2-C of [37]. (c) Logarithm of the equilibrium constant K_{eq} (computed using the time traces in (a)) as a function of f. The line is a fit with $\log K_{eq} = 10.4 + 1.79f$.

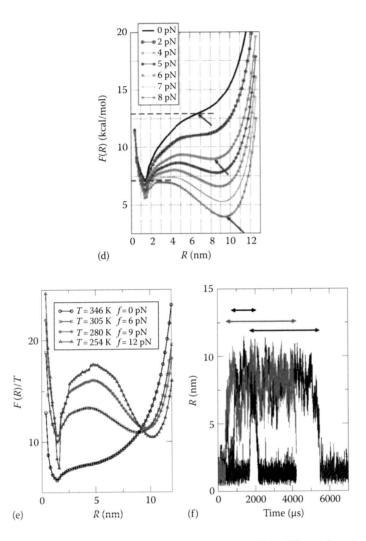

Figure 5.3 (continued)   (d) Equilibrium free-energy profiles $F(R)$ as a function of R at $T = 305$ K. The colors represent different f values that are displayed in the inset. The arrows give the location of the unfolded basin of attraction. Note that the transition state moves as a function f in accord with the Hammond postulate. (e, f) Hopping transitions along the phase boundary. (e) Free-energy profiles $F(R)$ along the phase boundary (T_m, f_m) (see a). The barrier separating NBA and UBA increases at low T_m and high f_m values. (f) Time traces of R obtained by using Brownian dynamics simulations. The values of T and f are 305 K and 6 pN, respectively. The arrows on the top of time traces indicate the residence times in the NBA for three trajectories.

To estimate the variations in τ_h along the (T_m, f_m) boundary, we performed three very long overdamped Langevin simulations at $T_m = 305$ K and $f_m = 6$ pN. The unfolding/refolding time is observed to be 1–4 ms (Figure 5.3f). From the free-energy profile, we find $\Delta F^{\ddagger}/T \sim 3$, so that $\tau_0 = 0.05$–0.2 ms. Consequently, τ_{h} at $T = 254$ K and $f = 12$ pN is estimated to be 1–4 s, which is three orders of magnitude greater than at the higher T_m and lower f_m. These simulations showed that only by probing the dynamics over a wide range of (T, f) values can the entire energy landscape be constructed.

To probe the structural transitions in the hairpin, we performed Brownian dynamics simulations at a constant force with $T = 254$ K. From the phase diagram, the equilibrium unfolding force at this temperature is 12 pN (Figure 5.3). To monitor the complete unfolding of P5GA, in the time course of the simulations, we applied $f = 42$ pN to one end of the hairpin with the other end fixed. In contrast to thermal unfolding (or refolding), the initially closed hairpin unzips from the end to the loop region. The unzipping dynamics, monitored by the time dependence of R, shows *quantized staircase-like jumps* with substantial variations in step length, which depend on the initial conditions. The lifetimes associated with the *intermediates* vary greatly. The large dispersion reflects the heterogeneity of the mechanical unfolding pathways. Approach to the stretched state occurs in a stepwise *quantized* manner [38], which was first shown in lattice models of proteins [38].

5.3.3 Force-Quench Refolding [32]

To monitor the dynamics of approach to the NBA, we initiated refolding from extended conformations with $R = 13.5$ nm, prepared by stretching at $T = 290$ K and $f = 90$ pN. Subsequently, we quench the force to $f = 0$, and the approach to the native state was monitored. From the distribution of first passage times, the refolding kinetics follow exponential kinetics with the mean folding time of ≈ 191 μs, compared with 12.4 μs in the temperature quench. It is remarkable that even though the final conditions ($T = 290$ K and $f = 0$) are the same as in thermal refolding, the time scale for hairpin formation upon force quench is significantly large than thermal refolding.

The large difference arises because the molecules that are fully stretched with $f \gg f_m$ and those that are generated at high T have vastly different initial conformations. Hence, they can navigate entirely different regions of the energy landscape in the approach to the native conformation. The distribution of R in the thermally denatured conformations is $P(R) \propto e^{-\beta V_{tot}(R)/k_B T_0}$ (T_0 is the initial temperature), whereas in the ensemble of the stretched conformation, $P(R) \propto \delta(R - R_s)$ where R_s is the value of R when the hairpin is fully extended. The initially stretched conformations ($R_s = 13.5$ nm) do not overlap with the accessible regions of the canonical ensemble of thermally denatured conformations [39].

As a consequence, the regions of the free-energy landscape from which folding commences in force-jump folding are vastly different from those corresponding to the initial population of thermally equilibrated ensemble.

The pathways explored by the hairpins en route to the NBA are heterogeneous. Different molecules reach the hairpin conformation by vastly different routes. Nevertheless, the time dependence of R shows that the approach to the native conformation occurs in stages. Upon release of force, there is a rapid initial decrease in R that results in the collapse of the hairpin. Surprisingly, this process takes an average of several microseconds, which is much longer than expectations based on theories of collapse kinetics of polymer coils [40,41]. In the second stage, the hairpin fluctuates in relatively compact state with R in the broad range (25–75 Å) for prolonged time periods. On these time scales, which vary considerably depending on the molecules, conformational search occurs among compact structures. The final stage is characterized by a further decrease in R that takes the molecules to the NBA. The last stage is the most cooperative and abrupt, whereas the first two stages appear to be much more continuous. Interestingly, similar relaxation patterns characterized by heterogeneous pathways and continuous collapse in the early stages have been observed in force-quench refolding of ubiquitin [42]. The simulations showed that the complexity of the folding landscape observed in ribozyme experiments was already reflected in the formation of simple RNA hairpin [32,43] just as β-hairpin formation captures much of the complexity of protein folding [44]. The multistage approach to the native stage is reminiscent of the three-stage refolding by Camacho–Thirumalai for protein refolding [45].

5.3.4 Complexity of Hairpin Formation

When viewed on length scales that span several bps, folding of a small RNA (or DNA) hairpin is remarkably simple. However, when probed on short times (ns–μs range), the formation of a small hairpin involving turn formation and base stacking is remarkably complex. Recent experiments show that the kinetics of hairpin formation in RNA (or ssDNA) deviates from the classical two-state kinetics and is best described as a multistep process [46]. Additional facets of hairpin formation have been revealed in single-molecule experiments that use mechanical force (f). These experiments prompted simulations that vary both T and f. The equilibrium-phase diagram showed two basins of attraction (folded and unfolded) at the locus of critical points (T_m, f_m). At T_m and f_m, the probability of being unfolded and folded is the same. The free-energy surface obtained from simulations explained the sharp bimodal transition between the folded and unfolded states when the RNA hairpin is subject to f [32,47]. Thus, from thermodynamic considerations, hairpin formation could be approximated to be two-state like.

5.3.5   Self-Organized Polymer (SOP) Model for RNA Folding

Although the TIS interaction model is one of the simplest possible representations of RNA molecules, we can further simplify the representation of RNA when the number of nucleotides is large and our concern lies in the dynamical features described by low-frequency modes. Instead of representing each nucleotide by three beads, we can represent each nucleotide by a single bead by taking the center of each nucleotide as the interaction center. The interactions stabilizing the native conformation are taken to be uniform. However, variations of this model are required for accurate modeling of RNA structures that have a subtle interplay between secondary and tertiary interactions.

One of the computational bottlenecks of TIS model simulations is the computation of the dihedral angle potential, largely because of the calculation of the trigonometric function in the energy function. The repeated calculation of the dihedral angle potential term is sufficiently burdensome that some choose to use lookup tables so that its calculation is done only at the beginning of the program run. If the configuration of the dihedral angle potential is not required then in simulation efficiency, an appreciable increase would be achieved, making such an approach attractive if it is reasonable. These arguments were the basis for the construction of the SOP model. In this very simple model, a single bead represents each nucleotide. Local interactions are defined by bond potentials and native contacts determine favorable long-range interactions. The Hamiltonian for the SOP model is identical for proteins except that the values of the parameters are different (see Table 1 in Hyeon et al. [48]).

5.3.6   Stretching Azoarcus Ribozyme

The structure of the (195 nt) Azoarcus ribozyme [49] (PDB code: 1u6b) is similar to the catalytic core of the Tetrahymena thermophila ribozyme, including the presence of a pseudoknot. The size of this system in terms of the number of nucleotides allows exploration of the forced unfolding over a wide range of loading conditions. For the Azoarcus ribozyme, mechanical unfolding trajectories generated at three different loading rates using SOP model reveal distinct unfolding pathways. At the highest loading rate, the FEC has six conspicuous rips (red FEC in Figure 5.4b), whereas at the lower r_f, the number of peaks is reduced to between two and four. The structures in each rip were identified by comparing the FECs (Figure 5.4b) with the history of rupture of contacts (Figure 5.4c). At the highest loading rate, the dominant unfolding pathway of the Azoarcus ribozyme is N → [P5] → [P6] → [P2] → [P4] → [P3] → [P1]. At medium loading rates, the ribozyme unfolds via N → [P1, P5, P6] → [P2] → [P4] → [P3], which leads to four rips in the FECs. At the lowest loading rate, the number of rips is further reduced to two, which we

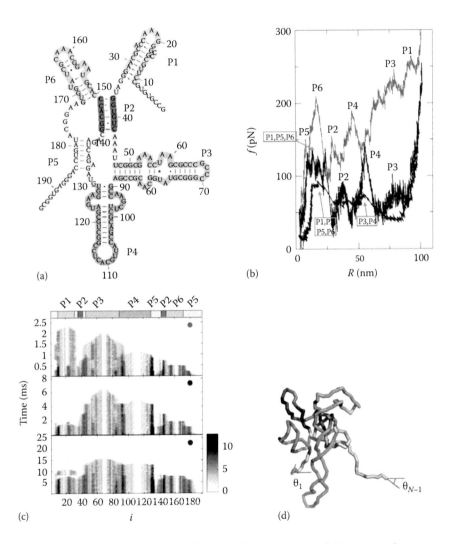

(a)

(b)

(c)

(d)

Figure 5.4 (See color insert.) (a) Secondary structure of Azoarcus ribozyme. (b) FECs of Azoarcus ribozyme at three r_f's (v = 43 µm/s, k_s = 28 pN/nm in red; v = 12.9 µm/s, k_s = 28 pN/nm in green; and v = 5.4 µm/s, k_s = 3.5 pN/nm in blue) obtained using the SOP model. (c) Contact rupture dynamics at three loading rates. The rips, resolved at the nucleotide level, are explicitly labeled. (d) Topology of Azoarcus ribozyme in the SOP representation. The first and the last alignment angles between the bond vectors and the force direction are specified.

(*continued*)

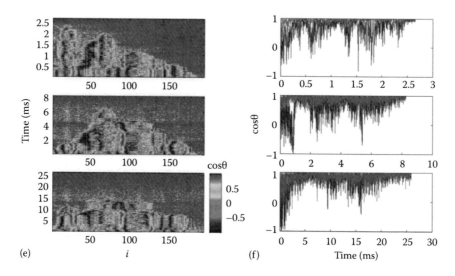

Figure 5.4 (continued)   **(See color insert.)** (e) Time evolutions of cos Θ_i ($i = 1, 2, ...,$ $N-1$) at three loading rates are shown. The values of cos Θ_i are color coded as indicated on the scale shown on the right of bottom panel. (f) Comparisons of the time evolution of cos Θ_i (blue) and cos Θ_{N-1} (red) at three loading rates show that the differences in the f_c values at the opposite ends of the ribozyme are greater as r_f increases.

identify with N → [P1, P2, P5, P6] → [P3, P4]. Unambiguously identifying the underlying pulling speed-dependent conformational changes requires not only the FECs but also the history of rupture of contacts (Figure 5.4c). The simulations using the SOP model also show that unfolding pathways can be altered by varying the loading rate.

To understand the profound changes in the unfolding pathways as r_f is varied, it is necessary to compare r_f with r_T, the rate at which the applied force propagates along RNA (or proteins) [48]. In both atomic force microscope (AFM) and laser optical tweezers (LOT) experiments, force is applied to one end of the chain (3′ end), while the other end is fixed. The initially applied tension propagates over time in a nonuniform fashion through a network of interactions that stabilize the native conformation. The variable $\lambda = r_T/r_f$ determines the rupture history of the biomolecules. If $\lambda \gg 1$, then the applied tension at the 5′ end of the RNA propagates rapidly so that, even prior to the realization of the first rip, force along the chain is uniform. This situation pertains to the LOT experiments (low r_f). In the opposite limit, $\lambda = 1$, the force is nonuniformly felt along the chain. In such a situation, unraveling of RNA begins in regions in which the value of local force exceeds the tertiary interactions. Such an event occurs close to the end at which the force is applied.

The intuitive arguments given earlier were made precise by computing the rate of propagation of force along the Azoarcus ribozyme. To visualize the propagation of force, we computed the dynamics of alignment of the angles between the bond segment vector $(r_{i,i+1})$ and the force direction during the unfolding process (Figure 5.4d through f). The nonuniformity in the local segmental alignment along the force direction, which results in a heterogeneous distribution of times in which segment vectors approximately align along the force direction, is most evident at the highest loading rate (Figure 5.4e). Interestingly, the dynamics of the force propagation occurs sequentially from one end of the chain to the other at high r_f. Direct comparison of the differences in the alignment dynamics between the first (θ_i) and last angles (θ_{N-1}) (see Figure 5.4d) illustrates the discrepancy in the force values between the 3′ and 5′ ends (Figure 5.4f). There is nonuniformity in the force values at the highest r_f whereas there is a more homogeneous alignment at low r_f. The microscopic variations in the dynamics of tension propagation are reflected in the rupture kinetics of tertiary contacts (Figure 5.4c) and, hence, in the dynamics of the rips (Figure 5.4b).

These results highlight an important prediction of the SOP model that the very nature of the unfolding pathways can drastically change depending on the loading rate, r_f. The dominant unfolding rate depends on r_f, suggesting that the outcomes of unfolding by LOT and AFM experiments can be dramatically different. In addition, predictions of forced unfolding based on all-atom MD simulations should also be treated with caution unless, for topological reasons (as in the Ig27 domain from muscle protein titin), the unfolding pathways are robust to large variations in the loading rates.

5.4 CONCLUDING REMARKS

We have provided a rationale for building a coarse-grained model for describing the dynamics of nucleic acids and have presented a handful of applications to show the power of using simple coarse-grained structure-based models in the context of folding and functions of RNA and proteins. It is remarkable that such simple models can capture the complexity of self-assembly.

There are several avenues that are likely to be explored both in the context of protein and RNA folding using coarse-grained models of increasing sophistication. First, experiments are starting to provide detailed information on the structures of unfolded states of proteins in the presence of denaturants such as urea and guanidinium hydrochloride. Direct simulations, therefore, require models of denaturants within the context of the CG models. Preliminary studies that tackle this challenging problem have already appeared [50]. Similarly, there is a challenge to model the counterion-dependent nature of unfolded states of ribozymes. This will require incorporating in an effective way counterion size and shape within the CG models. Second, it is increasingly clear that functions require interactions between biomolecules. Thus, the CG models will have to be expanded to include scales ranging from microns (DNA) to nanometers (RNA and proteins). Third, applications to molecular machines accounting for the complex relationship between the mechanochemical cycles and functions have also been initiated. Explaining the linkage between the conformational changes for biological machines will require progress in establishing the validity of the CG models as well as further developments in refining them. These and other challenges and progress to date show that the next 10 years will witness an explosion in routinely using CG models to quantitatively understand many phenomena ranging from folding to function.

ACKNOWLEDGMENTS

C. H. thanks Korea Institute for Advanced Study (KIAS) for providing computing resources. The work of DT was supported by the National Science Foundation.

REFERENCES

1. Fisher, M. E. (1974) Renormalization group in theory of critical behavior. *Rev. Mod. Phys.* 46, 597–616.
2. de Gennes, P. G. (1979) *Scaling Concepts in Polymer Physics.* Cornell University Press, Ithaca, NY.
3. Bustamante, C., Marko, J. F., Siggia, E. D., and Smith, S. (1994) Entropic elasticity of λ-phase DNA. *Science* 265, 1599–1600.
4. Valle, F., Favre, M., de Los Rios, P., Rosa, A., and Dietler, G. (2005) Scaling exponents and probability distribution of DNA end-to-end distance. *Phys. Rev. Lett.* 95, 158105.
5. Pastor, R. W., Zwanzig, R., and Szabo, A. (1996) Diffusion limited first contact of the ends of a polymer: Comparison of theory with simulation. *J. Chem. Phys.* 105, 3878–3882.
6. Toan, N., Greg Morrison, P., Hyeon, C., Thirumalai, D. et al. (2008) Kinetics of loop formation in polymer chains? *J. Phys. Chem. B* 112, 6094–6106.
7. Podtelezhnikov, A. A. and Vologodskii, A. V. (2000) Dynamics of small loops in DNA molecules. *Macromolecules* 33, 2767–2771.
8. Vologodskii, A., Levene, S., Klenin, K., Frank-Kamenetskii, M., and Cozzarelli, N. (1992) Conformational and thermodynamic properties of supercoiled DNA. *J. Mol. Biol.* 227, 1224–1243.
9. Podtelezhnikov, A. and Vologodskii, A. (1997) Simulations of polymer cyclization by Brownian dynamics. *Macromolecules* 30, 6668–6673.

10. Jun, S., Bechhoefer, J., and Ha, B.-Y. (2003) Diffusion-limited loop formation of semi-flexible polymers: Kramers theory and the interwined time scales of chain relaxation and closing. *Europhys. Lett.* 64, 420–426.

11. Hyeon, C. and Thirumalai, D. (2006) Kinetics of interior loop formation in semiflexible chains. *J. Chem. Phys.* 124, 104905.

12. Cloutier, T. and Widom, J. (2004) Spontaneous sharp bending of double-stranded DNA. *Mol. Cell.* 14, 355–362.

13. Du, Q., Smith, C., Shiffeldrim, N., Vologodskaia, M., and Vologodskii, A. (2005) Cyclization of short DNA fragments and bending fluctuations of the double helix. *Proc. Natl. Acad. Sci. USA* 102, 5397–5402.

14. Savelyev, A. and Papoian, G. (2010) Chemically accurate coarse graining of double-stranded DNA. *Proc. Natl. Acad. Sci. USA* 107(47), 20340–20345.

15. Vafabakhsh, R. and Ha, T. (2012) Extreme bendability of DNA less than 100 base pairs long revealed by single-molecule cyclization. *Science* 337, 1097–1101.

16. Smith, S. B., Finzi, L., and Bustamante, C. (1992) Direct mechanical measurements of the elasticity of single DNA molecules by using magnetic beads. *Science* 258, 1122–1126.

17. Jun, S. and Mulder, B. (2006) Entropy-driven spatial organization of highly confined polymers: Lessons for the bacterial chromosome. *Proc. Natl. Acad. Sci. USA* 103, 12388.

18. Jun, S. and Wright, A. (2010) Entropy as the driver of chromosome segregation. *Nat. Rev. Microbiol.* 8, 600–607.

19. Cremer, T. and Cremer, C. (2001) Chromosome territories, nuclear architecture and gene regulation in mammalian cells. *Nat. Rev. Genet.* 2, 292–301.

20. Grosberg, A., Rabin, Y., Havlin, S., and Neer, A. (1993) Crumpled globule model of the three-dimensional structure of DNA. *Europhys. Lett.* 23, 373.

21. Lieberman-Aiden, E., van Berkum, N., Williams, L., Imakaev, M. et al. (2009) Comprehensive mapping of long-range interactions reveals folding principles of the human genome. *Science* 326, 289.

22. Grosberg, A., Nechaev, S., and Shakhnovich, E. (1988) The role of topological constraints in the kinetics of collapse of macromolecules. *J. Phys.* 49, 2095–2100.

23. Thirumalai, D., Lee, N., Woodson, S. A., and Klimov, D. K. (2001) Early events in RNA folding. *Annu. Rev. Phys. Chem.* 52, 751–762.

24. Tinoco Jr., I. and Bustamante, C. (1999) How RNA folds. *J. Mol. Biol.* 293, 271–281.

25. Dima, R. I., Hyeon, C., and Thirumalai, D. (2005) Extracting stacking interaction parameters for RNA from the data set of native structures. *J. Mol. Biol.* 347, 53–69.

26. Heilman-Miller, S. L., Pan, J., Thirumalai, D., and Woodson, S. A. (2001) Role of counterion condensation in folding of tetrahymena ribozyme II. Counterion-dependence of folding kinetics. *J. Mol. Biol.* 309, 57–68.

27. Thirumalai, D. and Hyeon, C. (2005) RNA and protein folding: Common themes and variations. *Biochemistry* 44, 4957–4970.

28. Koculi, E., Lee, N. K., Thirumalai, D., and Woodson, S. A. (2004) Folding of the tetrahymena ribozyme by polyamines: Importance of counterion valence and size. *J. Mol. Biol.* 341, 27–36.

29. Koculi, E., Hyeon, C., Thirumalai, D., and Woodson, S. A. (2007) Charge density of divalent metal cations determines RNA stability. *J. Am. Chem. Soc.* 129, 2676–2682.

30. Heilman-Miller, S. L., Pan, J., Thirumalai, D., and Woodson, S. A. (2001) Role of counterion condensation in folding of tetrahymena ribozyme I. Equilibrium stabilization by cations. *J. Mol. Biol.* 306, 1157–1166.

31. Pan, J., Thirumalai, D., and Woodson, S. A. (1999) Magnesium-dependent folding of self-splicing RNA: Exploring the link between cooperativity, thermodynamics, and kinetics. *Proc. Natl. Acad. Sci. USA* 96, 6149–6154.

32. Hyeon, C. and Thirumalai, D. (2005) Mechanical unfolding of RNA hairpins. *Proc. Natl. Acad. Sci. USA* 102, 6789–6794.

33. Klimov, D. K., Betancourt, M. R., and Thirumalai, D. (1998) Virtual atom representation of hydrogen bonds in minimal off-lattice models of alpha helices: Effect on stability, cooperativity and kinetics. *Folding Des.* 3, 481–498.

34. Mathews, D., Sabina, J., Zuker, M., and Turner, D. (1999) Expanded sequence dependence of thermodynamic parameters improves prediction of RNA secondary structure. *J. Mol. Biol.* 288, 911–940.

35. Walter, A. E., Turner, D. H., Kim, J., Lyttle, M. H., Muller, P., Mathews, D. H., and Zuker, M. (1994) Coaxial stacking of helices enhances binding of oligoribonucleotides and improves predictions of RNA folding. *Proc. Natl. Acad. Sci. USA* 91, 9218–9222.

36. Onoa, B. and Tinoco, Jr, I. (2004) RNA folding and unfolding. *Curr. Opin. Struct. Biol.* 14, 374–379.

37. Liphardt, J., Dumont, S., Smith, S. B., Tinoco, Jr., I., and Bustamante, C. (2002) Equilibrium information from nonequilibrium measurements in an experimental test of Jarzynski's equality. *Science* 296, 1832–1835.

38. Klimov, D. K. and Thirumalai, D. (1999) Stretching single-domain proteins: Phase diagram and kinetics of force-induced unfolding. *Proc. Natl. Acad. Sci. USA* 96, 6166–6170.

39. Hyeon, C. and Thirumalai, D. (2006) Forced-unfolding and force-quench refolding of RNA hairpins. *Biophys. J.* 90, 3410–3427.

40. Thirumalai, D. (1995) From minimal models to real proteins: Time scales for protein folding kinetics. *J. Phys. I* (Fr.) 5, 1457–1467.

41. Pitard, E. and Orland, H. (1998) Dynamics of the swelling or collapse of a homopolymer. *Europhys. Lett.* 41, 467–472.

42. Fernandez, J. M. and Li, H. (2004) Force-clamp spectroscopy monitors the folding trajectory of a single protein. *Science* 303, 1674–1678.

43. Chen, S. J. and Dill, K. A. (2000) RNA folding energy landscapes. *Proc. Natl. Acad. Sci. USA* 97, 646–651.

44. Thirumalai, D. and Klimov, D. K. (1999) Deciphering the time scales and mechanisms of protein folding using minimal off-lattice models. *Curr. Opin. Struct. Biol.* 9, 197–207.

45. Camacho, C. J. and Thirumalai, D. (1993) Kinetics and thermodynamics of folding in model proteins. *Proc. Natl. Acad. Sci. USA* 90, 6369–6372.

46. Ma, H., Proctor, D. J., Kierzek, E., Kierzek, R., Bevilacqua, P. C., and Gruebele, M. (2006) Exploring the energy landscape of a small RNA hairpin. *J. Am. Chem. Soc.* 128, 1523–1530.

47. Hyeon, C. and Thirumalai, D. (2008) Multiple probes are required to explore and control the rugged energy landscape of RNA hairpins. *J. Am. Chem. Soc.* 130, 1538–1539.

48. Hyeon, C., Dima, R. I., and Thirumalai, D. (2006) Pathways and kinetic barriers in mechanical unfolding and refolding of RNA and proteins. *Structure* 14, 1633–1645.
49. Rangan, P., Masquida, B., Westhof, E., and Woodson, S. A. (2003) Assembly of core helices and rapid tertiary folding of a small bacterial group i ribozyme. *Proc. Natl. Acad. Sci. USA* 100, 1574–1579.
50. O'Brien, E., Dima, R., Brooks, B., and Thirumalai, D. (2007) Interactions between hydrophobic and ionic solutes in aqueous guanidinium chloride and urea solutions: Lessons for protein denaturation mechanism. *J. Am. Chem. Soc.* 129, 7346–7353.

Chapter 6

Amyloid Peptide Aggregation

*Computational Techniques to Deal with
Multiple Time and Length Scales*

Luca Larini and Joan-Emma Shea

CONTENTS

6.1　INTRODUCTION

A number of diseases are associated with the pathological aggregation of normally soluble proteins and the deposition of these aggregates in the form of amyloid plaques on various organs on the body. More than 20 such amyloid diseases, including Parkinson's and Alzheimer's diseases, have been identified [1,2]. The proteins involved in amyloid disorders are varied in terms of sequence and native fold, yet they self-assemble via similar nucleation–growth kinetics to a similar end product, a cross-β fibril with characteristic properties, including the ability to bind to amyloidogenic dyes. Importantly, even proteins that are not naturally associated with amyloidosis can form fibers under conditions that destabilize the native fold. A number of short peptides have also been shown to aggregate readily under concentrated conditions, and certain organisms utilize the aggregation process for functional purposes, such as scaffolding in bacteria [3]. These observations suggest that the formation of amyloid fibers is a common feature of all proteins [1].

The aggregation process spans several time (ms to hours) and length scales (nm to 100 nm). During the nucleation process, which involves relatively few proteins, interactions specific to each protein sequence drive the initial aggregation. The size and nature of the nucleus will be influenced by sequence details. However, at longer time and length scales, more generic properties of the proteins start to dominate, resulting in full-fledged fibers that share similar structure independently of the protein sequence. In order to tackle the multiple time and length scales associated with the aggregation from a computational standpoint, it is necessary to resort to a hierarchy of computational models of different levels of resolution.

In this chapter, we discuss three classes of models: fully atomic simulations in explicit solvent, fully atomic simulations in implicit solvent, and coarse-grained models, as well as the computational techniques relevant for each model. We focus primarily on one major technique per model, complemented by a *case study* example. At the end of each section, we provide a brief overview of other notable techniques used in the literature.

6.2 REPLICA EXCHANGE MOLECULAR DYNAMICS AND ENHANCED SAMPLING TECHNIQUES

6.2.1 Introduction

Molecular dynamics (MD) simulations are routinely used to understand the behavior of complex systems at the atomistic scale. For this purpose, several atomistic force fields have been developed to date, such as AMBER [4–10], CHARMM [11–14], and OPLS-AA [15–21]. These simulations have provided important insight into the chemistry and physics of biological molecules and their phase transitions at the atomic scale.

Standard fully atomic MD simulations are unable to access the time scales required to study aggregation. As a consequence, to properly study the slow reorganization time associated with protein dynamics (~seconds), techniques to enhance sampling are necessary. Generally, in order to enhance the sampling efficiency, these enhanced sampling methods apply a bias to the dynamics of the system. We can picture the effect of the bias as forcing the system to evolve on a distorted free-energy surface. To be effective, the new free-energy landscape must be distorted in such a way that barriers among different minima are decreased. As a result, the molecules can more easily explore the different minima and avoid being trapped in one minimum for the length of the simulation. A drawback of these sampling schemes, however, is that since the system is now evolving on a different free-energy landscape, the dynamics of the process are altered.

Because of the way the distorted landscape is constructed, the weight of each conformation is increased with respect to the original one. It is very important to choose a bias whose effect can be easily removed, a process called reweighting. Reweighting consists in inputting the biased and enhanced probability of each conformation of the distorted free-energy landscape to obtain unbiased probabilities in the original landscape. Additionally, care must be taken in choosing a bias that does not excessively distort the free-energy landscape. In fact, if the distorted free-energy landscape is not selective enough, the risk is that it samples too many conformations that when reweighted have a negligible weight in the original landscape. Or, in the worst case scenario, it can lead to out-of-equilibrium conformations.

In the following sections, we will describe different methods that aim at enhancing the sampling efficiency of fully atomistic simulations. All of these methods make a good choice for the bias and have simple ways of reweighting the biased conformations.

6.2.2   Method Details

Details about the replica exchange molecular dynamics (REMD) protocol [22–24] are given in Chapter 1. In this section, we will briefly discuss the idea behind this methodology.

To enhance the sampling of the conformation on a rugged landscape, the REMD method takes advantage of the fact that at high temperature, the system can more easily cross the free-energy barriers that separate minima. REMD employs multiple temperatures, starting around the temperature of interest (usually around room temperature, a temperature at which the protein is folded) and ending at high temperature (a temperature at which the protein unfolds) suitable to cross the free-energy barriers to other minima.

Multiple simulations (called replicas) are performed at the same time, each at a different temperature. At regular intervals (typically a few ps), conformations from different replicas of the system are compared and susceptible to exchange based on a Metropolis criterion (the latter provides the proper reweighting of each conformation). We can picture the REMD method as a process where a conformation that is trapped in one minimum at low temperature is slowly heated at a high temperature, where it can easily overcome free-energy barriers and change conformation. The new conformation is then slowly cooled to the original temperature. The Metropolis criterion guarantees that this conformation is not overheated or supercooled, which would lead to out of equilibrium conformations. In addition, as the preceding statement holds for each replica, REMD provides enhanced sampling at all the temperatures employed in the study.

6.2.3   Case Study: Amyloid-β [25–35]

We illustrate the REMD approach for a fully atomic, solvated system by considering a case study of the self-assembly of the amyloid-β [25–35]. The amyloid-β [25–35] is the smallest naturally occurring by-product of enzymatic cleavage of the amyloid precursor protein (APP). This peptide forms toxic oligomers and provides a computationally tractable, yet biologically relevant system to study the formation of oligomers. In a series of papers [25–27], we used REMD simulations, with an atomistically detailed description of the peptide (OPLS-AA force field [28,29]) and explicit solvent molecules (TIP3P [30]), to investigate the conformational space populated by monomer, dimers, trimers, and tetramers. Simulations were run in the canonical NVT ensemble, with electrostatic interactions accounted for using the particle mesh Ewald method [31] with a real-space cutoff of 1.2 nm. The same cutoff was used to compute van der Waals' interactions as well as long-range dispersion corrections for both energy and pressure. The temperature was kept constant by a Nosé–Hoover thermostat, and the equations of motion were integrated by means of the leapfrog algorithm with a time step of 2 fs. Constraints were used for the heavy atoms connected to hydrogen atoms, with the LINCS algorithm [32] for the protein and the SETTLE algorithm [33] for water. We used periodic boundary conditions, with a box of water edge of 5.4 nm. The temperature ranges used in the replica exchange simulations were 290–497.4 K (monomer), 290–356 K (dimer), and 290–411 K (trimer and tetramer). To achieve proper sampling and statistics, simulations were run for 206 ns (monomer), 330 ns (dimer), 502 ns (trimer), and 517 ns (tetramer). The initial 107 ns (monomer), 123 ns (dimer), 207 ns (trimer), and 208 ns (tetramer) were discarded as equilibration data. The output of the simulations was clustered based on mutual root-mean-square deviation of the peptide backbone using the Daura algorithm [34]. All simulations were performed using the GROMACS package [35].

Our simulations showed that the monomer sampled a variety of compact conformations, including β-hairpin structures. The monomer conformations are selected through a competition between electrostatic and hydrophobic interactions. The dimers populated primarily compact conformations consisting of associated β-hairpins and a small minority of extended dimer structures. In contrast to the monomer case, the conformations adopted by the dimers are mainly dictated by electrostatics. A significant conformational change was observed between dimer and trimer, with the trimers now favoring extended conformations over compact ones. Tetramers show an even more pronounced favoring of extended conformations. Nonetheless, both trimers and tetramers show a heterogeneous distribution of conformations, reflecting a competition between electrostatic interactions (which drive the chains to extend) and hydrophobic interactions (which favor compaction). A particularly interesting observation from our simulations is that the extended conformers tend to have a hairpin present at the edge of the sheet. Our simulations suggest that this hairpin plays a dual role in (1) stabilizing the β-sheet in a flat conformation and (2) serving as the growing end of the sheet. We find that the hairpin must possess sufficient flexibility to rearrange its structure once bound to the sheet. These observations led to a new picture for the initial stages of aggregation that proceeds as follows (see Figure 6.1): In the first step, two monomers in β-hairpin conformations associate to form a compact dimer. In the second step, one of the chains is destabilized and adopts an extended conformation, thus enabling a new monomer to be added to the growing front. The β-hairpin plays a critical role in keeping the growing sheet extended. A summary of this scheme is given in Figure 6.1.

6.2.4 Other Sampling Techniques

6.2.4.1 Metadynamics

In metadynamics [36–38], the biasing potential consists of a penalty to the system for exploring regions of the space already visited. For a system of n particles, one must first define a collective variable (CV) $C(\vec{r}^{(n)})$, which in

Fibril

Figure 6.1 Schematic of the aggregation pathway for Aβ [25–35], highlighting the crucial role of the β-hairpin. On- and off-pathways structures are shown. (Adapted from Larini, L. and Shea, J.E., *Biophys. J.*, 103, 576, 2012.)

principle can be a function of all the position $\vec{r}$ of each particle. At regular intervals, the method checks the value of $C(\vec{r}^{(n)})$ at time t and adds a biasing potential $V'(C,t)$ that is different from zero only around C and is chosen to push the system away from it (Figure 6.2, panel b). In this way, the overall potential of the system is composed of a time-independent part (the atomistic potential) and a time-dependent part V'. A pictorial way to understand the idea behind metadynamics is to think of the atomistic potential as a set of wells (minima of the potential) that over time are filled by the potential V' (see Figure 6.2).

Initially, the effect of V' consists of reducing the barrier between the minima as a result of filling the bottom of each minimum (Figure 6.2c and d). However, after a certain period, all the wells will be completely filled so that there will no longer be any barrier present (Figure 6.2e). At this stage, all the information about the shape and depth of the atomistic potential is contained in the filling potential V'. In this way, the potential V' can be used to reconstruct the original atomistic potential.

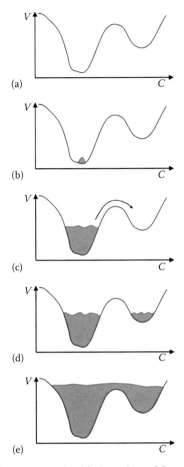

Figure 6.2   How metadynamics works. (a) The unbiased free-energy landscape is composed of several minima (in this example, only two are depicted). Due to the height of the
barriers, the system would very likely be trapped within one minimum for a very long
time, which in principle can exceed the typical time scale of a standard MD simulation.
(b) For this reason, metadynamics adds a biasing potential that pushes the system away
from its current position. In the example, the system was located at the bottom of one
minimum. As a consequence, a repulsive potential (orange area) is added to the original
force field to push the system into other regions of the free-energy landscape. (c) As the
system evolves, metadynamics keeps adding repulsive regions to the current location of
the system, until the bottom of one minimum has been partially filled. In this way, the
effective barrier between the two minima keeps decreasing over time, until the system is
able to travel to another minimum. (d) At this point, metadynamics will start filling the
second minimum as well. (e) The process is repeated until the whole free-energy surface
has been filled. At this point, the known biasing potential (orange area) that has been
added over time has adopted the shape of the original, unknown free-energy landscape.

6.2.5 Umbrella Sampling

Another very popular methodology to enhance sampling is umbrella sampling [39,40]. As discussed in the case of metadynamics, a CV $C(\vec{r}^{(n)})$ is used, but in contrast to metadynamics, the system is now forced to fluctuate around a value C_i of the CV (usually the application of a harmonic potential $K_i(C(\vec{r}^{(n)})-C_i)^2$ to the whole system). Like in REMD, multiple simulations at different value of C_i are performed, taking care that distributions of $C(\vec{r}^{(n)})$ around nearby C_i show some overlap. In contrast to REMD, however, no attempt at exchanging conformations is performed.

Once all of the simulations have reached equilibrium, the whole free-energy landscape (which is assumed to be a function of $C(\vec{r}^{(n)})$) can be reconstructed through the weighted histogram analysis method (WHAM) [41].

6.2.5.1 Hamiltonian REMD

Hamiltonian replica exchange molecular dynamics (HREMD) [42,43] is a modified version of REMD in which different replicas of the system have different Hamiltonians describing their evolution in time. The idea behind this method is that a system is trapped in a minimum because of strong electrostatic or hydrophobic interactions. As a consequence, if these interactions are weakened (e.g., changing the charges or the Lennard-Jones parameters), the system can more easily move between minima.

As in the case of standard REMD, the HREMD method requires to perform multiple simulations at the same time, exchanging between them at regular intervals. All the simulations are at the same temperature, with the Hamiltonians usually differing by a scaling factor of their parameters.

6.3 IMPLICIT SOLVENT MODEL

6.3.1 Introduction

As discussed in the previous section, atomistic modeling of molecules is a powerful technique that allows a study of the dynamics of the system on time scales that can reach µs. However, as the size of the system increases (>20 residues), these simulations become prohibitively expensive in large part because of the presence of a large number of solvent molecules surrounding the protein of interest. In this case, approaches where the explicit solvent is removed are extremely valuable. Implicit solvent models are suitable for studying protein conformations as well as for free-energy calculations and are a standard tool for studying protein–ligand interactions as well as protein aggregation.

6.3.2 Method Details

In explicit solvent models, the system that is being studied is represented as an ensemble of atoms interacting with each other (solvent and protein). In implicit solvent models on the other hand, the atomistic description is retained only for the protein(s), whereas the solvent is approximated by a continuum dielectric medium. As a result, the free energy G of the system can be written down as [44–47]

$$G = E_{MM} + G_{solv}^{el} + G_{solv}^{surf} - TS = \langle H^{IS} \rangle - TS, \tag{6.1}$$

where

E_{MM} is the atomistic energy from the fully atomistic portion of the system

G_{solv}^{el} and G_{solv}^{surf} are the electrostatic and nonelectrostatic portions of the solvation free energy, respectively (namely, the free energy of transferring one molecule from vacuum into the solution)

S is the entropic contribution to the free energy at the temperature T

In the NVT ensemble (constant temperature and volume), Equation 6.1 can be seen as the free energy associated with a system that evolves according to the Hamiltonian H^{IS} (a more formal definition for the Hamiltonian can be found in [48]). In this picture, the molecules under investigation evolve in the same way as all-atom simulations, whereas the solvent is kept at an equilibrium conformation through the implicit $G_{solv}^{el} + G_{solv}^{surf}$ term. In fact, in explicit solvent models, when one protein changes its conformation, it is expected that the solvent molecules would require some time to relax around the new conformation of the protein. In the case of implicit solvent models, on the other hand, the solvent relaxes instantaneously around the molecule.

Different implicit solvent models are based on different approximations of H^{IS}. In the following sections, we review the most common approaches.

6.3.2.1 Poisson–Boltzmann's Methods

In the Poisson–Boltzmann (PB) methods, the solvent is assumed to be a continuum medium described by Poisson's equation:

$$-\nabla \cdot \varepsilon(\vec{r}) \nabla \phi(\vec{r}) = 4\pi\rho(\vec{r}), \tag{6.2}$$

where $\phi(\vec{r})$ is the potential at position $\vec{r}$ of an inhomogeneous medium of relative permittivity $\varepsilon(\vec{r})$. $\rho(\vec{r})$ represents the distribution of charges in the medium and takes into account both the explicit charges of the molecules under investigation $\rho_M(\vec{r})$ and the mean field generated by m ions in solution $\rho_m(\vec{r})$ defined as

$$\rho_m(\vec{r}) = \sum_{j=1}^{m} \bar{n}_j q_j \exp\left[\frac{-q_j\phi(\vec{r})}{k_B T} - \frac{V_j(\vec{r})}{k_B T}\right], \tag{6.3}$$

where

$\bar{n}_j$ is the concentration of ion j in solution

q_j is its charge

$V_j(\vec{r})$ takes into account the excluded volume between the molecule under investigation and the ion

The set of equation described so far represents the most general form of the (nonlinear) PB equation. For practical applications, the PB equation derived previously is usually simplified further through assumptions on the nature of the ions in solution as well as linearized in the case that $\phi(\vec{r}) \ll 1$. The interested reader can find more details about these different approaches in [46]. Depending on the approximation employed, different free-energy functional G_{solv}^{el} can be constructed to be used in association with Equation 6.1.

The major drawback of the PB methodologies is that the free-energy functional is generally very expensive to evaluate. For this reason, they are usually not used in MD simulations, where this functional should in principle be evaluated at each time step. On the other hand, these methods are very useful to evaluate the free energy of a single *snapshot*. In this case, we talk about molecular mechanics simulations (MM), instead of MD, because no dynamics are computed. A very popular methodology for computing free energies in this way is the MM-PBSA method [44–47].

6.3.2.2 Generalized Born's Model

One of the most common approximations to the PB theory discussed earlier is the generalized Born (GB) model [46,49].

In the case of spherical ion of radius a and charge q, G_{solv}^{el} can be computed analytically and gives the Born formula:

$$G_{solv}^{el} = -\frac{q^2}{2a}\left(1 - \frac{1}{\varepsilon_{sol}}\right). \tag{6.4}$$

If the molecule under investigation is described by a set of N charges q_j with radius a_j and the distance r_{ij} between charges is large enough, then

$$G_{solv}^{el} \approx -\sum_i^N \frac{q_i^2}{2a_i}\left(1 - \frac{1}{\varepsilon_{sol}}\right) + \frac{1}{2}\sum_i^N\sum_{j\neq i}^N \frac{q_i q_j}{r_{ij}}\left(\frac{1}{\varepsilon_{sol}} - 1\right). \tag{6.5}$$

One of the advantages of this formulation when compared to PB is that now the quantity ε refers only to the solvent. In addition, the PB approaches deal with the functional of the free energy, whereas in the GB methods, the functional can be solved analytically.

However, in order to properly capture the properties of real molecules, Equation 6.5 is generalized to the form

$$G_{solv}^{el} \approx \left(1 - \frac{1}{\varepsilon_{sol}}\right)\frac{1}{2}\sum_{i}^{N}\sum_{j \neq i}^{N}\frac{q_i q_j}{f_{ij}^{GB}}. \tag{6.6}$$

where the radius of each ion is now converted into an effective Born radius f_{ij}^{GB}, which is generally defined as

$$f_{ij}^{GB}(r_{ij}) = \left[r_{ij}^2 + R_i R_j \exp\left(\frac{-r_{ij}^2}{4R_i R_j}\right)\right]^{1/2} \tag{6.7}$$

and

$$\frac{1}{R_i} = \frac{1}{a_i} - \frac{1}{4\pi}\int_{in,r>a_i}\frac{1}{r^4}d\vec{x} \tag{6.8}$$

where the integral is computed in the region inside the molecule, excluding a volume of radius a_i around the origin.

In the literature, different GB methods differ in the way they deal with Equation 6.8. In the case study discussed later, we will use the formulation proposed by [50].

6.3.2.3 Surface Area

Until this point, we have focused on the evaluation of the electrostatic contribution G_{solv}^{el} to the free energy. In this section, we will analyze the contribution due to the molecular surface exposed to the solvent G_{solv}^{surf}. This term describes the free energy required to build a cavity in the solvent that can host the molecule under examination as well as the nonelectrostatic interactions between molecule and solvent.

Generally, this contribution is assumed proportional to the solvent-accessible surface area A_{SASA} [51]:

$$G_{solv}^{surf} = \gamma A_{SASA} + b, \tag{6.9}$$

where γ and b are adjustable parameters. An alternative form of Equation 6.9 is [51]

$$G_{solv}^{surf} = \gamma A_{MSA} + H_{ss}^{vdW}, \qquad (6.10)$$

where

A_{MSA} is the molecular surface area

H_{ss}^{vdW} is the dispersion interaction between solvent and solute

In the case study analyzed in the following, we will use the model proposed by [52].

6.3.3   Case Study: Islet Amyloid Polypeptide

We illustrate REMD simulations with implicit solvent with a case study of the islet amyloid polypeptide (IAPP) implicated in type 2 diabetes. This peptide is 37 residues long, more than three times longer than the Aβ [25–35] peptide discussed in the preceding section on explicit solvent simulations. Due to its size, simulations in explicit solvent, while possible [53–55], are very costly from a computational standpoint.

The IAPP, also known as amylin, is cosecreted with insulin by the β-cells of the pancreas. This peptide plays a beneficial role in controlling blood sugar, but its aggregation has been associated with β-cell death and with the emergence of type 2 diabetes [56,57]. The IAPP is present in a number of species, with a few point mutation differences between species [58–60]. Interestingly, not all forms of IAPP aggregate. For instance, while the human form aggregates, the rat form does not, and rats do not develop type 2 diabetes associated with aggregate formation. Moreover, transgenic mice expressing human IAPP can develop type 2 diabetes [61,62]. The sequence of the human and rat forms of IAPP is shown in Figure 6.3.

We performed REMD simulations of both the rat and human variants of IAPP to investigate whether conformational differences were already apparent in the monomeric forms of these two peptides. Simulations were performed using the Amber ff96 force field, and solvation effects were captured with the IGB = 5 implicit solvent model augmented by a surface term [63]. Sixteen replicas were used, exponentially spaced between 270 and 465 K and exchanges were attempted every 3 ps, with an exchange rate of 20%. Simulations were run for 200 ns/replica. Structures were clustered by mutual RMSD over the C_α atoms, with a cutoff of 3 Å, and a representative structure for the two most populated clusters for the human and rat forms is shown in Figure 6.3. The conformational differences between the two peptides are striking. While the rat adopts primarily collapsed coil and helical structures, notably with helical structure in the N-terminus, the human form populates β-hairpin structures as well as structures containing a helix at the N-terminus and β-structure at

Human IAPP   KCNTATCATQRLANFLVHSSNNFGAILSSTNVGSNTY-NH$_2$

Rat IAPP   KCNTATCATQRLANFLV**R**SSNNLG**P**V**LPP**TNVGSNTY-NH$_2$

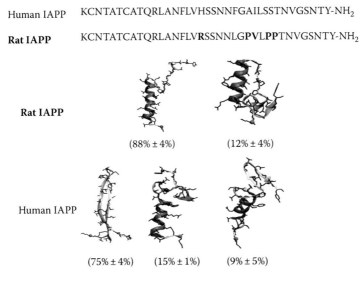

Figure 6.3 Sequence and representative structures from the most populated clusters for rat and human IAPP. (Adapted from Dupuis, N.F. et al., *J. Am. Chem. Soc.*, 131, 18283, 2009.)

the C-terminus. A solved NMR structure for the rat form of IAPP shows helicity in the N-terminal region, consistent with the REMD simulations [64]. Additionally, the structures from simulation are in good agreement with collision cross sections obtained from ion-mobility mass spectrometry experiments [65]. The human form of IAPP, on the other hand, is considered an intrinsically disordered peptide and there is no solved NMR structure for this peptide [64]. However, NMR data are consistent with some helical structuring in the N-terminal region. Kayed et al. suggested based on a series of experiments involving circular dichroism, filtration assays, AFM, and EM that the human form of IAPP populates two distinct conformations, one corresponding to an amyloidogenic form and the other to a nonamyloidogenic form [66]. Ion-mobility mass spectrometry experiments on human IAPP report two families of structures, one with larger collision cross section that matches theoretical collision cross sections for the hairpin structure seen in simulations and a more compact family, with collision cross sections consistent with the helical-rich structures [65]. The simulations support a picture in which the conformations of human IAPP that share the structured N-terminal helix with the rat IAPP are the *physiological* conformations that bind to the AMY receptors and play a role in glucose regulation. The hairpin conformation, on the other hand, would be a *pathological* conformation, a precursor to aggregation. The notion of a hairpin as an amyloid precursor is supported by 2D IR experiments [67,68] and by biochemical experiments [67,68] that identify the interpeptide interaction regions of human IAPP as corresponding to the β-strands of the hairpin seen in simulation. This example illustrates how simulation can complement experiment and, in the case of the intrinsically disordered human IAPP peptide, even provide structural information at an atomistic resolution that cannot be achieved experimentally.

6.3.4 Other Methods

6.3.4.1 Langevin' Dipoles
In the case of PB and GB methods, the solvent is depicted as a continuum dielectric medium. In this approach, all the molecular details regarding the solvent are lost. An alternative approach proposed by Florián and Warshel [69–71] tries to recover the discrete nature of the solvent surrounding the protein by means of point dipoles fixed on a cubic grid. A schematic of the method is presented in Figure 6.4, showing a protein surrounded by water. The system under examination can be separated into four regions. The first region

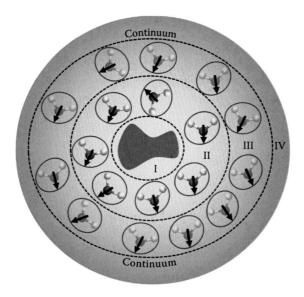

Figure 6.4   Langevin' dipoles. In the Langevin dipole model, the protein under investigation (gray, region I) is the source of the electric field that affects the solvent. The response of the solvent to the electric field is described by a lattice of polarizable dipoles that can rotate but not translate (regions II and III). At long distances, the solvent can be described by a continuum representation of the solvent (region IV).

(region I) is composed of the protein and represents the origin of the electric field $\vec{\xi}$. The second region (region II) is represented by a lattice of point dipole. These dipoles can change their orientation and their strength (namely, it is a polarizable dipole) as they interact with the electric field $\vec{\xi}$. Region III is the region where the interaction with the dipole becomes weaker, before the electric field wears off and the solvent is seen as a continuum (region IV). The difference between regions II and III is the way dipoles are treated. In region II, dipoles are updated at every time step, whereas the treatment of the dipoles in region III is bound to the strength of $\vec{\xi}$.

6.3.4.2   Reference Interaction Site Model

The methods described so far model the solvent as an electric medium augmented by contribution that takes into account the presence of a molecular surface between a molecule under examination and the solvent. On the other hand, the 3D reference interaction site model (3D-RISM) focuses on the structural properties of the solvent [46,72,73]. Assuming that the solute s is surrounded by the solvent S, the structure of the solvent can be described by

$$\rho h_\gamma^{Ss}(\vec{r}) = \sum_\alpha \int c_\alpha^{Ss}(\vec{r} - \vec{r}') \chi_{\alpha\gamma}^{SS}(\vec{r}')d\vec{r}', \qquad (6.11)$$

where

  α and γ represent solvent sites (usually, atoms)
  $\vec{r}$ is the position where the function is evaluated (usually on a grid)
  ρ is the average density
  $h_\gamma^{Ss}(\vec{r})$ is the correlation function
  $c_\alpha^{Ss}(\vec{r} - \vec{r}')$ is the direct correlation function
  $\chi_{\alpha\gamma}^{SS}(\vec{r})$ is the susceptibility

This equation is usually solved self-consistently with another equation that provides a closure and keeps into account the atomistic potential of the solute.

As a consequence, the 3D-RISM approach provides not only a way to compute the effect of the solvent on the solute, but it computes correlation functions that describe the structure of the solvent as well. For example, analysis of the correlation function of the hydrogen atoms of the water can be used to identify regions of the solute that form hydrogen bonds with the solvent [72].

6.4   COARSE-GRAINING MODELING

For the study of large aggregates, composed of tens or hundreds of proteins, it becomes necessary not only to simplify the representation of the water but to coarse grain the protein as well.

6.4.1   Example of Coarse-Grained Models

Figure 6.5 shows a compilation of recent coarse-grained models that have been used to simulate the aggregation process. They range from simple models (cuboids and tube models), to single-bead lattice models, to multiple-bead off-lattice models. Simulation techniques include the Monte Carlo, Langevin dynamics, and discrete MD.

We illustrate in the following a case study using an intermediate-range coarse-grained model in which the peptide is represented by three beads, two for the backbone and one for the side chain [74–77]. The side chains are either hydrophobic (H), polar (P), or charged. The sequence considered here consists of alternating hydrophobic and polar side chains, capped by oppositely charged residues. The sequence of the seven amino acid peptide and the force field used is shown in Figure 6.6. The bonded parameters are obtained from geometric data in the protein data bank, while the nonbonded hydrophobic–hydrophobic energy terms are taken to be consistent with free-energy transfer of hydrophobic amino acids from nonpolar to polar solvent. The β-strand propensity is modulated through a torsional term involving two side-chain residues and two backbone (X) atoms, with near-trans conformations corresponding to extended, stiff, high β-strand peptides. Peptides with a lower energy constant K in this torsional term correspond to more flexible peptides, with lower β-strand propensity. The torsional potential involves residues $S_k - X_k - X_{k+1} - S_k$, where S corresponds to one of the side chains and k is an index corresponding to a given residue. The functional form of this torsional potential is $K\cos(\alpha + \delta)$, with higher K corresponding to stiffer peptides.

A system of 27 peptides was considered, and simulations are performed using the Langevin dynamics, both at constant temperature to study the kinetics of assembly (100 simulations, each 500 ns in length) and replica exchange (30 replicas, spaced between 300 and 388 K, each run for 200 ns, with an acceptance ratio between 10% and 30%) to fully characterize the thermodynamics. The degree of β-strand propensity was modulated from $K = 1.0$ (very flexible) to $K = 3.0$ (stiff) and was seen to have a dramatic effect on the kinetics and thermodynamics of aggregation, as illustrated in Figure 6.7. Peptides with low β-strand propensity tend to follow an aggregation pathway in which they first coalesce into an amorphous assembly, from which β-sheets then emerge. Peptides with high β-strand propensity, on the other hand, assemble via ordered β-sheet aggregates. These simulations were able to rationalize the different aggregation behaviors of a number of experimental systems, for instance, the sup35 peptide [78] (corresponding to a low K in our simulations) and the KFFE peptide [79] (corresponding to a high K case). An additional important outcome of these simulations is the suggestion that peptides with greater β-strand propensity may be less harmful to cells than peptides with lower β-strand propensity that have a greater tendency to populate nonfibrillar aggregates.

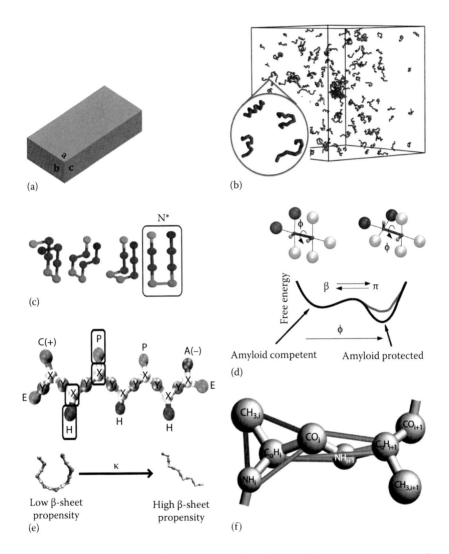

Figure 6.5   A compilation of coarse-grained models used in recent years to study aggregation. (Adapted from Wu, C. and Shea, J.E., *Curr. Opin. Struct. Biol.*, 21, 209, 2011.) (a) Cuboid model. (From Zhang, J.N. and Muthukumar, M., *J. Chem. Phys.*, 130, 035102, 2008.) (b) Tube model. (From Auer, S. et al., *PLoS Comput. Biol.*, 5, e1000458, 2009.) (c) Lattice model. (From Li, M.S. et al., *J. Chem. Phys.*, 129, 175101, 2008.) (d) Caflisch model. (From Pellarin, R. and Caflisch, A., *J. Mol. Biol.*, 360, 882, 2006.) (e) Off-lattice Shea model. (From Bellesia, G. and Shea, J.E., *J. Chem. Phys.*, 130, 145103, 2009; Bellesia, G. and Shea, J.-E., *J. Chem. Phys.*, 126, 245104, 2007.) (f) Off-lattice discrete MD Hall model. (From Nguyen, H.D. and Hall, C.K., *J. Am. Chem. Soc.*, 128, 1890, 2006.)

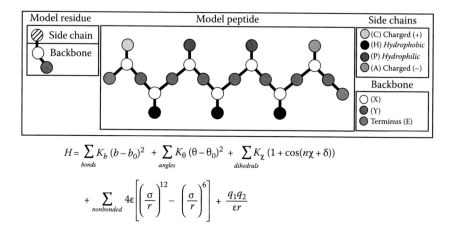

$$H = \sum_{bonds} K_b\,(b - b_0)^2 \;+\; \sum_{angles} K_\theta\,(\theta - \theta_0)^2 \;+\; \sum_{dihedrals} K_\chi\,(1 + \cos(n\chi + \delta))$$

$$+ \sum_{nonbonded} 4\varepsilon\left[\left(\frac{\sigma}{r}\right)^{12} - \left(\frac{\sigma}{r}\right)^6\right] + \frac{q_1 q_2}{\varepsilon r}$$

Figure 6.6 Force field and geometry for the model peptide. (From Bellesia, G. and Shea, J.E., *J. Chem. Phys.*, 130, 145103, 2009; Bellesia, G. and Shea, J.-E., *J. Chem. Phys.*, 126, 245104, 2007.)

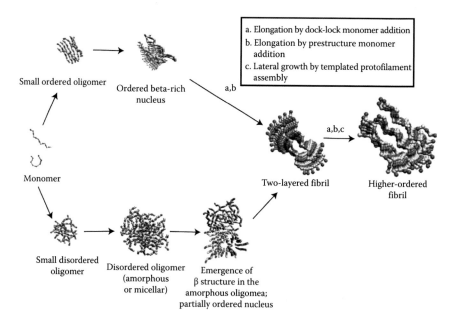

Figure 6.7 A schematic of the aggregation pathways for peptides with high and low β-strand propensities. (Adapted from Wu, C. and Shea, J.E., *Curr. Opin. Struct. Biol.*, 21, 209, 2011.)

6.4.2 Systematic Approaches

The coarse-graining (CG) approaches illustrated in Figure 6.5 are phenomenological. In this section, we focus on systematic approaches to CG. These methodologies aim at constructing coarse-grained potentials starting from available experimental or simulation data.

The first step in all these methods consists in defining a mapping operator $M(\vec{r}^{(n)})$ that maps the atomistic coordinates $\vec{r}$ of a system of n atoms into a system of $N < n$ CG coordinates $\vec{R}$. An example is the operator that maps the atoms into their center of mass. The second step consists in computing the effective potential $U(\vec{R}^{(N)})$ among the newly defined CG sites. The latter is usually achieved by defining some target quantity (generally different for different CG approaches) and then constructing an algorithmic procedure that aims at its optimal approximation. Here, we review the most popular methods. Except for the relative entropy (RE) methods, the other methodology is freely available in the VOTCA toolkit [80].

6.4.2.1 Iterative Boltzmann's Inversion

The iterative Boltzmann inversion [81] (IBI) aims at reproducing the distribution functions associated with a specific potential (e.g., the radial distribution function for nonbonded interactions or angular distribution for angular potentials). In this way, a system is constructed where each equation deals with a different distribution. This set of equations is then solved in an iterative way until the difference between the CG distribution P_{CG} and the target distribution P_{AA} falls below a certain threshold.

The iterative scheme for the iteration $k + 1$ is defined as [80]

$$U^{k+1}\left(\vec{R}^{(N)}\right)=U^{k}\left(\vec{R}^{(N)}\right)+\Delta U^{k}\left(\vec{R}^{(N)}\right)$$

$$\Delta U^{k}\left(\vec{R}^{(N)}\right)=k_{B}T\ln\frac{P_{CG}^{k}(\vec{R}^{(N)})}{P_{AA}\left(M\left(\vec{r}^{(n)}\right)\right)},$$

where
T is the temperature
k_{B} is Boltzmann's constant

6.4.2.2 Inverse Monte Carlo

Inverse Monte Carlo [82,83] (IMC) differs from IBI for the way $\Delta U(\vec{R}^{(N)})$ is computed. The potential of the system is written as $U(\vec{R}^{(N)})=\sum_{\alpha}K_{\alpha}S_{\alpha}(\vec{R}^{(N)})$, where $S_{\alpha}(\vec{R}^{(N)})$ is a function of the coordinates and K_{α} a set of parameters. In the original paper [82], it was shown that a suitable choice for $S_{\alpha}(\vec{R}^{(N)})$ is such that $\langle S_\alpha(\vec{R}^{(N)})\rangle$ is the radial distribution function. At the same time, the radial distribution function depends on the potential $U(\vec{R}^{(N)})$. As a consequence, the IMC method requires an iterative method so that the parameters K_α are chosen in such a way that the correct radial distribution function of the system is reproduced correctly. Formally, this is equivalent to state that

$$\left\langle S_\alpha\left(\vec{R}^{(N)}\right)\right\rangle_{CG} - \left\langle S_\alpha\left(M\left(\vec{r}^{(n)}\right)\right)\right\rangle_{AA} = \sum_\gamma \frac{\partial\left\langle S_\alpha\left(\vec{R}^{(N)}\right)\right\rangle_{CG}}{\partial K_\alpha}\Delta K_\gamma.$$

6.4.2.3 Multiscale Coarse Graining

In the multiscale coarse-graining [84–86] (MS-CG) approach, the coarse-grained potentials are constructed using data from MD simulations. In standard MD simulations, the forces $\vec{f}_i$ acting on each atom i of the system are calculated and used to propagate the dynamics of the system.

When coarse graining such a system, atoms are grouped together into coarse-grained interaction sites I, which can represent a whole molecule as well as portions of it. To propagate the dynamics of this new set of CG sites, a CG potential is defined so that the force $\vec{F}_I$ acting on the Ith CG site is defined as $\vec{F}_I = \sum_{i\in I} \vec{f}_i$, namely, the force acting on each CG site is the sum of the atomistic forces acting on each atom belonging to that CG site. Generally, the functional form of these CG potentials is extremely complicated and involves many-body contributions. As a consequence, the forces $\vec{F}_I$ must be approximated. In the MS-CG methods, the approximations are achieved through a mean least-square fit of the original force $\vec{F}_I$ computed from atomistic MD simulations, and a new set of approximated forces $\vec{F}_I^{CG}$ is obtained as a result. These new forces $\vec{F}_I^{CG}$ are then used to perform MD simulations of the CG system.

6.4.2.4 Relative Entropy

The RE [87] method is derived from the information theory framework. In this framework, CG is seen as a loss of information, and the RE method optimizes the potential in such a way as to minimize this loss. Formally, the method minimizes the function

$$S_{rel} = \sum_i p_{AA}(i)\ln\frac{p_{AA}(i)}{p_{CG}(i)},$$

where

S_{rel} is the RE

$p(i)$ is the probability of finding the configuration I in an ensemble

6.5 CONCLUSIONS

Protein aggregation is a complex problem, which spans multiple time and length scales. To fully describe this process from a computational standpoint, it becomes necessary to resort to a hierarchy of modeling techniques, from fully atomistic in explicit solvent, to atomistic in implicit solvent, to coarse-grained modeling. In this chapter, we have highlighted the theoretical foundations behind these approaches and given a few illustrative examples of the application of these simulation techniques to protein aggregation.

REFERENCES

1. Chiti, F. and C. M. Dobson. 2006. Protein misfolding, functional amyloid, and human disease. *Annu. Rev. Biochem.* 75:333–366.
2. Chiti, F. and C. M. Dobson. 2009. Amyloid formation by globular proteins under native conditions. *Nat. Chem. Biol.* 5:15–22.
3. Fowler, D. M., A. V. Koulov, W. E. Balch, and J. W. Kelly. 2007. Functional amyloid—From bacteria to humans. *Trends Biochem. Sci.* 32:217–224.
4. Cornell, W. D., P. Cieplak, C. I. Bayly, I. R. Gould, K. M. Merz, D. M. Ferguson, D. C. Spellmeyer, T. Fox, J. W. Caldwell, and P. A. Kollman. 1995. A second generation force field for the simulation of proteins, nucleic acids, and organic molecules. *J. Am. Chem. Soc.* 117:5179–5197.
5. Duan, Y., C. Wu, S. Chowdhury, M. C. Lee, G. Xiong, W. Zhang, R. Yang et al. 2003. A point-charge force field for molecular mechanics simulations of proteins based on condensed-phase quantum mechanical calculations. *J. Comput. Chem.* 24:1999–2012.
6. García, A. E. and K. Y. Sanbonmatsu. 2002. α-Helical stabilization by side chain shielding of backbone hydrogen bonds. *Proc. Natl. Acad. Sci. USA* 99:2782–2787.
7. Hornak, V., R. Abel, A. Okur, B. Strockbine, A. Roitberg, and C. Simmerling. 2006. Comparison of multiple Amber force fields and development of improved protein backbone parameters. *Proteins* 65:712–725.
8. Kollman, P. A. 1996. Advances and continuing challenges in achieving realistic and predictive simulations of the properties of organic and biological molecules. *Acc. Chem. Res.* 29:461–469.
9. Lindorff-Larsen, K., S. Piana, K. Palmo, P. Maragakis, J. L. Klepeis, R. O. Dror, and D. E. Shaw. 2010. Improved side-chain torsion potentials for the Amber ff99SB protein force field. *Proteins* 78:1950–1958.
10. Wang, J., P. Cieplak, and P. A. Kollman. 2000. How well does a restrained electrostatic potential (RESP) model perform in calculating conformational energies of organic and biological molecules? *J. Comput. Chem.* 21:1049–1074.
11. Feller, S. E. and A. D. MacKerell. 2000. An improved empirical potential energy function for molecular simulations of phospholipids. *J. Phys. Chem. B* 104:7510–7515.
12. Foloppe, N. and J. A. D. MacKerell. 2000. All-atom empirical force field for nucleic acids: I. Parameter optimization based on small molecule and condensed phase macromolecular target data. *J. Comput. Chem.* 21:86–104.

13. MacKerell, A. D., D. Bashford, M. Bellott, R. L. Dunbrack, J. D. Evanseck, M. J. Field, S. Fischer et al. 1998. All-atom empirical potential for molecular modeling and dynamics studies of proteins. *J. Phys. Chem. B* 102:3586–3616.

14. Mackerell, A. D., M. Feig, and C. L. Brooks. 2004. Extending the treatment of backbone energetics in protein force fields: Limitations of gas-phase quantum mechanics in reproducing protein conformational distributions in molecular dynamics simulations. *J. Comput. Chem.* 25:1400–1415.

15. Jorgensen, W. L., D. S. Maxwell, and J. Tirado-Rives. 1996. Development and testing of the OPLS all-atom force energetics and properties of organic liquids. *J. Am. Chem. Soc.* 118:11225–11236.

16. Jorgensen, W. L. and N. A. McDonald. 1998. Development of an all-atom force field for heterocycles. Properties of liquid pyridine and diazenes. *J. Mol. Struct.: THEOCHEM* 424:145–155.

17. Kaminski, G. A., R. Friesner, J. Tirado-Rives, and W. Jorgensen. 2001. Evaluation and reparametrization of the OPLS-AA force field for proteins via comparison with accurate quantum chemical calculations on peptides. *J. Phys. Chem. B* 105:6474–6487.

18. McDonald, N. A. and W. L. Jorgensen. 1998. Development of an all-atom force field for heterocycles. Properties of liquid pyrrole, furan, diazoles, and oxazoles. *J. Phys. Chem. B* 102:8049–8059.

19. Price, M. L. P., D. Ostrovsky, and W. L. Jorgensen. 2001. Gas-phase and liquid-state properties of esters, nitriles, and nitro compounds with the OPLS-AA force field. *J. Comput. Chem.* 22:1340–1352.

20. Rizzo, R. C. and W. L. Jorgensen. 1999. OPLS all-atom model foramines: Resolution of the amine hydration problem. *J. Am. Chem. Soc.* 121:4827–4836.

21. Watkins, E. K. and W. L. Jorgensen. 2001. Perfluoroalkanes: Conformational analysis and liquid-state properties from ab initio and Monte Carlo calculations. *J. Phys. Chem. A* 105:4118–4125.

22. Hukushima, K. and K. Nemoto. 1996. Exchange Monte Carlo method and application to spin glass simulations. *J. Phys. Soc. Jpn.* 65:1604–1608.

23. Sugita, Y. and Y. Okamoto. 1999. Replica-exchange molecular dynamics method for protein folding. *Chem. Phys. Lett.* 314:141–151.

24. Lei, H. and Y. Duan. 2007. Improved sampling methods for molecular simulation. *Curr. Opin. Struct. Biol.* 17:187–191.

25. Wei, G. H. and J.-E. Shea. 2006. Effects of solvent on the structure of the Alzheimer amyloid-β(25–35) peptide. *Biophys. J.* 91:1638–1647.

26. Wei, G. H., A. I. Jewett, and J.-E. Shea. 2010. Structural diversity of the dimer of the Alzheimer amyloid-β(25–35) peptide: Insights into the molecular nature of polymorphism. *Phys. Chem. Chem. Phys.* 12:3622–3629.

27. Larini, L. and J. E. Shea. 2012. Role of β-hairpin formation in aggregation: The self-assembly of the amyloid-β(25–35) peptide. *Biophys. J.* 103:576–586.

28. Kaminski, G. A., R. A. Friesner, J. Tirado-Rives, and W. L. Jorgensen. 2001. Evaluation and reparametrization of the OPLS-AA force field for proteins via comparison with accurate quantum chemical calculations on peptides. *J. Phys. Chem. B* 105:6474–6487.

29. Jorgensen, W. L., D. S. Maxwell, and J. Tirado-Rives. 1996. Development and testing of the OPLS all-atom force field on conformational energetics and properties of organic liquids. *J. Am. Chem. Soc.* 118:11225–11236.

30. Jorgensen, W. L., J. Chandrasekhar, J. D. Madura, R. W. Impey, and M. L. Klein. 1983. Comparison of simple potential functions for simulating liquid water *J. Chem. Phys.* 79:926–935.

31. Darden, T., D. York, and L. Pedersen. 1993. Particle Mesh Ewald—An N·Log(N) method for Ewald sums in large systems. *J. Chem. Phys.* 98:10089–10092.

32. Hess, B., H. Bekker, H. J. C. Berendsen, and J. G. E. M. Fraaije. 1997. LINCS: A linear constraint solver for molecular simulations. *J. Comput. Chem.* 18:1463–1472.

33. Miyamoto, S. and P. A. Kollman. 1992. Settle—An analytical version of the shake and rattle algorithm for rigid water models. *J. Comput. Chem.* 13:952–962.

34. Daura, X., K. Gademann, B. Jaun, D. Seebach, W. F. van Gunsteren, and A. E. Mark. 1999. Peptide folding: When simulation meets experiment. *Angew. Chem. Int. Ed.* 38:236–240.

35. Van der Spoel, D., E. Lindahl, B. Hess, G. Groenhof, A. E. Mark, and H. J. C. Berendsen. 2005. Gromacs: Fast, flexible, and free. *J. Comput. Chem.* 26:1701–1718.

36. Laio, A. and F. L. Gervasio. 2008. Metadynamics: A method to simulate rare events and reconstruct the free energy in biophysics, chemistry and material science. *Rep. Prog. Phys.* 71:126601.

37. Barducci, A., M. Bonomi, and M. Parrinello. 2011. Metadynamics. *Wiley Interdiscip. Rev.: Comput. Mol. Sci.* 1:826–843.

38. Laio, A. and M. Parrinello. 2002. Escaping free-energy minima. *Proc. Natl. Acad. Sci. USA* 99:12562–12566.

39. Torrie, G. M. and J. P. Valleau. 1974. Monte Carlo free energy estimates using non-Boltzmann sampling: Application to the sub-critical Lennard-Jones fluid. *Chem. Phys. Lett.* 28:578–581.

40. Kästner, J. 2011. Umbrella sampling. *Wiley Interdiscip. Rev.: Comput. Mol. Sci.* 1:932–942.

41. Kumar, S., J. M. Rosenberg, D. Bouzida, R. H. Swendsen, and P. A. Kollman. 1992. The weighted histogram analysis method for free-energy calculations on biomolecules. I. The method. *J. Comput. Chem.* 13:1011–1021.

42. Affentranger, R., I. Tavernelli, and E. E. Di Iorio. 2006. A novel Hamiltonian replica exchange MD protocol to enhance protein conformational space sampling. *J. Chem. Theory Comput.* 2:217–228.

43. Fukunishi, H., O. Watanabe, and S. Takada. 2002. On the Hamiltonian replica exchange method for efficient sampling of biomolecular systems: Application to protein structure prediction. *J. Chem. Phys.* 116:9058–9067.

44. Homeyer, N. and H. Gohlke. 2012. Free energy calculations by the molecular mechanics Poisson–Boltzmann surface area method. *Mol. Inform.* 31:114–122.

45. Srinivasan, J., T. E. Cheatham, P. Cieplak, P. A. Kollman, and D. A. Case. 1998. Continuum solvent studies of the stability of DNA, RNA, and phosphoramidate–DNA helices. *J. Am. Chem. Soc.* 120:9401–9409.

46. Luchko, T. and D. A. Case. 2012. Implicit solvent models and electrostatics in molecular recognition. In *Protein-Ligand Interactions*, H. Gohlke (ed.). Wiley-VCH Verlag GmbH & Co. KGaA, Weinheim, Germany, pp. 171–189.

47. Kollman, P. A., I. Massova, C. Reyes, B. Kuhn, S. Huo, L. Chong, M. Lee et al. 2000. Calculating structures and free energies of complex molecules: Combining molecular mechanics and continuum models. *Acc. Chem. Res.* 33:889–897.

48. Roux, B. and T. Simonson. 1999. Implicit solvent models. *Biophys. Chem.* 78:1–20.

49. Bashford, D. and D. A. Case. 2000. Generalized Born models of macromolecular solvation effects. *Annu. Rev. Phys. Chem.* 51:129–152.

50. Onufriev, A., D. Bashford, and D. A. Case. 2004. Exploring protein native states and large-scale conformational changes with a modified generalized born model. *Proteins* 55:383–394.

51. Gohlke, H. and D. A. Case. 2004. Converging free energy estimates: MM-PB(GB)SA studies on the protein–protein complex Ras–Raf. *J. Comput. Chem.* 25:238–250.

52. Weiser, J., P. S. Shenkin, and W. C. Still. 1999. Approximate atomic surfaces from linear combinations of pairwise overlaps (LCPO). *J. Comput. Chem.* 20:217–230.

53. Reddy, A. S., L. Wang, S. Singh, Y. Ling, L. Buchanan, M. T. Zanni, J. L. Skinner, and J. J. De Pablo. 2010. Stable and metastable states of human amylin in solution. *Biophys. J.* 99:2208–2216.

54. Reddy, A. S., L. Wang, Y. S. Lin, Y. Ling, M. Chopra, M. T. Zanni, J. L. Skinner, and J. J. De Pablo. 2010. Solution structures of rat amylin peptide: Simulation, theory, and experiment. *Biophys. J.* 98:443–451.

55. Andrews, M. N. and R. Winter. 2011. Comparing the structural properties of human and rat islet amyloid polypeptide by MD computer simulations. *Biophys. Chem.* 156:43–50.

56. Pittner, R. A., K. Albrandt, K. Beaumont, L. S. L. Gaeta, J. E. Koda, C. X. Moore, J. Rittenhouse, and T. J. Rink. 1994. Molecular physiology of amylin. *J. Cell. Biochem.* 55:19–28.

57. Kruger, D. F., P. M. Gatcomb, and S. K. Owen. 1999. Clinical implications of amylin and amylin deficiency. *Diabetes Educ.* 25:389–397.

58. Betsholtz, C., L. Christmanson, U. Engstrom, F. Rorsman, K. Jordan, T. D. Obrien, M. Murtaugh, K. H. Johnson, and P. Westermark. 1990. Structure of cat islet amyloid polypeptide and identification of amino-acid-residues of potential significance for islet amyloid formation. *Diabetes* 39:118–122.

59. Betsholtz, C., L. Christmansson, U. Engstrom, F. Rorsman, V. Svensson, K. H. Johnson, and P. Westermark. 1989. Sequence divergence in a specific region of islet amyloid polypeptide (IAPP) explains differences in islet amyloid formation between species. *FEBS Lett.* 251:261–264.

60. Christmanson, L., C. Betsholtz, A. Leckstrom, U. Engstrom, C. Cortie, K. H. Johnson, T. E. Adrian, and P. Westermark. 1993. Islet amyloid polypeptide in the rabbit and European hare—Studies on its relationship to amyloidogenesis. *Diabetologia* 36:183–188.

61. Matveyenko, A. V. and P. C. Butler. 2006. Islet amyloid polypeptide (IAPP) transgenic rodents as models for type 2 diabetes. *ILAR J.* 47:225–233.

62. Soeller, W. C., J. Janson, S. E. Hart, J. C. Parker, M. D. Carty, R. W. Stevenson, D. K. Kreutter, and P. C. Butler. 1998. Islet amyloid-associated diabetes in obese A(vy)/a mice expressing human islet amyloid polypeptide. *Diabetes* 47:743–750.

63. Onufriev, A., D. Bashford, and D. A. Case. 2000. Modification of the generalized Born model suitable for macromolecules. *J. Phys. Chem. B* 104:3712–3720.

64. Williamson, J. A. and A. D. Miranker. 2007. Direct detection of transient α-helical states in islet amyloid polypeptide. *Protein Sci.* 16:110–117.

65. Dupuis, N. F., C. Wu, J.-E. Shea, and M. T. Bowers. 2009. Human islet amyloid poly-peptide monomers form ordered β-hairpins: A possible direct amyloidogenic pre-cursor. *J. Am. Chem. Soc.* 131:18283–18292.

66. Kayed, R., J. Bernhagen, N. Greenfield, K. Sweimeh, H. Brunner, W. Voelter, and A. Kapurniotu. 1999. Conformational transitions of islet amyloid polypeptide (IAPP) in amyloid formation in vitro. *J. Mol. Biol.* 287:781–796.

67. Wang, L., C. T. Middleton, S. Singh, A. S. Reddy, A. M. Woys, D. B. Strasfeld, P. Marek et al. 2011. 2DIR spectroscopy of human amylin fibrils reflects stable β-sheet struc-ture. *J. Am. Chem. Soc.* 133:16062–16071.

68. Shim, S. H., R. Gupta, Y. L. Ling, D. B. Strasfeld, D. P. Raleigh, and M. T. Zanni. 2009. Two-dimensional IR spectroscopy and isotope labeling defines the pathway of amyloid formation with residue-specific resolution. *Proc. Natl. Acad. Sci. USA* 106:6614–6619.

69. Florián, J. and A. Warshel. 1997. Langevin dipoles model for ab initio calculations of chemical processes in solution: Parametrization and application to hydration free energies of neutral and ionic solutes and conformational analysis in aqueous solu-tion. *J. Phys. Chem. B* 101:5583–5595.

70. Florián, J. and A. Warshel. 1999. Calculations of hydration entropies of hydropho-bic, polar, and ionic solutes in the framework of the Langevin dipoles solvation model. *J. Phys. Chem. B* 103:10282–10288.

71. Warshel, A. 1979. Calculations of chemical processes in solutions. *J. Phys. Chem.* 83:1640–1652.

72. Beglov, D. and B. Roux. 1997. An integral equation to describe the solvation of polar molecules in liquid water. *J. Phys. Chem. B* 101:7821–7826.

73. Kovalenko, A. and F. Hirata. 2000. Potentials of mean force of simple ions in ambient aqueous solution. I. Three-dimensional reference interaction site model approach. *J. Chem. Phys.* 112:10391–10402.

74. Bellesia, G. and J. E. Shea. 2009. Diversity of kinetic pathways in amyloid fibril formation. *J. Chem. Phys.* 131:111102.

75. Bellesia, G. and J. E. Shea. 2009. Effect of β-sheet propensity on peptide aggregation. *J. Chem. Phys.* 130:145103.

76. Bellesia, G. and J.-E. Shea. 2007. Self-assembly of β-sheet forming peptides into chiral fibrillar aggregates. *J. Chem. Phys.* 126:245104.

77. Morriss-Andrews, A., G. Bellesia, and J. E. Shea. 2012. β-sheet propensity controls the kinetic pathways and morphologies of seeded peptide aggregation. *J. Chem. Phys.* 137:145104.

78. Lindquist, S., S. K. DebBurman, J. R. Glover, A. S. Kowal, J. J. Liu, E. C. Schirmer, and T. R. Serio. 1998. Amyloid fibres of Sup35 support a prion-like mechanism of inheri-tance in yeast. *Biochem. Soc. Trans.* 26:486–490.

79. Tjernberg, L., W. Hosia, N. Bark, J. Thyberg, and J. Johansson. 2002. Charge attraction and beta propensity are necessary for amyloid fibril formation from tetrapeptides. *J. Biol. Chem.* 277:43243–43246.

80. Rühle, V., C. Junghans, A. Lukyanov, K. Kremer, and D. Andrienko. 2009. Versatile object-oriented toolkit for coarse-graining applications. *J. Chem. Theory Comput.* 5:3211–3223.

81. Reith, D., M. Pütz, and F. Müller-Plathe. 2003. Deriving effective mesoscale poten-tials from atomistic simulations. *J. Comput. Chem.* 24:1624–1636.

82. Lyubartsev, A. P. and A. Laaksonen. 1995. Calculation of effective interaction potentials from radial distribution functions: A reverse Monte Carlo approach. *Phys. Rev. E* 52:3730–3737.

83. Müller-Plathe, F. 2002. Coarse-graining in polymer simulation: From the atomistic to the mesoscopic scale and back. *ChemPhysChem* 3:754–769.

84. Izvekov, S. and G. A. Voth. 2005. A multiscale coarse-graining method for biomolecular systems. *J. Phys. Chem. B* 109:2469–2473.

85. Izvekov, S. and G. A. Voth. 2005. Multiscale coarse graining of liquid-state systems. *J. Chem. Phys.* 123:134105–134113.

86. Noid, W. G., J.-W. Chu, G. S. Ayton, V. Krishna, S. Izvekov, G. A. Voth, A. Das, and H. C. Andersen. 2008. The multiscale coarse-graining method. I. A rigorous bridge between atomistic and coarse-grained models. *J. Chem. Phys.* 128:244114.

87. Shell, M. S. 2008. The relative entropy is fundamental to multiscale and inverse thermodynamic problems. *J. Chem. Phys.* 129:144108.

88. Wu, C. and J. E. Shea. 2011. Coarse-grained models for protein aggregation. *Curr. Opin. Struct. Biol.* 21:209–220.

89. Zhang, J. N. and M. Muthukumar. 2008. Simulations of nucleation and elongation of amyloid fibrils. *J. Chem. Phys.* 130:035102.

90. Auer, S., A. Trovato, and M. Vendruscolo. 2009. A condensation-ordering mechanism in nanoparticle-catalyzed peptide aggregation. *PLoS Comput. Biol.* 5:e1000458.

91. Li, M. S., D. K. Klimov, J. E. Straub, and D. Thirumalai. 2008. Probing the mechanisms of fibril formation using lattice models. *J. Chem. Phys.* 129:175101.

92. Pellarin, R. and A. Caflisch. 2006. Interpreting the aggregation kinetics of amyloid peptides. *J. Mol. Biol.* 360:882–892.

93. Nguyen, H. D. and C. K. Hall. 2006. Spontaneous fibril formation by polyalanines; discontinuous molecular dynamics simulations. *J. Am. Chem. Soc.* 128:1890–1901.

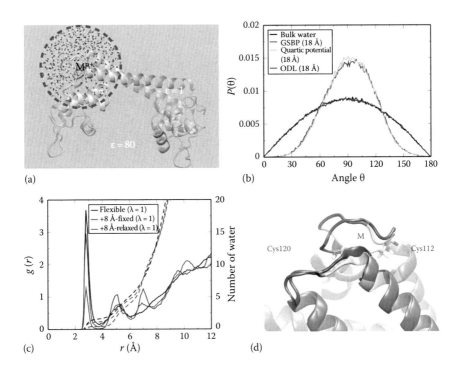

(a)

(b)

(c)

(d)

Figure 2.3 Properties of PBC and GSBP simulations with a partially or fully flexible CueR. (a) A snapshot that illustrates the GSBP (20 Å) setup; protein atoms beyond 20 Å of M are fixed (cyan), those within 20 Å of M are flexible (yellow), and the bulk continuum is indicated with a gray background. (b) The orientation distribution of water molecules in bulk (from PBC calculations), an orientationally disordered liquid (ODL) ensemble, and two other water droplet simulations; for the last three cases, orientations for water molecules within 1.5 Å from the droplet surface are shown. The orientation is defined as the angle between the water dipole and the normal vector of the local surface. (c) Radial distribution function of water oxygen around M (oxidized state, $\lambda = 1$) in PBC simulations with different parts of the CueR protein held fixed. The integrated RDFs are shown as dashed lines. (d) Overlay of +8 Å-fixed and +8 Å-relaxed structures for the oxidize state after 1 ns production run. The flexible protein region from the +8 Å-fixed structure is shown in yellow, and that in the +8 Å-relaxed structure is shown in purple; the fixed protein region is colored in blue. Clearly, the metal binding loop responds to the metal oxidation and therefore should be allowed to relax for reliable oxidation free energy calculations (see Table 2.2 and text for discussion).

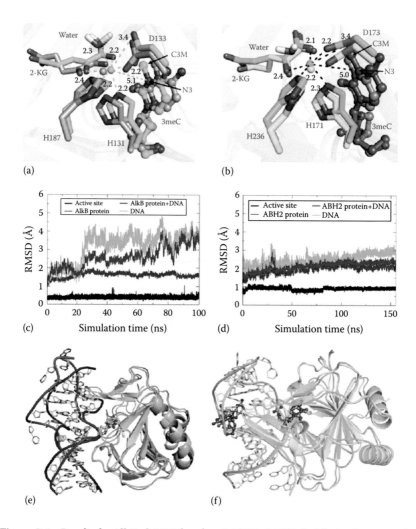

Figure 2.4 Results for AlkB–dsDNA (a,c,e) and ABH2–dsDNA (b,d,f) simulations with a simple MM model for the iron site. (a, b) Overlay of the active site between crystal structure and the snapshot at the end of the ~100 ns MD simulation. For AlkB, the crystal structure is colored in cyan, and the MD snapshot is colored in pink. Residues in the active site are shown in sticks, lesion bases (3meC) in sticks and spheres, and the distances in the crystal structure (in Å) are shown in black. For ABH2, the crystal structure is colored in green while the lesion base colored in marine, and the snapshot from MD simulation is colored in yellow with the lesion base colored in magenta. (c, d) RMSD of the active site, protein, DNA, and the entire protein–DNA complex between the MD simulation and crystal structure. All nonhydrogen atoms are included in the calculation. (e, f) Superposition of the AlkB–dsDNA/ABH2–dsDNA crystal structure and the snapshot after 100 ns of MD simulation to illustrate conformational changes in the dsDNA and protein–DNA interactions; much larger changes are observed for AlkB.

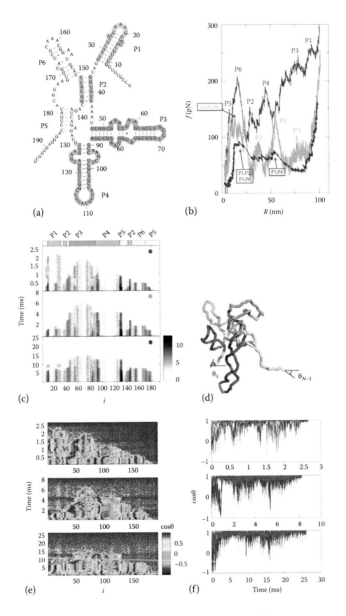

Figure 5.4 (a) Secondary structure of Azoarcus ribozyme. (b) FECs of Azoarcus ribozyme at three r_f's ($v = 43\,\mu m/s$, $k_s = 28\,pN/nm$ in red; $v = 12.9\,\mu m/s$, $k_s = 28\,pN/nm$ in green; and $v = 5.4\,\mu m/s$, $k_s = 3.5\,pN/nm$ in blue) obtained using the SOP model. (c) Contact rupture dynamics at three loading rates. The rips, resolved at the nucleotide level, are explicitly labeled. (d) Topology of Azoarcus ribozyme in the SOP representation. The first and the last alignment angles between the bond vectors and the force direction are specified. (e) Time evolutions of $\cos\Theta_i$ ($i = 1, 2, ..., N-1$) at three loading rates are shown. The values of $\cos\Theta_i$ are color coded as indicated on the scale shown on the right of bottom panel. (f) Comparisons of the time evolution of $\cos\Theta_i$ (blue) and $\cos\Theta_{N-1}$ (red) at three loading rates show that the differences in the f_c values at the opposite ends of the ribozyme are greater as r_f increases.

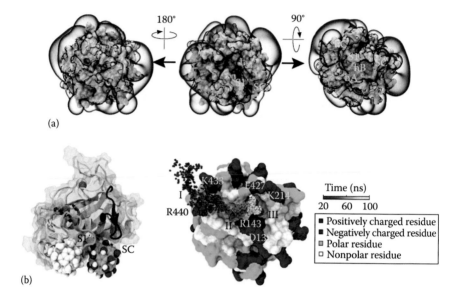

(a)

(b)

Figure 10.4 Electrostatic potential map and representative inhibitory pathway. (a) While a large negative potential prevails around the active site of MMP-9 (center), a hydrophobic patch with positive hot spots is stretched out over the area encompassing the ligand specificity S1′ and SC loops (right). The electrostatic potential was computed by using APBS 1.3 tool (Adaptive Poisson-Boltzmann Solver) [72]. Blue and red lobes represent isoelectric potential surfaces of +0.5 and –0.5 kT, respectively. (b, left) A representative binding mode shows that Gd@C$_{82}$(OH)$_{22}$ (a solid ball) binds between the S1′ ligand specificity loop (green ribbon) and the SC loop (purple ribbon) leading to the ligand binding groove. Alternatively, Gd@C$_{82}$(OH)$_{22}$ (a gray ball) can bind at the back entrance of the S1′ cavity leading into the active site (ball and stick for active sites and orange ball for the catalytic Zn^{2+}). (b, right) A possible binding pathway—depending on major driving forces, Gd@C$_{82}$(OH)$_{22}$ binds on MMP-9 along with three different phases: phase I, a diffusion-controlled nonspecific electrostatic interaction; phase II, a transient nonspecific hydrophobic interaction; and phase III, a specific hydrophobic and hydrogen-bonded stable binding.

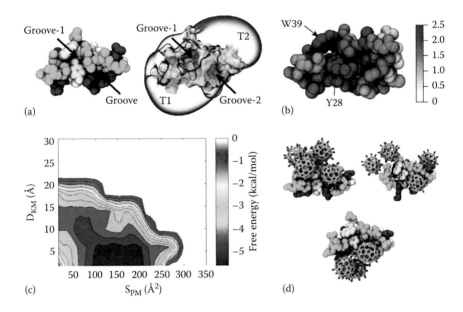

Figure 10.6 Residue-specific contacts and binding free energy surface from the binary system of Gd@C$_{82}$(OH)$_{22}$ and the WW domain. (a) The WW domain is colored according to the residue types with Y28 and W38 highlighted in yellow: negatively charged, positively charged, and polar and nonpolar residues are colored in red, blue, green, and white, respectively. (b) Site-specific contact ratio of the WW domain, where residues in PRM binding site have high contact ratio to Gd@C$_{82}$(OH)$_{22}$. (c) The binding free energy surface, where D_{KM} is the minimum distance between Gd@C$_{82}$(OH)$_{22}$ and the signature residues (Y28 and W39) of the WW domain and S_{PM} is the contacting surface area. (d) Representative binding modes found in the global minimum. Yellow, key residues; white, hydrophobic; green, noncharged polar; red, negatively charged; and blue, positively charged residues.

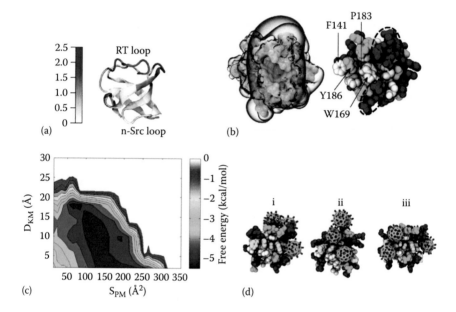

Figure 10.8 Residue-specific contacts and binding free energy surface from the binary system of Gd@C_{82}(OH)$_{22}$ and the WW domain. (a) Residue-specific contacts of Gd@C_{82}(OH)$_{22}$ on the SH3 domain. (b) Electrostatic potential surface of the SH3 domain, where a large negative electrostatic field is formed around the PRM binding site by negative residue clusters at the n-Src and RT loops. (c) Binding free energy landscape in the binary system and (d) representative structures in the low-energy basin, where residues are colored with residue types as described in Figure 10.2. Key residues are marked in yellow.

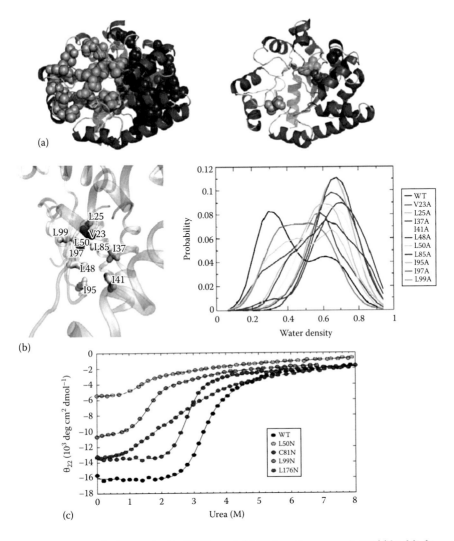

(a)

(b)

(c)

Figure 12.1 Dehydration inside TIM barrel. (a) Ribbon diagrams of αTS (a) highlighting the three hydrophobic clusters formed by the ILV residues: Cluster 1 (blue), Cluster 2 (orange), and Cluster 3 (green) obtained using a 4.2 Å cutoff distance between pairs of ILV side chains and (b) showing the locations of hydration mutations in the crystal structure, Leu50 in β2 (purple), Cys81 in α2 (blue), Leu99 in β3 (green), and Leu176 in β6 (orange). Coordinates of αTS from *Salmonella typhimurium* were used to generate the figure from a refined version of PDB file 1BKS. (b) Ribbon diagram of the interior of Cluster 1, portraying the ILV side chains selected for alanine-scanning mutagenesis, in space filling format (left). Histograms of water density inside Cluster 1 for the wild-type protein and the 10 alanine variants (right). (c) Experimental urea-induced equilibrium unfolding profiles for wild type and the hydration variants, L50N, C81N, L99N, and L176N. The continuous lines represent the fit of the data to a three-state model. (Reproduced from Das, P. et al., *J. Am. Chem. Soc.*, 135, 1882, 2013. With permission.)

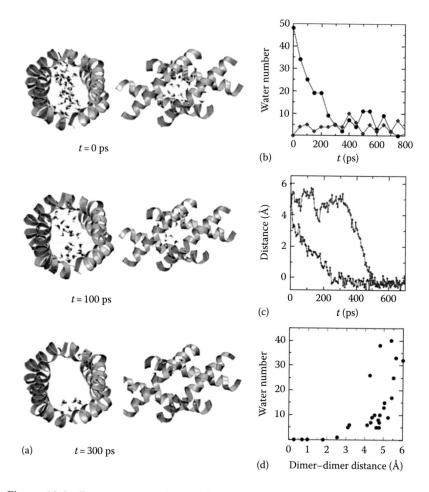

Figure 12.2 Dewetting in melittin. (a) Snapshots of water molecules inside the gap of the melittin tetramer (the protein is shown as ribbons and water as sticks). Only water molecules near the center of the channel are plotted, which is defined as the region with a spherical radius of 10 Å from the center of the enlarged tetramer. (b) Plot of the number of water molecules inside the channel against the MD time for *dewetting* (black) and *wetting* (red) simulations. (c) The kinetics of the *folding* simulation starting from two different initial separations, $d = 6$ Å (black) and $d = 4.5$ Å (red). (d) Plot of the number of water molecules inside the channel against the dimer–dimer distance for one folding trajectory starting at an initial separation of 6 Å, which indicates a drying-induced collapse. (Reproduced from Liu, P. et al., *Nature*, 437, 159, 2005. With permission.)

Chapter 7

Lipid Bilayers

Structure, Dynamics, and Interactions
with Antimicrobial Peptides

Yi Wang and J. Andrew McCammon

CONTENTS

7.1 INTRODUCTION

Lipids are the primary components of biological membranes. Arranged in bilayers, they not only provide a means for cellular compartmentalization but are also crucial for the function of many proteins embedded within them. For instance, the mitochondrial ADP/ATP carrier, which imports ADP into the mitochondria and exports synthesized ATP, requires the presence of the −2 charged cardiolipins for its proper function [1]. Rhodopsin, a G-protein-coupled receptor, can only function in a membrane with the correct balance of pressures in the headgroup and tail region [2]. Given the close interaction of lipids and membrane proteins, the study of bilayers not only is important to our understanding of their own properties but also provides the basis for investigating the function of many membrane proteins.

Molecular dynamics (MD) simulation [3–5], which provides both high spatial and temporal resolutions of the system under investigation, has been widely applied on the study of lipid bilayers. The first MD simulation of lipid bilayers was reported by van der Ploeg and Berendsen in the early 1980s [6]. This simulation was performed using a simplified lipid model with the headgroup represented by a single bead. In the early 1990s, simulations of phospholipid bilayers were reported (see [7] and references therein). Since these early studies, simulations of bilayers have significantly improved our understanding in their structure, dynamics, and interaction with membrane proteins. In this chapter, we review some of the recent lipid simulation work, with a focus on pure bilayer studies as well as those involving antimicrobial peptides (AMPs). Due to space limitation, we cannot discuss all the important work in these areas, and we refer the readers to a number of review articles for more details [8–11]. This chapter is organized as the following: In Section 7.2, we review the calculation of various structural and dynamic properties of a bilayer; in Section 7.3, we discuss methods to expedite lipid simulations; in Section 7.4, we comment on the challenges in bilayer simulations and discuss recent progress in lipid force field (FF) development; and in Section 7.5, we review bilayer simulations involving AMPs. These discussions are followed by a summary and an outlook for future bilayer simulations.

7.2 STRUCTURAL, ELASTIC, AND DYNAMIC PROPERTIES OF BILAYERS

Unlike proteins, lipid bilayers are intrinsically disordered in the fully hydrated, fluid (L_α or liquid crystalline) phase (Figure 7.1), which is most biologically relevant. As a result, bilayer structures are characterized by statistical distribution functions of atomic positions. A combination of techniques, including computer modeling as well as experimental measurements, such as x-ray scattering,

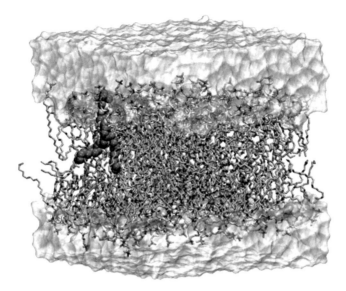

Figure 7.1   MD simulation system of a POPC bilayer in the L_α phase. Lipids are shown in sticks, with one molecule highlighted in vdW representation. Water molecules are shown as a transparent surface.

neutron diffraction, and nuclear magnetic resonance (NMR) are used to determine the structure of a lipid bilayer. The past couple of decades have witnessed considerable progress toward the development of these techniques [9–12]. Important structural parameters, such as area per lipid and hydrophobic thickness, can now be measured with high precision for the L_α-phase bilayers. In this process, MD simulations have played an important role, serving as a tool to both interpret experimental results and generate structural models [8,13–15].

Apart from structural properties, MD simulations are increasingly used to study the elastic and dynamic properties of lipid bilayers [14,16–19]. For instance, the lateral diffusion of lipids, the collective undulatory motions, and the permeability of small molecules have all been investigated in recent MD studies [20–23]. Table 7.1 lists a number of bilayer structural and dynamic properties that are frequently calculated from simulations and/or measured from experiments. In the following, we briefly review each property, with a focus on their calculation through MD simulations as well as the comparison with experimental results.

7.2.1   Area per Lipid

The synergy between MD simulations and experiments is particularly evident in the measurement of the area per lipid (A). While it may appear trivial,

TABLE 7.1   LIST OF BILAYER PROPERTIES FREQUENTLY CALCULATED FROM MD SIMULATIONS AND/OR MEASURED FROM EXPERIMENTS

Bilayer Properties	Calculation
Area per lipid A [24]	$A = 2A_{tot}/N$
Deuterium order parameter S_{CD} [8]	$S_{CD} = \left\langle \dfrac{3}{2}\cos^2\beta - \dfrac{1}{2} \right\rangle$
Molecular tilt S_{tilt} [8]	$S_{tilt} = \left\langle \dfrac{3}{2}\cos^2\theta - \dfrac{1}{2} \right\rangle$
Electron density profile $\rho(z)$ [24]	$\rho(z) = \sum\limits_{i \in slice} e_i/\Delta V$
Surface tension γ [24]	$\gamma = \left\langle L_z \times \left(P_{zz} - \left(\dfrac{1}{2}\right)\left(P_{xx} + P_{yy}\right) \right) \right\rangle$
Pressure profile $P_{local}(z)$ [26–28]	$P_{local}(z) = \sum\limits_{i \in slice} m_i v_i \otimes v_i$ $-\left(\dfrac{1}{\Delta V}\right)\sum\limits_{i<j} F_{ij} \otimes r_{ij} f(z, z_i, z_j)$
Area compressibility K_A [20,29]	$K_A = A_0 \left(\dfrac{\partial \gamma}{\partial A}\right)$
Bending modulus K_c [20,29]	$\langle u_{und}^2(q) \rangle = \dfrac{kT}{A\left(K_c q^4 + \gamma q^2\right)}$
Lateral diffusion coefficient D [30]	$D = \dfrac{1}{N}\sum\limits_{i}^{N} \dfrac{1}{4t}\langle |\mathbf{r}_i(t) - \mathbf{r}_i(0)|^2 \rangle$
Water permeability p_f [31]	$\dfrac{1}{p_f} = \int\limits_{z_1}^{z_2} \dfrac{\exp(\Delta G_w(z)/RT)}{D_w(z)} dz$

Note: $\langle ... \rangle$ indicates average over the relevant atoms and over a simulation trajectory. References are given for each property.

measurement of A represented a considerable challenge, as reflected in the large uncertainty of results of the model lipid 1,2-dipalmitoyl-*sn*-glycero-3-phosphocholine (DPPC), which ranged from 58 to 71 Å in early studies [9,24]. Interested readers are referred to a recent review [9], which summarized the development of experimental techniques measuring the area per lipid of bilayers. From MD simulations, the calculation of A is straightforward—given the small size of most simulation systems, little undulation effect is present, and the area per lipid is simply obtained as the total area of the system along the membrane plane divided by half of the total number of lipids (Table 7.1). For lipid mixtures, more complex algorithms can be applied to calculate individual A for each lipid species [25].

Combining simulation and experiment data, a commonly used strategy to determine A is to simulate the bilayers at different area per lipid and calculate their form factors $F(q)$ or electron density profiles (EDPs) $\rho(z)$. Through comparison with experimental $F(q)$ or $\rho(z)$, the area per lipid that produces the best fit can then be obtained. Feller et al. [24] used this strategy to determine A for DPPC. While their study was performed more than a decade ago, the result (62.9 ± 1.3 Å) is remarkably close to those obtained in later studies with newer technologies [9].

7.2.2   Order Parameters

The deuterium order parameter (S_{CD}) is obtained from NMR experiment with selected lipid tail hydrogens deuterated. S_{CD} is routinely used to characterize the disorder of lipid tails—higher absolute values indicate more order and lower absolute values indicate more disorder. Computationally, S_{CD} is obtained according to the equation shown in Table 7.1, where β stands for the angle between the carbon–deuterium (CD) vector and the bilayer normal. When the lipid tails are in a perfect *trans* conformation, $\beta = 90°$ and $\cos(\beta) = 0$. As a result, $S_{CD} = 0.5$ for perfectly *trans* lipids. When the lipid molecules are in an isotropic state, the CD vector is uniformly distributed. Through integration in the spherical coordinates, where φ represents the azimuthal angle, it can be shown that $\langle \cos^2\beta \rangle = 1/3$:

$$
\begin{aligned}
\langle \cos^2\beta \rangle &= \frac{1}{2\pi} \int_0^{2\pi} \int_0^{\pi/2} \cos^2\beta \sin\beta \, d\beta \, d\phi \\
&= \int_0^{\pi/2} \cos^2\beta \sin\beta \, d\beta \\
&= \int_0^{\pi/2} -\cos^2\beta \, d\cos\beta \\
&= \int_1^0 -t^2 \, dt = \frac{1}{3}.
\end{aligned}
$$

As a result, $S_{CD} = 0$ for molecules in an isotropic state. In practice, when lipids are arranged in a L_α-phase bilayer, the absolute values of S_{CD} generally range from 0.1 to 0.3. Higher order is often found near the lipid headgroups, while the terminal methyl groups tend to have the lowest $|S_{CD}|$.

Apart from S_{CD}, another order parameter, S_{tilt}, is used to describe the overall molecular tilt of the lipids. The equation for calculating S_{tilt} is very similar to that of S_{CD} (Table 7.1), with β replaced by θ, which represents the angle between the long axis of the lipid and the bilayer normal. By calculating S_{tilt} from simulations and comparing the result with NMR experiments, Pastor et al. [8,15] studied the details of lipid motions: They showed that the internal motion, which accounts for the fast relaxation, rapidly average a lipid into a *cylinder*; the *wobbling* motion of the cylinder in a cone-like region is then found to be the primary component of the *slow* relaxation on the 10 ns timescale [15].

7.2.3 Electron Density Profile

An important structural parameter of the bilayer and a key result from Neutron scattering and x-ray diffraction experiments is the EDP. The overall scattering density profile or equivalently the observed structure factors [32] are obtained directly from the aforementioned experiments. Fourier reconstruction from these structure factors then yields the 1D EDP. Based on MD simulations, EDP can be calculated in several ways: the most straightforward is to divide the simulation cell into slabs of approximately 0.1 Å and then assign all the electrons of an atom to its nucleus [24]. The time-averaged electrons in each slab yields the EDP (Figure 7.2). Alternatively, one may assume a Gaussian distribution of the electrons on each atomic center with a standard deviation equal to the van der Waals radius [33]. Finally, a third method was proposed by White and colleagues [32]: Following the same protocol used in experiments, the authors first determined the discrete structure factors and then used the Fourier reconstruction to obtain the EDP. The advantage of their model-independent method is that no assumption in the shape of the electron distribution is made and no artificial smoothing of the EDP is performed. However, despite the differences in the aforementioned three methods, qualitatively similar results are obtained. In particular, the similarity between results of methods 1 and 3 supports the usage of the former method to generate a fast approximation of the EDP.

7.2.4 Surface Tension and Pressure Profile

Experimentally, a patch of lipid bilayer that is allowed to freely adjust its volume and area has zero surface tension. In simulations, the finite size of the simulated lipid patch has led to arguments that the surface tension should be

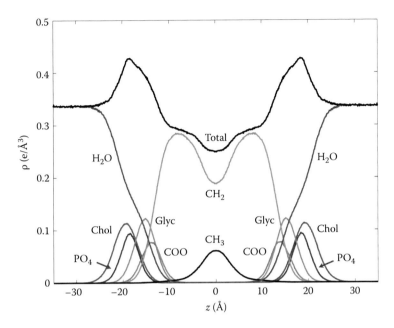

Figure 7.2 EDP of a POPC bilayer.

nonzero (see [34] and references therein), although counter arguments have also been presented (see [6] and references therein). As summarized in a recent review by Pastor and MacKerell [35], the consensus now is that the surface tension in a simulation should be zero for a bilayer at its equilibrium surface area. A small γ of a few dyn/cm may be present if undulations are taken into account.

The study of surface tension in bilayer simulations has been complicated by a long-standing issue in lipid force fields. For instance, until recently, the widely used CHARMM FF failed to reproduce experimentally obtained area per lipid under zero surface tension [36]. For this reason, many simulations were performed under constant pressure and constant area (NPAT) conditions, or under a fixed surface tension (NPγT). With the latest update of CHARMM [36], bilayer simulations can be performed under zero surface tension in a *semi-isotropic* NPT ensemble ($\gamma = 0$, the x and y dimensions scale uniformly, while the z dimension scales independently). We will discuss more about this issue and lipid FF development in later sections.

Apart from surface tension, more detailed analysis of the internal pressure in a bilayer can be obtained from MD simulations. Lindahl and Edholm [26] first reported the spatial decomposition of surface tension in a DPPC bilayer. Such decomposition (Figure 7.3), later referred to as a pressure profile (PP), reveals the bilayer surface tension to be a sum of very large terms with opposite signs,

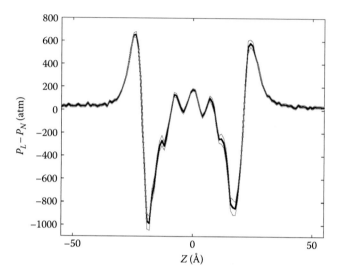

Figure 7.3 PP of a POPC bilayer. Thin lines represent the data plus and minus the standard error. The calculation was performed using methods described in Gullingsrud and Schulten. (From Gullingsrud, J. and Schulten, K., *Biophys. J.*, 86, 3496, 2004.)

thereby explaining its sensitivity to FFs and simulation details. PPs provide a means to study the effect of lipid environment on membrane proteins. For instance, Gullingsrud and Schulten [27] showed that PPs affect the gating of the MscL channel through its second moment in a tension-independent manner. Their calculations also revealed a pressure trough below the lipid headgroups in each monolayer, which was suggested to play a role in positioning small molecules within the bilayer [27].

Although PP calculations provide detailed information on local pressure within a bilayer, there exist certain methodological issues that should be taken into consideration. Sonne et al. [28] pointed out that PP is not uniquely defined, since the expression of local pressure involves an arbitrary choice of an integration pathway. Nevertheless, similar results were obtained for DPPC using two different expressions (Irving–Kirkwood and Harasima) of the local pressure tensor [28]. Therefore, this issue is likely not a major concern for similar lipid bilayers. In comparison, a more disconcerting issue is the dependence of the PPs on the MD integrator: the multiple-timestepping (MTS) algorithm, frequently used to speed up MD simulations, is found to significantly affect the shape and value of PPs, especially in the hydrophobic tail region [27]. As MTS is widely used to expedite MD calculations, this issue is difficult to avoid. Fortunately, comparison of two PPs obtained using the same simulation protocol may allow some cancellation of errors, thereby alleviating, if not eliminating, the aforementioned problem.

7.2.5 Area Compressibility and Bending Modulus

The area compressibility modulus, K_A, and the bending modulus, K_c, characterize the elasticity of a bilayer. These mesoscopic properties can only be calculated from systems of much larger dimension than the membrane thickness [20,29]. Such expensive calculations only became possible in the past decade: Simulating a fully hydrated DPPC bilayer with 1024 lipids, Lindahl and Edholm analyzed both undulatory and peristaltic (fluctuation in membrane thickness) motions of the bilayer and found a K_c of 4×10^{-20} J and a K_A of 250–300 mN/m [20]. A separation of scales is clearly observed in their spectral analysis: at length scales larger than the membrane thickness ($q < 1.5$ nm^{-1}), mesoscopic undulation dominates; on the intermediate length scale ($1.5 < q < 4$ nm^{-1}), collective lipid protrusion is observed; finally, on small length scales comparable to interlipid distances ($q > 4$ nm^{-1}), the spectrum disappears into noise due to the statistical nature of lipid atomic distributions.

Marrink and Mark [29] performed a similar analysis to address the question of whether a surface tension should be applied in simulations of a finite-sized lipid bilayer. As mentioned earlier, this issue was discussed in multiple studies, some of which argued that a surface tension was needed to compensate for the suppressed undulations in finite-sized bilayers [37]. The undulation data of Marrink and Mark revealed a strong dependence of the surface tension on surface area but also indicated that for stress-free simulations, the equilibrium area did not have a strong dependence on system size. As a result, the authors concluded that no surface tension should be applied to compensate for the system size in a bilayer simulation. The same conclusion is reached in others' work and summarized in a recent review by Pastor and MacKerell [35]. Currently, with significant improvement in lipid FFs, the majority of bilayer simulations can be performed under zero surface tension in the semi-isotropic NPT conditions described earlier. We will discuss some of these lipid FF developments in the following section.

7.2.6 Lateral Diffusion

From MD simulations, the lateral diffusion coefficient of lipids (D) can be calculated according to the Einstein relation (Table 7.1). In this formula, N is the number of lipids in the system and $\mathbf{r}_i(t)$ and $\mathbf{r}_i(0)$ are the coordinates of a lipid molecule in the membrane plane at time t and time 0, respectively. To improve statistics and make use of all data points, the time origin can be shifted along the simulation trajectory. Note that the aforementioned equation is only valid at large t—when t is small, the velocities of a molecule are highly correlated and the calculation tends to overestimate D.

Apart from the diffusion of lipid molecules within a bilayer, the upper and the lower monolayers also move in a diffusion-like manner. Therefore, a natural question is whether we should remove the COM movement of the two monolayers before calculating D. If so, the obtained D reflects the *net* diffusion of a lipid within its monolayer. Recently, Roark and Feller [23] showed that for a small system, the correlation length of monolayer COM motions was comparable with the dimension of the simulation unit cell. As a result, a small system may produce an artificially large D if the COM motion is not removed [14]. Others have also argued that the diffusion of monolayers is purely spurious and should therefore be excluded from the calculation of D. In a recent study performed in our lab [30], we chose to remove the monolayer COM in the calculation of D. Our major concern was the relatively small size of the simulation system (78 lipid molecules) and the poor statistical quality of the monolayer COM data. Although the aforementioned treatment suffers from the finite size effect described by Yeh and Hummer [38], it significantly improves the statistical quality of the result. Furthermore, when comparing two simulation systems, the impact of removing COM motion is significantly reduced, as long as both systems are treated using the same method.

The lateral diffusion coefficient of lipids reflects their fluidity within the bilayer plane. Multiple simulations have found that D is on the order of 10^{-7} cm^2/s. Therefore, on average, a lipid molecule only travels ~10 Å in the bilayer plane during a 25 ns simulation. To put these numbers in perspective, a water molecule can travel ~200 Å in the same amount of simulation time. Clearly, the slow diffusion of lipids renders it time consuming to study dynamic behaviors at the lipid–water interface. To expedite this process, various methods have been designed and we will discuss some of them in the following section.

7.2.7 Water Permeability

Owing to the hydrophobic nature of lipids, the bilayer presents an energetic barrier against the transmembrane permeation of a large number of molecules. This barrier is essential for living cells—it enables them to establish a chemical gradient across cellular membranes and achieve selective transport through specialized membrane proteins [39,40]. For instance, ions, amino acids, sugars, and many drug molecules rely on specific membrane channels or transporters to cross the membranes. The bilayer itself primarily serves as a pathway for very small and hydrophobic molecules. An interesting exception is water: Although water molecules are polar, given their small sizes and extremely high concentration in the human body, water transport was long deemed to be primarily mediated by lipid bilayers. The discovery of dedicated water channels, namely, aquaporins [41], has altered this viewpoint. Nevertheless, water permeability of bilayers sets the baseline for its transport and, therefore, remains an important subject of research.

Using MD simulations, water permeability of a DPPC bilayer was first calculated by Marrink and Berendsen [31]. The authors developed an *inhomogeneous solubility and diffusion* model, where the free energy profile of water across the bilayer ($\Delta G_w(z)$) and the location-dependent diffusion coefficient ($D_w(z)$) are used to calculate the permeability of water (p_f) according to the equation shown in Table 7.1. This model can be also used to calculate permeability of other small molecules, as long as their thermodynamic gradients across the membrane remain small [42].

7.3 EXPEDITING LIPID BILAYER SIMULATIONS

Due to the slow diffusion of lipids within a bilayer, the dynamic properties of a membrane can be prohibitively expensive to study with conventional all-atom simulations. Two commonly used strategies to expedite these simulations are (1) reducing the complexity of the system with coarse-grained (CG) models and (2) speeding up the calculation using enhanced-sampling methods. In the following, we briefly review each strategy, with focus on two representative methods. The readers are referred to recent reviews for details on other methods [43–48].

7.3.1 Coarse-Grained Models

Along with the development in all-atom simulations of lipid bilayers, CG modeling has been an active area of research in the past decade. As an extreme of coarse graining, bilayers can be treated as continuum models, with its elastic properties characterized by a number of parameters such as K_A and K_c (Table 7.1). In between the continuum and all-atom representations, models with reduced complexity and resolution have been developed [43,44]. For instance, MARTINI, first developed for lipid bilayers [49,50], is one of the most widely used CG models and has been subsequently expanded to include parameters for proteins [51] and carbohydrates [52]. The particles (termed *interaction sites*) in MARTINI are based on a four-to-one mapping of their all-atom models. They are classified into four main types: polar (P), nonpolar (N), apolar (C), and charged (Q). Each type has a number of subtypes to better characterize the chemical nature of the underlying atomic structure. Parameterization of the particles is achieved through calibration against thermodynamic data, in particular, the oil/water partition coefficients of small chemicals. As validation of the model, free energies of vaporization, hydration, and partition of ~30 small compounds were calculated and compared with experimental data [50]. Additionally, a number of structural, elastic, and dynamic properties of the CG DPPC bilayer were compared with all-atom simulation results [50].

The speedup in MARTINI CG simulations is partly achieved through a smoothened potential energy surface, which, in turn, allows the usage of a much larger timestep (20 fs instead of 2 fs) in the integration of equations of motion. The reduced number of particles in the system also saves considerable computational resource and the effective diffusive dynamics of water is usually considered to be four times faster than in all-atom simulations [50]. As a result, at least two orders of magnitude speedup can be expected from CG simulations performed with MARTINI. Such significant increase in computational efficiency has allowed the method to be used in a large number of studies (see [53] and references therein).

Despite its successful application, the original MARTINI FF used an over-simplified model of water. As a result, the effects of polarization and proper screening of charged interactions are absent [50]. This problem has been addressed recently with a polarizable water model in the MARTINI FF [54]. With such a model, interactions between charged and polar groups in a low-dielectric environment can be described more realistically. Consequently, processes such as electroporation of an octane slab and translocation of ions across a membrane can now be modeled with comparable accuracy to that of all-atom simulations [54]. Although it is computationally more expensive, given the importance of polarizability in biomolecule simulations, the new water model will see wide application in future CG studies.

7.3.2 Accelerated Molecular Dynamics

Enhanced-sampling methods are a class of methods used to speed up the sampling of conformational space in MD simulations. In theory, any such method suitable for protein simulations should also be applicable on lipids; in practice, however, different simulation protocols may be required to ensure the proper simulation of lipid bilayers. An example is the accelerated molecular dynamics (aMD) method, which was initially developed for proteins [55] and later benchmarked against the 1-palmitoyl-2-oleoyl-*sn*-glycero-3-phosphocholine (POPC) and 1,2-dimyristoyl-*sn*-glycero-3-phosphocholine (DMPC) bilayers [30].

In essence, aMD is an enhanced-sampling method that improves the conformational space sampling by reducing energy barriers separating different states of a system. Since it was first proposed by Hamelberg et al. [55], the aMD method has been applied to study a variety of biological systems [56]. During an aMD simulation, the potential energy landscape of a system is modified such that when the system's energy falls below a threshold energy,

E, a bias potential is added, and the modified potential, $V^*(\mathbf{r})$, is related to the original potential, $V(\mathbf{r})$, via

$$V^*(\mathbf{r}) = \begin{cases} V(\mathbf{r}) + \Delta V(\mathbf{r}) & V(\mathbf{r}) < E \\ V(\mathbf{r}) & V(\mathbf{r}) \geq E \end{cases}$$

where $\Delta V(\mathbf{r})$ is the bias potential:

$$\Delta V(\mathbf{r}) = \frac{(E - V(\mathbf{r}))^2}{\alpha + E - V(\mathbf{r})}.$$

In the aforementioned equation, α is a positive number that controls the roughness of the modified potential surface. Together with E, it determines the *amount* of acceleration applied to the system. When α becomes very small, the modified potential below E approaches a constant energy surface, in which case all the enthalpic barriers are removed and sampling is only limited by entropic contributions as well as diffusion in the conformational space.

In a recent study, we applied the aMD method to enhance the lateral diffusion and mixing of POPC and DMPC molecules [30]. Comparison with conventional MD (cMD) simulations revealed moderate speedup in lipid mixing in the aMD simulation of the POPC:DMPC bilayer (Figure 7.4). Specifically, analysis of the

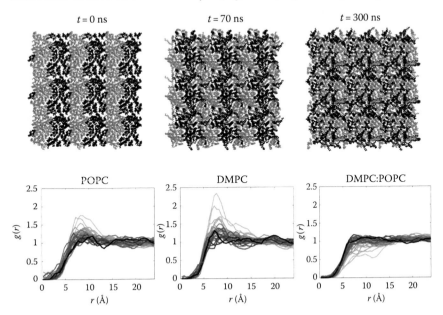

Figure 7.4 Mixing of POPC and DMPC lipids in aMD simulations. (Data from Wang, Y. et al., *J. Chem. Theory Comput.*, 7, 3199, 2011.)

radial distribution function $g(r)$ shows that lipids are well mixed at $t = 50$ ns in the aMD simulation, whereas at least 130 ns is required by the cMD simulation to reach a similar state (Figure 7.4). Based on these results, aMD is estimated to provide approximately twofold to threefold speedup compared to cMD. Such a moderate speedup may be increased with higher boost energy, although the artificial *trans/gauche* transition of the unsaturated oleoyl tail in POPC sets an upper bound for the boost energy [30]. Selective aMD may be a promising direction for avoiding such artifacts—while it has been used in protein simulations [57], no bilayer simulation using selective aMD has been reported. Therefore, future efforts on aMD simulation of bilayers may focus on applying selective boost to further speed up lipid diffusion while preserving its structural integrity.

7.4   CHALLENGES IN BILAYER SIMULATIONS AND RECENT DEVELOPMENT OF LIPID FORCE FIELD

Unlike neighboring amino acids in a protein, neighboring lipids in a bilayer are not covalently bonded to each other. As a result, nonbonded interactions play a crucial role in determining the structure and dynamics of a bilayer. These interactions generally manifest into two opposing tendencies, namely, the tendency of the hydrophobic tails to minimize the surface area and the tendency of the hydrophilic headgroups to maximize it. In order to reproduce the structural and dynamic properties of a bilayer, a proper balance between these two opposing tendencies must be maintained.

The challenges in bilayer simulations are reflected in the development of lipid FFs. Lipid parameters have been developed for all of the four most widely used protein FFs, namely, AMBER [58,59], CHARMM [60], GROMOS [61], and OPLS [62,63]. Among them, the most widely used lipid FFs have been GROMOS [64,65] and CHARMM [36]. Since 2009, important updates have been reported for both FFs. Additionally, the AMBER lipid FF has also been updated in recent studies [66–68]. All of these developments have been aimed at reproducing important bilayer structural and thermodynamic properties described earlier. In particular, the latest FFs all report the correct area per lipid at zero surface tension.

Recently, Piggot et al. [69] conducted a series of simulations to compare the performance of four GROMOS FFs, 43A1-S3 [70], 53A6$_L$ [65], Berger et al. [64], 53A6 Kukol [71], as well as the latest CHARMM c36 FF [36]. As discussed in their work [69], each lipid FF is found to possess its own strength and weakness, although the combination of certain FF and simulation protocols appears to be *disastrous* and should be strictly avoided. Such combinations include GROMOS 53A6 POPC parameters of Kukol [71] and the use of 1.0 nm cutoffs for DPPC of Berger et al. [64]. These results reflect again the intricacy of bilayer simulations, since both FF and simulation protocols may have a significant effect on the result of a bilayer simulation. Indeed, similar effects have also been reviewed by Pastor and MacKerell in their review [35]. Focusing on the latest CHARMM FF, they stressed the importance of simulating bilayers with the same van der Waals (vdW) cutoff and water models as those used in the parameterization of the FF. Furthermore, it has also been pointed out that a FF that works for one lipid species may not work well for other lipids. An example is the widely used Berger FF [64], which works well for DPPC but fails to produce results consistent with experimental data for POPG and POPE (see [69] and references therein). Clearly, the development of lipid FFs remains an ongoing task. In addition to further refinement of current FFs, polarizable lipid models will likely provide a more accurate description of bilayer structural, elastic, and dynamic properties.

7.5   SIMULATIONS OF BILAYER–PEPTIDE INTERACTIONS

Simulations of pure lipid bilayers provide the basis for studying their interactions with peptides and proteins. An important class of molecules in the former category is AMPs. AMPs are small, cationic, and amphipathic peptides with a wide spectrum of antimicrobial activity [72,73]. As part of the innate immune system, AMPs form the first line of defense against pathogenic invasion in a large number of species [74]. While some AMPs act on intracellular targets, it is generally accepted that the majority of them target the cellular membranes through disrupting the physical integrity of the bilayers. In the latter case, many AMPs induce pore formation in the bilayer, destroying the transmembrane electrochemical gradient, and, thereby, causing osmolysis, cell swelling, and eventually, cell death [75].

Since their mode of action differs significantly from most existing antibiotics, AMPs are considered promising candidates for new antibiotic drugs. However, from thousands of natural and synthetic AMPs, only a few have actually proceeded into clinical trials (see [10,11] and references therein). One of the difficulties in the development of AMP antibiotics is the lack of understanding of their selectivities and toxicities over different biomembranes—apart from antimicrobial activity, some AMPs also cause lysis of human red blood cells (hemolytic), making them unsuitable for therapeutic use. Therefore, designing peptides with high antimicrobial and low hemolytic activities is critical to AMP drug development. Since MD simulations provide an atomic-resolution model of the system under investigation, they have been used extensively to investigate AMP–lipid interactions [76–82].

7.5.1   Binding, Insertion, and Pore Formation of AMPs

The initial association of AMPs with bilayers, that is, binding and insertion, has been studied extensively using all-atom simulations [10,11]. Most AMPs are

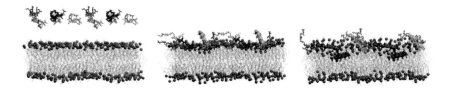

Figure 7.5   MD simulation of CM15 insertion into a mixed POPC:POPG bilayer.

found to preferentially reside near the lipid–water interface, with the hydrophobic side facing the lipid tails and the hydrophilic side interacting with water molecules and polar groups of the lipids (Figure 7.5). Compared with peptide binding and insertion, pore formation is more challenging to study due to its stochastic nature and high computational cost. In a 250 ns simulation performed by Marrink and colleagues [83], the spontaneous pore formation induced by an AMP was observed for the first time. Four copies of the AMP magainin-H2, placed in the bulk solution at the beginning of the simulation, bound to a DPPC bilayer surface and subsequently formed a toroidal pore. During this process, the peptide is inserted deeply into one monolayer of the membrane, thereby imposing a large curvature stress. A continuous pore was then formed when solvent molecules established stable interactions with hydrophilic residues of the peptide buried inside the membrane [83].

The aforementioned simulation provided great detail regarding the structure and dynamics of the toroidal pore formed by an AMP. For instance, it was shown that contrary to experimental models, the toroidal pore formed by magainin-H2 is highly disordered—the peptides never adopted fully perpendicular orientation in the simulation; instead, they stabilize the pore by binding close to the rim. In a follow-up study of the peptide melittin and DPPC, the authors further showed that only one or two peptides line the pore, in strong contrast to the traditional models [84]. Peptide aggregation, either prior to or after binding to the membrane surface, is found to be a prerequisite to pore formation [84]. Subsequent simulations using CG models or multiscale approaches [85,86] agreed well with the all-atom studies. For more comprehensive discussions of these simulation studies, we refer the readers to recent reviews [10,11].

7.5.2   Sensitivity of AMP Simulations to Force Field and Computation Protocols

As discussed in Section 7.4, due to an intricate balance between two large, opposing tendencies in the membrane, bilayer simulations can be sensitive to both FF and simulation protocols. A natural question is whether such sensitivity is also present in bilayer-AMP simulations. Although the peptide binding and insertion processes have been successfully studied with multiple FFs and simulation protocols, to the best of our knowledge, the only all-atom simulations that reported spontaneous pore formation [83,84,87] were based on the Berger FF and adopted the reaction field (RF) approach for electrostatic computation. This combination is commonly used with the simulation program GROMACS. In comparison, the majority of simulations performed with the program NAMD employ the CHARMM lipid FF and particle mesh ewald (PME) method for electrostatic calculation.

Following the magainin-H2 work, Sengupta et al. [84] performed a series of simulations of the AMP melittin in a DPPC bilayer. By varying the peptide/lipid ratio and the number of counter ions in the system, the authors systematically studied the conditions for pore formation, especially the effect of counter ions. Interestingly, they found that the presence of counter ions slows down the formation of melittin-induced pores. Along the same line, Kandasamy and Larson [77] studied the effect of salt on the interactions of the AMP magainin with a POPC bilayer. Through calculating the peptide insertion depth and analyzing its hydrophobic contact with the lipids, they showed that peptide binding to lipids is stronger at lower concentrations of salt. However, the effect of electrostatic calculation protocols was not examined in the aforementioned studies. In the work by Leontiadou et al. [88], RF and PME methods were found to have little impact on the stability of a tension-induced pore. However, this study used preexisting pores, a condition significantly different from spontaneous pore formation induced by AMPs. Therefore, it is unclear from the evidences given earlier how much effect (if any) electrostatic calculation protocols have on the spontaneous pore formation induced by AMPs. Similarly, the effect of FFs on AMP simulations remains to be fully elucidated. As in the case of pure bilayer simulations, it may be difficult to examine the independent effect of FF, electrostatics, and other simulation protocols in bilayer-AMPs simulations. However, such studies are required to fully understand the behavior of these small peptides, the limits of current lipid FFs, and how we can further improve the accuracy of bilayer and membrane protein simulations.

7.6 SUMMARY AND OUTLOOK

The past couple of decades have witnessed the rapid development in MD simulations of lipid bilayers. In the early 1990s, a 170 ps simulation of a 72 lipid–water system required 6 months of computer time [8]. Today, simulations five orders of magnitude longer can be completed within a couple of days on specialized hardwares [89]. Additionally, MD simulations can be performed in full atomistic details on bilayers with up to ~1000 lipids [20]. These studies have significantly improved our understanding in the structural, elastic, and dynamic properties of bilayers and have laid the foundation for more complex simulations involving peptides and membrane proteins (membrane protein simulations are reviewed in Chapter 9 of this book). Indeed, significant progress has been made in membrane protein simulations in the past few years, largely thanks to the recent improvement in computer hardware and software [90]. For instance, in a landmark simulation performed on the machine Anton, Dror et al. [91] studied the spontaneous binding of beta-blockers to the G-protein-coupled receptor b2-AR. Initially placed in the bulk water, the beta-blockers are found to bind to the protein in remarkably similar conformations to those revealed by x-ray crystallographic studies. This result points at an exciting direction for future simulation work, namely, using *brute force* MD simulations to study protein–ligand binding in computer-aided drug discovery. Given such recent progress and the continued development in computer hardware and software, we can expect MD simulations to play an ever more important role in our study of lipid bilayers and membrane proteins.

ACKNOWLEDGMENTS

Work at UCSD is supported in part by the National Science Foundation, the National Institutes of Health, Howard Hughes Medical Institute, Center for Theoretical Biological Physics, the National Biomedical Computation Resource, and the NSF supercomputer centers. Work at CUHK is supported by a research startup fund from the Chinese University of Hong Kong.

REFERENCES

1. K. Beyer and M. Klingenberg. ADP/ATP carrier protein from beef heart mitochondria has high amounts of tightly bound cardiolipin, as revealed by phosphorus-31 nuclear magnetic resonance. *Biochemistry*, 24:3821–3826, 1985.
2. A.V. Botelho, N.J. Gibson, R.L. Thurmond, Y. Wang, and M.F. Brown. Conformational energetics of rhodopsin modulated by nonlamellar-forming lipids. *Biochemistry*, 41:6354–6368, 2002.
3. M.P. Allen and D.J. Tildesley. *Computer Simulation of Liquids*. Oxford University Press, New York, 1987.
4. D. Frenkel and B. Smit. *Understanding Molecular Simulation from Algorithms to Applications*. Academic Press, San Diego, CA, 2002.
5. Y. Wang and J.A. McCammon. Introduction to molecular dynamics: Theory and applications in biomolecular modeling. In N.V. Dokholyan, ed., *Computational Modeling of Biological Systems: From Molecules to Pathways*, pp. 3–30. Springer, New York, 2012.
6. H.J.C. Berendsen and D.P. Tieleman. Molecular dynamics: Studies of lipid bilayers. In P. von Rague Schleyer ed., *Encyclopedia of Computational Chemistry*, vol. 3, pp. 1639–1650, John Wiley & Sons, New York, 1998.

7. R.W. Pastor. Molecular dynamics and Monte Carlo simulations of lipid bilayers. *Curr. Opin. Struct. Biol.*, 4:486–492, 1994.

8. R.W. Pastor, R.M. Venable, and S.E. Feller. Lipid bilayers, NMR relaxation, and computer simulations. *Acc. Chem. Res.*, 35:438–446, 2002.

9. S. Tristram-Nagle and J.F. Nagle. Lipid bilayers: Thermodynamics, structure, fluctuations and interactions. *Chem. Phys. Lipids*, 127:3–14, 2004.

10. E. Mátyus, C. Kandt, and D.P. Tieleman. Computer simulation of antimicrobial peptides. *Curr. Med. Chem.*, 14:2789–2798, 2007.

11. P.J. Bond and S. Khalid. Antimicrobial and cell-penetrating peptides: Structure, assembly and mechanisms of membrane lysis via atomistic and coarse-grained molecular dynamics simulations. *Protein Pept. Lett.*, 17:1313–1327, 2010.

12. J.F. Nagle and S. Tristram-Nagle. Lipid bilayer structure. *Curr. Opin. Struct. Biol.*, 10:474–480, 2000.

13. H.I. Petrache, K. Tu, and J.F. Nagle. Analysis of simulated NMR order parameter for lipid bilayer structure determination. *Biophys. J.*, 76:2479–2487, 1999.

14. J.B. Klauda, N. Kucerka, B.R. Brooks, R.W. Pastor, and J.F. Nagle. Simulation-based methods for interpreting x-ray data from lipid bilayers. *Biophys. J.*, 90:2798–2807, 2006.

15. J.B. Klauda, M.F. Roberts, A.G. Redfield, B.R. Brooks, and R.W. Pastor. Rotation of lipids in membranes: Molecular dynamics simulation, 31P spin-lattice relaxation, and rigid-body dynamics. *Biophys. J.*, 97:3074–3083, 2008.

16. S.E. Feller. Molecular dynamics simulations of lipid bilayers. *Curr. Opin. Colloid Interface Sci.*, 5(3–4):217–223, 2000.

17. L. Saiz, S. Bandyopadhyay, and M.L. Klein. Towards an understanding of complex biological membranes from atomistic molecular dynamics simulations. BSR, 22(2):151–173, 2002.

18. G.S. Ayton and G.A. Voth. Mesoscopic lateral diffusion in lipid bilayers. *Biophys. J.*, 87:3299–3311, 2004.

19. E. Falck, T. Róg, M. Karttunen, and I. Vattulainen. Lateral diffusion in lipid membranes through collective flows. *J. Am. Chem. Soc.*, 130:44–45, 2008.

20. E. Lindahl and O. Edholm. Mesoscopic undulations and thickness fluctuations in lipid bilayers from molecular dynamics simulations. *Biophys. J.*, 79:426–433, 2000.

21. L. Saiz and M.L. Klein. Computer simulation studies of model biological membranes. *Acc. Chem. Res.*, 35:482–489, 2002.

22. M.C. Pitman, F. Suits, K. Gawrisch, and S.E. Feller. Molecular dynamics investigation of dynamical properties of phosphatidylethanolamine lipid bilayers. *J. Chem. Phys.*, 122:244715, 2005.

23. M. Roark and S.E. Feller. Molecular dynamics simulation study of correlated motions in phospholipid bilayer membranes. *J. Phys. Chem. B*, 113:13229–13234, 2009.

24. S.E. Feller, R.M. Venable, and R.W. Pastor. Computer simulation of a DPPC phospholipid bilayer: Structural changes as a function of molecular surface area. *Langmuir*, 13:6555–6561, 1997.

25. O. Edholm and J.F. Nagle. Areas of molecules in membranes consisting of mixtures. *Biophys. J.*, 89:1827–1832, 2005.

26. E. Lindahl and O. Edholm. Spatial and energetic-entropic decomposition of surface tension in lipid bilayers from molecular dynamics simulations. *J. Chem. Phys.*, 113(9):3882–3893, 2000.

27. J. Gullingsrud and K. Schulten. Lipid bilayer pressure profiles and mechanosensitive channel gating. *Biophys. J.*, 86:3496–3509, 2004.

28. J. Sonne, F.Y. Hansen, and G.H. Peters. Methodological problems in pressure profile calculations for lipid bilayers. *J. Chem. Phys.*, 122:124903, 2005.

29. S.J. Marrink and A.E. Mark. Effect of undulations on surface tension in simulated bilayers. *J. Phys. Chem. B*, 105:6122–6127, 2001.

30. Y. Wang, P.R.L. Markwick, C.A.F. de Oliveira, and J. Andrew McCammon. Enhanced lipid diffusion and mixing in accelerated molecular dynamics. *J. Chem. Theory Comput.*, 7:3199–3207, 2011.

31. S.J. Marrink and H.J.C. Berendsen. Simulation of water transport through a lipid membrane. *J. Phys. Chem.*, 98:4155–4168, 1994.

32. R.W. Benz, F. Castro-Romn, D.J. Tobias, and S.H. White. Experimental validation of molecular dynamics simulations of lipid bilayers: A new approach. *Biophys. J.*, 88:805–817, 2005.

33. K. Tu, D.J. Tobias, and M.L. Klein. Constant pressure and temperature molecular dynamics simulation of a fully hydrated liquid crystal phase dipalmitoylphosphatidylcholine bilayer. *Biophys. J.*, 69:2558–2562, 1995.

34. S.E. Feller and R.W. Pastor. Constant surface tension simulations of lipid bilayers: The sensitivity of surface areas and compressibilities. *J. Chem. Phys.*, 111(3):1281–1287, 1999.

35. R.W. Pastor and A.D. MacKerell. Development of the CHARMM force field for lipids. *J. Phys. Chem. Lett.*, 2:1526–1532, 2011.

36. J.B. Klauda, R.M. Venable, J.A. Freites, J.W. O'Connor, D.J. Tobias, C. Mondragon-Ramirez, I. Vorobyov, A.D. MacKerell Jr., and R.W. Pastor. Update of the CHARMM all-atom additive force field for lipids: Validation on six lipid types. *J. Phys. Chem. B*, 114(23):7830–7843, 2010.

37. S.E. Feller and R.W. Pastor. On simulating lipid bilayers with an applied surface tension: Periodic boundary conditions and undulations. *Biophys. J.*, 71:1350–1355, 1996.

38. I.-C. Yeh and G. Hummer. System-size dependence of diffusion coefficients and viscosities from molecular dynamics simulations with periodic boundary conditions. *J. Phys. Chem. B*, 108:15873–15879, 2004.

39. F. Khalili-Araghi, J. Gumbart, P.-C. Wen, M. Sotomayor, E. Tajkhorshid, and K. Schulten. Molecular dynamics simulations of membrane channels and transporters. *Curr. Opin. Struct. Biol.*, 19:128–137, 2009.

40. Y. Wang, S.A. Shaikh, and E. Tajkhorshid. Exploring transmembrane diffusion pathways with molecular dynamics. *Physiology*, 25:142–154, 2010.

41. P. Agre. The aquaporin water channels. *Proc. Am. Thorac. Soc.*, 3:5–13, 2006.

42. S.J. Marrink and H.J.C. Berendsen. Permeation process of small molecules across lipid membranes studied by molecular dynamics simulations. *J. Phys. Chem.*, 100:16729–16738, 1996.

43. M. Venturoli, M.M. Sperotto, M. Kranenburg, and B. Smit. Mesoscopic models of biological membranes. *Phys. Rep.*, 437:1–54, 2006.

44. G.A. Both. *Coarse-Graining of Condensed Phase and Bimolecular Systems*. CRC Press, Boca-Raton, FL, 2008.

45. Y.Z. Ohkubo, T.V. Pogorelov, M.J. Arcario, G. Christensen, and E. Tajkhorshid. Accelerating membrane insertion of peripheral proteins with a novel membrane mimetic model. *Biophys. J.*, 102:2130–2139, 2012.

46. M.R. Shirts, D.L. Mobley, and J.D. Chodera. Alchemical free energy calculations: Ready for prime time? *Annu. Rep. Comput. Chem.*, 3:41–59, 2007.

47. A. Pohorille, C. Jarzynski, and C. Chipot. Good practices in free-energy calculations. *J. Phys. Chem. B*, 114:10235–10253, 2010.

48. J. Wereszczynski and J.A. McCammon. Statistical mechanics and molecular dynamics in evaluating thermodynamic properties of biomolecular recognition. *Q. Rev. Biophys.*, 45(1):1–25, 2012.

49. S.J. Marrink, A.H. de Vries, and A.E. Mark. Coarse grained model for semiquantitative lipid simulations. *J. Phys. Chem. B*, 108:750–760, 2004.

50. S.J. Marrink, H.J. Risselada, S. Yefimov, D.P. Tieleman, and A.H. de Vries. The MARTINI force field: Coarse grained model for biomolecular simulations. *J. Phys. Chem. B*, 111:7812–7824, 2007.

51. L. Monticelli, S.K. Kandasamy, X. Periole, R.G. Larson, D.P. Tieleman, and S.J. Marrink. The MARTINI coarse grained force field: Extension to proteins. *J. Chem. Theory Comput.*, 4:819–834, 2008.

52. C.A. Lopez, A.J. Rzepiela, A.H. de Vries, L. Dijkhuizen, P.H. Hünenberger, and S.J. Marrink. Martini coarse-grained force field: Extension to carbohydrates. *J. Chem. Theory Comput.*, 5:3195–3210, 2009.

53. X. Periole and S.J. Marrink. The martini coarse-grained force field. In L. Monticelli and E. Salonen, eds., *Methods in Molecular Biology*, pp. 533–565. Springer, New York, 2013.

54. S.O. Yesylevskyy, L.V. Schäfer, D. Sengupta, and S.J. Marrink. Polarizable water model for the coarse-grained MARTINI force field. *PLoS Comput. Biol.*, 6:e1000810, 2010.

55. D. Hamelberg, J. Mongan, and J.A. McCammon. Accelerated molecular dynamics: A promising and efficient simulation method for biomolecules. *J. Chem. Phys.*, 120(24):11919–11929, June 2004.

56. Y. Wang and J.A. McCammon. Accelerated molecular dynamics: Theory, implementation and applications. *AIP Conf. Proc.*, 1456:165, 2012.

57. J. Wereszczynski and J.A. McCammon. Using selectively applied accelerated molecular dynamics to enhance free energy calculations. *J. Chem. Theory Comput.*, 6:3285–3292, 2010.

58. J. Wang, P. Cieplak, and P.A. Kollman. How well does a restrained electrostatic potential (RESP) model perform in calculating conformational energies of organic and biological molecules? *J. Comput. Chem.*, 21:1049–1074, 2000.

59. V. Hornak, R. Abel, A. Okur, B. Strockbine, A. Roitberg, and C. Simmerling. Comparison of multiple Amber force fields and development of improved protein backbone parameters. *Proteins*, 65:712–725, 2006.

60. A.D. MacKerell Jr., M. Feig, and C.L. Brooks III. Extending the treatment of backbone energetics in protein force fields: Limitations of gas-phase quantum mechanics in reproducing protein conformational distributions in molecular dynamics simulations. *J. Comput. Chem.*, 25:1400–1415, 2004.

61. C. Oostenbrink, A. Villa, A.E. Mark, and W.F. Van Gunsteren. A biomolecular force field based on the free enthalpy of hydration and solvation: The GROMOS force-field parameter sets 53A5 and 53A6. *J. Comput. Chem.*, 25:1656–1676, 2004.

62. W.L. Jorgensen, D.S. Maxwell, and J. Tirado-Rives. Development and testing of the OPLS all-atom force field on conformational energetics and properties of organic liquids. *J. Am. Chem. Soc.*, 118:11225–11236, 1996.

63. M.L.P. Price, D. Ostrovsky, and W. L. Jorgensen. Gas-phase and liquid-state properties of esters, nitriles, and nitro compounds with the OPLS-AA force field. *J. Comput. Chem.*, 22:1340–1352, 2001.

64. O. Berger, O. Edholm, and F. Jähnig. Molecular dynamics simulations of a fluid bilayer of dipalmitoylphosphatidylcholine at full hydration, constant pressure, and constant temperature. *Biophys. J.*, 72:2002–2013, 1997.

65. D. Poger, W.F. van Gunsteren, and A.E. Mark. A new force field for simulating phosphatidylcholine bilayers. *J. Comput. Chem.*, 31:1117–1125, 2010.

66. J.P.M. Jambeck and A.P. Lyubartsev. An extension and further validation of an all-atomistic force field for biological membranes. *J. Chem. Theory Comput.*, 8:2938–2948, 2012.

67. C.J. Dickson, L. Rosso, R.M. Betz, R.C. Walker, and I.R. Gould. Gafflipid: A general amber force field for the accurate molecular dynamics simulation of phospholipid. *Soft Mater.*, 8:9617–9627, 2012.

68. A.A. Skjevik, B.D. Madej, R.C. Walker, and K. Teigen. LIPID11: A modular framework for lipid simulations using amber. *J. Phys. Chem. B*, 116:11124–11136, 2012.

69. T.J. Piggot, A. Pineiro, and S. Khalid. Molecular dynamics simulations of phosphatidylcholine membranes: A comparative force field study. *J. Chem. Theory Comput.*, 8:4593–4609, 2012.

70. S.-W. Chiu, S.A. Pandit, H.L. Scott, and E. Jakobsson. An improved united atom force field for simulation of mixed lipid bilayers. *J. Phys. Chem. B*, 113:2748–2763, 2009.

71. A. Kukol. Lipid models for united-atom molecular dynamics simulations of proteins. *J. Chem. Theory Comput.*, 5:615–626, 2009.

72. K. Matsuzaki. Control of cell selectivity of antimicrobial peptides. *Biochim. Biophys. Acta*, 1788:1687–1692, 2009.

73. L.T. Nguyen, E.F. Haney, and H.J. Vogel. The expanding scope of antimicrobial peptide structures and their modes of action. *Trends Biotechnol.*, 29:464–472, 2011.

74. K.A. Brogden. Antimicrobial peptides: Pore formers or metabolic inhibitors in bacteria? *Nat. Rev. Mol. Cell Biol.*, 3:238–250, 2005.

75. S. Qian, L. Yang, W. Wang, and H.W. Huang. Structure of transmembrane pore induced by bax-derived peptide: Evidence for lipidic pores. *Proc. Natl. Acad. Sci. USA*, 105:17379–17383, 2008.

76. D.P. Tieleman and M.S.P. Sansom. Molecular dynamics simulations of antimicrobial peptides: From membrane binding to transmembrane channels. *Int. J. Quantum Chem.*, 83:166–179, 2001.

77. S.K. Kandasamy and R.G. Larson. Effect of salt on the interactions of antimicrobial peptides with zwitterionic lipid bilayers. *Biochim. Biophys. Acta Biomembr.*, 1758(9):1274–1284, 2006.

78. J.C. Hsu and C.M. Yip. Molecular dynamics simulations of indolicidin association with model lipid bilayers. *Biophys. J.*, 92:L100–L102, 2007.

79. H.D. Herce and A.E. Garcia. Molecular dynamics simulations suggest a mechanism for translocation of the hiv-1 tat peptide across lipid membranes. *Proc. Natl. Acad. Sci. USA*, 104:20805–20810, 2007.

80. H. Khandelia, J.H. Ipsen, and O.G. Mouritsen. Impact of peptides on lipid membranes. *Biochim. Biophys. Acta Biomembr.*, 1778:1528–1536, 2008.

81. J. Pan, D.P. Tieleman, J.F. Nagle, N. Kucerka, and S. Tristram-Nagle. Alamethicin in lipid bilayers: Combined use of x-ray scattering and MD simulations. *Biochim. Biophys. Acta Biomembr.*, 1788:1387–1397, 2009.

82. C.M. Dunkin, A. Pokorny, P.F. Almeida, and H.S. Lee. Molecular dynamics studies of transportan 10 (tp10) interacting with a POPC lipid bilayer. *J. Phys. Chem. B*, 115:1188–1198, 2011.

83. H. Leontiadou, A.E. Mark, and S.J. Marrink. Antimicrobial peptides in action. *J. Am. Chem. Soc.*, 128:12156–12161, 2006.

84. D. Sengupta, H. Leontiadou, A.E. Mark, and S.J. Marrink. Toroidal pores formed by antimicrobial peptides show significant disorder. *Biochim. Biophys. Acta*, 1778:2308–2317, 2008.

85. L. Thøgersen, B. Schiøtt, T. Vosegaard, N.C. Nielsen, and E. Tajkhorshid. Peptide aggregation and pore formation in a lipid bilayer: A combined coarse-grained and all atom molecular dynamics study. *Biophys. J.*, 95(9):4337–4347, 2008.

86. D.L. Parton, E.V. Akhmatskaya, and M.S.P. Sansom. Multiscale simulations of the antimicrobial peptide maculatin 1.1: Water permeation through disordered aggregates. *J. Phys. Chem. B*, 116:8584–8593, 2012.

87. F. Jean-François, J. Elezgaray, P. Berson, P. Vacher, and E.J. Dufourc. Pore formation induced by an antimicrobial peptide: Electrostatic effects. *Biophys. J.*, 95:5748–5756, 2008.

88. H. Leontiadou, A.E. Mark, and S.J. Marrink. Ion transport across transmembrane pores. *Biophys. J.*, 92:4209–4215, 2007.

89. D.E. Shaw, M.M. Deneroff, R.O. Dror, J.S. Kuskin, R.H. Larson, J.K. Salmon, C. Young et al. Anton, a special-purpose machine for molecular dynamics simulation. *Commun. ACM*, 51(7):91–97, 2008.

90. R.O. Dror, R.M. Dirks, J.P. Grossman, H. Xu, and D.E. Shaw. Biomolecular simulation: A computational microscope for molecular biology. *Annu. Rev. Biophys.*, 41:429–452, 2012.

91. R.O. Dror, A.C. Pan, D.H. Arlow, D.W. Borhani, P. Maragakis, Y. Shan, H. Xu, and D.E. Shaw. Pathway and mechanism of drug binding to g-protein-coupled receptors. *Proc. Natl. Acad. Sci. USA*, 108:13118–13123, 2011.

Biomolecular Interactions

Chapter 8

Protein and Nucleic Acid Interactions with Molecular Dynamics Simulations

Zhen Xia and Ruhong Zhou

CONTENTS

8.1   INTRODUCTION

The discovery of protein–nucleic acid interactions can be traced back to the nineteenth century, when the association of protein–DNA strands was first observed by scientists with the assistance of microscopes. Since then, proteins interacting with nucleic acids have been demonstrated to play a central role in a wide range of fundamental biological processes, including gene regulation of transcription, translation; DNA replication, repair and recombination; as well as RNA processing and translocation [1,2]. During the last decade, more than a thousand high-resolution structures of protein–nucleic acid complexes have been determined to advance our understanding of their biological functions and molecular mechanisms in atomistic details (Protein Data Bank; www.pdb.org) [3].

Proteins and nucleic acids can bind in different ways. In general, it can be divided into the sequence-specific binding and non-sequence-specific binding [4,5]. The nucleic acid binding regions of proteins are always located in the conserved domains, where multiple DNA- or RNA-binding domains (DBDs or RBDs) are found within their tertiary structures. Therefore, the identity of the individual domains and their relative arrangement are functionally important for the protein–nucleic acid binding. Several common DBDs have been discovered, including zinc finger [6], helix-turn-helix [7], helix-loop-helix [8], winged helix [9], and leucine zipper [10]. RNA-binding specificity and function are determined by the zinc finger [6], K homology [11], S1 [12], PAZ [13], PUF [14], PIWI [15], and RNA recognition motif (RRM) domains [16–18]. The binding affinity can be further increased through protein oligomerization or multidomain protein complex.

For certain target nucleic acid sequences, multiple nucleic acid binding domains can increase the specificity and affinity of the protein, mediate a conformational change in the target nucleic acid, properly position other nucleic acid sequences for recognition, and regulate the activity of enzymatic domains within the binding protein. In RNAs, their particular dynamic secondary and tertiary structures are extremely important for protein recognition and sequence-specific binding [19].

Although the details of interactions between protein and nucleic acid can vary widely by their binding sites, the general principles remain similar by several physical forces, such as base stacking (dispersion forces), hydrogen bonding (dipolar interactions), salt bridges (electrostatic interactions), and hydrophobic interactions. These noncovalent binding forces are relatively weak and the overall binding affinity is the summation of many interactions. Because DNA and RNA are highly charged polymers, perhaps the most obvious force is the electrostatic interactions, in which positively charged amino acids (e.g., lysine and arginine) are usually rich in their binding interface. Other polar interactions like hydrogen bonds are also widespread and always of importance to the sequence specificity. In some complexes, the hydrophobic stacking force is unexpectedly prevalent between the protein and nucleic acid bases.

There have been an increasing number of studies of protein–nucleic acid complexes that address the dynamics of protein–nucleic acid interactions computationally and experimentally over recent years [20]. Collaborations between experimental and computational researchers have been increasingly critical because they focus on the dynamics of protein–nucleic acid systems at different time and length scales. While experimental methods can capture the functional changes and dynamics in a macroscopic way, very few atomic details can be observed. Computational modeling serves as a powerful tool to remediate this problem. Techniques like molecular dynamics (MD) simulation can be used to investigate complex allostery, conformational change, induced fit, and the role of solvent molecules and ions at the interface of the protein–nucleic acid complex. By sharing complementary information with each other, experimental and computational researchers are able to address the biological significance with more evidence. In this chapter, we summarize the recent advances in protein–nucleic acid recognition and interaction using computational approaches. We start with the basic simulation methods and the progress of atomic force field development for the protein and nucleic acid complex and then introduce the studies of several important RNA–protein (and DNA–protein) complex systems using atomistic MD simulations. The molecular mechanisms of ribosome in RNA–protein interactions are particularly discussed. We conclude with remaining challenges and possible future directions for protein–nucleic acid simulations.

8.2 FORCE FIELDS

In molecular mechanics approaches, interactions among atoms are described by empirical potential energy functions that are often referred to as force fields. The classical molecular mechanics models treat atoms as rigid particles and typically utilize fixed atomic charges, point dispersion–repulsion, and simple empirical functions for valence interactions. Over the past two decades, many classical force fields, such as Amber [21], CFF [22], CHARMM [23], GROMOS [24], MM3 [25], and OPLS-AA [26], have been developed primarily for atomistic simulations of macromolecules. These developed force fields for protein simulations have gone through comprehensive refinements and validations, which have shown great success in a wide range of biological systems. An example of potential energy function in CHARMM force field is represented as

$$U = \sum_{bonds} k_b (b-b_0)^2 + \sum_{UB} k_{UB} (S-S_0)^2 + \sum_{angle} k_a (\theta-\theta_0)^2 + \sum_{dihedrals} k_\chi (1+\cos(n\chi-\chi_0))$$

$$+ \sum_{impropers} k_{imp} (\varphi-\varphi_0)^2 + \sum_{Vdw} \varepsilon_{ij} \left[\left(\frac{R_{ij}}{r_{ij}} \right)^{12} - \left(\frac{R_{ij}}{r_{ij}} \right)^{6} \right] + \frac{q_i q_j}{\varepsilon r_{ij}}$$

where
U is the totally potential energy
k_b is the bond force constant
b_0 is the equilibrium distance
k_{UB} is the Urey–Bradley force constant
S_0 is the equilibrium distance
k_a is the valence angle force constant
θ_0 is the equilibrium angle
k_χ is the dihedral force constant
n and χ_0 are the multiplicity and the phase angle, respectively
k_{imp} is the improper force constant
φ_0 is the equilibrium improper angle

The van der Waals (Vdw) interaction is represented as the Lennard-Jones function, where ε_{ij} is the well depth and R_{ij} is the minimum interaction radius; the electrostatic interaction is represented as Coulomb's law, where ε is the dielectric constant and the q is the partial charge for each atom.

The force field development for nucleic acids has gotten much more attention in the past decade, mainly due to the great discovery of noncoding RNAs in many crucial functions in gene expression and cell regulation [27–37]. Different from protein, nucleic acids are highly charged polymers with typical nonglobular shapes, whose structures highly rely on the concentration of salt. Therefore, more careful treatment to the solvent and electrostatic interactions is necessary during the parameterizations. Recently, Orozco and coworkers have improved the AMBER force field for nucleic acids by refining the representation of α/γ concerted rotation in the backbone [38], while Mackerell and coworkers have updated the CHARMM force field for RNA simulations by improving the conformational sampling of 2′-hydroxyl orientation in the ribose [39]. Other force fields may also potentially be used for nucleic acids simulations, although the number of studies is still limited.

8.3 RNA–PROTEIN INTERACTION

Overall, RNA has more flexible structure than DNA. Even for RNA itself, it can form complex secondary and tertiary structures, and the biological function highly relies on the dynamics and its interaction with the protein [40–49]. RNA–protein interactions are involved in almost every step in gene expression, such as mRNA splicing, protection, translation, and degradation.

8.3.1 mRNA Regulations

RNA–protein complexes play a pivotal role in the cellular machinery in passing genetic information, from transcription and splicing to translation of the mature mRNA. To form a mature mRNA in eukaryotes, a number of edits are processed from the pre-mRNA, such as the intron removal and the splicing of the coding regions. A number of simulation studies have been performed on the U1A RNA–protein complex, a well-characterized protein containing an RBD. Reyes and Kollman first performed MD simulations to probe the molecular basis of U1A protein binding to the RNA hairpin and its internal loop [50]. Meanwhile, Tang and Nilsson examined the sequence-specific interaction of the human U1A protein with the hairpin II of U1 snRNA in solution, a common sequence-specific interaction feature that was proposed for two RNA substrates bound to the U1A protein [51]. The influence of cations in the U1A RNA–protein recognition was further revealed by simulations under both low and high ionic strength in solution [52]. The high-concentration cations around the RNA were found to compete with the protein for binding sites and specifically destabilize the residues at the RNA–protein interface [52]. More recently, Baranger and coworkers investigated the contribution of highly conserved aromatic amino acids to RNA binding of the U1A protein [53]. Beveridge and coworkers calculated the positional cross-correlations of atomic fluctuations that contribute to the cooperativity and corresponding observed thermodynamic coupling [54,55]. Similar correlated motions in the U1 snRNA stem–loop and U1A protein complex were found by Showalter and Hall [56,57].

8.3.2 MicroRNA–Protein Interaction

The discovery of many new noncoding RNAs has led to a paradigm shift in molecular biology in the past decade. MicroRNAs (miRNAs) are one of the major small noncoding RNA molecules found in plants and animals, which function in transcriptional and posttranscriptional regulations of gene expression [58,59]. More than 20,000 miRNAs have been identified in over 193 species since 2012 (The microRNA Database; www.mirbase.org, release 19).

A series of enzymes (proteins) have been found during the miRNA-related gene regulation, such as DROSHA, Exportin 5 (XPO5), Dicer endonuclease, and Argonaute (Ago) [60,61]. MiRNAs have been shown to be involved in many fundamental processes in gene regulation [62–70].

The Ago protein is a critical component of the RNA-induced silencing complexes (RISC) that directly mediate the recognition and cleavage of the target mRNA [71–74]. Wang et al. examined the global dynamics of Ago protein by MD simulations, where the miRNA and the target mRNA were docked to the Ago protein and simulated based on the experimental x-ray structures [75]. Correlated domain (PAZ, Mid, and PIWI domains in Ago) movements were considered crucial to the interaction of Ago–RNA, and the presence of two Mg^{2+} ions was found to be important to the catalytic cleavage activity of the Ago complex [75].

In the Ago–RNA complex, the complete Watson–Crick pairing at position 2-7 of an miRNA's 5′ region (also called the *seed*) has been shown to play a central role in the recognition of mRNA target [61,76]. However, a number of target mRNAs with noncomplete Watson–Crick pairing at the *seed* were found in different miRNA-target identifications [77–82], and these nonconserved targets may be even more widespread than the conserved targets (complete Watson–Crick pairs on seed) [83–87].

Zhou and coworkers performed all-atom MD simulations of the Ago–miRNA–mRNA ternary complexes by introducing the G·U wobbles, mismatches, bulges, and combinations at various positions of the miRNA *seed* region (Figure 8.1) [88]. Simulations have shown only minor fluctuations on the Ago complexes with multiple G:C base pairs replaced by the G·U wobbles in the *seed*. Bulge insertions only caused slightly larger distortions on both guide miRNA and target mRNA sides. It is also interesting to see the combination of multiple G·U wobbles and a single bulge led to stable structures as well. These findings indicate that target sequences with imperfect matches at the seed are more diverse than the guide seed might suggest [88], which was consistent with several distinct publicly available high-throughput sequencing of RNAs isolated by cross-linking immune precipitation (HITS-CLIP) datasets [77].

In the same study, the ruinous G to C mutations with 1–4 C·C mismatches further revealed the inner dynamics of the Ago–miRNA–mRNA complex [88]. In the *damaging* four-position C·C mismatch mutant, the hydrogen bonds between the nucleic acid duplex and L1/L2 segments of the Ago protein are broken, introducing a bending motion of the PAZ domain along the L1/L2 hinge-like link. The detailed motions of the PAZ domain in the C·C mismatch mutant were further revisited by principle component analysis, resulting in an opening of the nucleic acid binding channel in the Ago–RNA complex. This finding was consistent with the experimental determined binary and ternary structures [89–91]. Furthermore, similar domain motions were found in MD

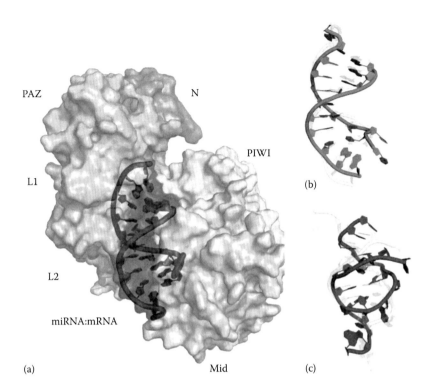

Figure 8.1 (a) Structural views of the Ago–miRNA–mRNA complex (PDB entry:3F73). The Ago protein is rendered as cartoon and molecular surface. The miRNA–mRNA heteroduplex is presented as cartoon. The structure of the guide miRNA and target mRNA duplex for the wild type (b) and the mutant (c) during the 100 ns MD simulation. The conformational change is shown by superimposing the final snapshot to the starting native structure.

simulations of the Ago–DNA–mRNA complex [92] and other independent simulation study [75]. Additional simulations with extra 3′-compensatory pairing of the longer miRNA showed a minor contribution to the stability of the complex in the mutants, which was again in good agreement with a recent finding that the complementary base pairing beyond the seed region was not relevant for the repression of the *cog-1* 3′UTR and other *Caenorhabditis elegans* 3′UTRs by the *lsy-6* miRNA [93,94].

8.3.3 Ribosome

Ribosome is one of the largest RNA–protein complexes in cells. It is a molecular machine that serves as the primary site of biological protein synthesis.

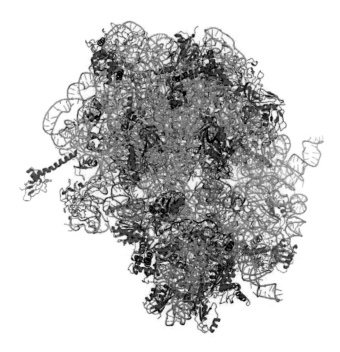

Figure 8.2 Illustrations of the structure of the 70S ribosome (*Thermus thermophilus*) complexed with mRNA, tRNA, and paromomycin (PDB entry: 2J00 and 2J01). The entire ribosome is represented as cartoon.

The size of the simulation system could reach as large as three million atoms for an intact ribosome with explicit waters (Figure 8.2). A set of large-scale MD simulations were performed to investigate the dynamic structures of the ribosome after high-resolution x-ray crystal structures of individual subunits were released [95]. Because of the large size of ribosome, trade-offs with various approximations were made between the level of molecular details and the simulation times. For example, in the early studies, the normal mode analysis was applied to determine the global motions of the ribosome [96–98]. Wriggers and coworkers first utilized normal mode analysis to reproduce the 30S rotation and stalk movement relative to 50S in the ribosomal complex [96].

The first explicit solvent MD simulation with intact ribosome was performed by Sanbonmatsu et al. to predict the regions of the large subunit that interact with the tRNA during this conformational change (also known as *accommodation corridor*) [99]. The large-scale conformational change was successfully predicted during the accommodation of tRNA from its partially bound state

(A/T) to its fully bound state (A/A). The in silico prediction was later verified by several recent experimental studies [100–104]. More recent simulations further studied the interactions of the ribosome exit tunnel with the initiator and elongator tRNAs in the large subunit of the ribosome [105–107]. The valuable results also assisted the production of all-atom fitting of the cryo-electron microscopy (cryo-EM) reconstructions. Microsecond sampling of the intact ribosome in explicit solvent was performed to calculate the diffusion coefficients of tRNA within the ribosome, aiming at connecting experimentally measured kinetic rates to free-energy barrier heights [108].

Because of the large size of the intact ribosome, some computational studies focus on critical functional regions of the ribosome in protein synthesis. Sanbonmatsu and Joseph examined the stability and dynamics of the hydrogen bond network at the decoding center with mRNA and tRNA anticodon stem–loops. It was interesting to find that ribosomal RNA has different effects to the stability of cognate codon–anticodon interactions with certain noncognate combinations [109]. The corresponding binding free energies of the codon–anticodon interactions were revisited by Aqvist and coworkers with a longer sampling time on different anticodon stem–loops [110,111]. The role of multihelix junctions in functional fluctuations of the ribosome has been examined by Sponer and coworkers [112]. MD simulations indicated that three-way junctions acts as the anisotropic flexible elements contributing to the dynamic of the ribosome in 5S rRNA, the A-site finger, and L7/L12 stalk region. In addition, Sponer and coworkers performed detailed studies at the large ribosome subunit to show the central role of the elbow-shaped kink-turn in the motions of helix 38 [113].

The RNA–protein interaction in the A-site of decoding center is also a popular targeting site for aminoglycosidic antibiotics, which fundamentally affects the fidelity of protein translation in bacteria. The drug–ribosome interactions were investigated by several MD simulations [114–117]. Vaiana et al. examined the conformational and hydration patterns by MD simulations of the aminoglycoside paromomycin bound to a eubacteria ribosomal decoding A-site with various mutations [116]. Later, Vaiana and Sanbonmatsu utilized advanced sampling method (e.g., replica exchange MD simulations) with microsecond simulations to show the decoding center bases (A1492 and A1493) flipping in and out of helix 44 in the absence of ligand gentamicin [114]. Overall, computational methods have made significant progress to bridge the static structures and dynamics of the ribosome.

8.3.4 Virus RNA–Protein Interaction

MD simulations have been applied to study the RNA–protein complex in virus assembly and virus–host interactions [118,119]. With the fast-growing high-performance computing, the current supercomputer could extend the simulation systems as large as millions of atoms or the simulation time as long as micro- to milliseconds [99,120]. A complete satellite tobacco mosaic virus was first simulated at an all-atom scale by Schulten and coworkers. With up to one million atoms and over 50 ns total simulation time [119], the authors indicated that the full virion and RNA genome are dynamically stable, but different components in the capsid showed a rich pattern of behaviors. Surprisingly, the simulated virus quickly becomes unstable and implodes in the absence of RNA, indicating the RNA could be of importance to the assembly of the entire capsid.

Large-scale simulations could provide atomic insights into the interactions and coevolution relationship between RNA virus proteins and their hosts. Zhou and coworkers explored the recognition and interaction mechanism between a virus protein p19 and a 21 nt short interfering RNA (siRNA) (Figure 8.3) [118]. The p19 protein is originally encoded from *Tombusvirus* that can bind to small RNAs to suppress RNA silencing in its host [121–125]. Later, p19 quickly becomes a useful tool to dissect the RNAi pathway in vivo or in vitro because of its ability to bind small RNAs in both plant and mammalian cells [36,126,127]. Regular MD simulation indicated that the p19–RNA complex becomes less stable when two key interacting residues are mutated. The detailed structural analysis showed that the hydrophobic and stacking interactions between amino acid tryptophan and the ribose bases are crucial for the binding, in good agreement with experimental data [128]. The steered MD and rigorous free-energy perturbation (FEP) simulations were then performed to quantitatively measure the binding affinity between the p19 protein and siRNA, which further confirmed that the strong stacking interaction is important to the p19–siRNA recognition [118].

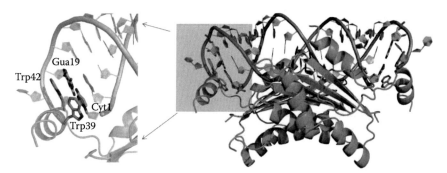

Figure 8.3 The structure of p19-siRNA complex (PDB entry: 1RPU). The siRNA and the p19 protein are rendered as cartoon. Residues Trp39 and Trp42 (sticks) in the N-terminal subdomain with the interaction of nucleic acids Gua19 and Cyt1 (sticks) in siRNA.

8.4 DNA–PROTEIN INTERACTION

DNA–protein interaction and recognition is crucial for many biological processes, including transcriptional regulation, DNA replication, and DNA damage repair. One of the pioneer works was done by Nilsson and coworkers in 1995 [129], where all-atom MD simulations were utilized to study the dynamic structures and interactions between the double-stranded DNA and dimer of the glucocorticoid receptor DNA-binding domain (GR DBD) complex. During the simulation, an increased bending of DNA helix axis was found at two half-sites responsible for the DNA–protein complex recognition. The bend of DNA actually facilitated the DNA–protein binding with increased number of interactions between the DNA and sequence specifically in the N-terminal region. A set of highly conserved residues within the family of nuclear receptors formed strong hydrogen-bonding networks to the DNA major groove, suggesting a common feature for the DBD recognition DNA helix.

Protein 53 (p53) is a cell cycle regulation protein that acts as the tumor suppressor and associates with many cancers [130–134]. Sequence-specific DNA and p53 binding can activate the expression of several genes including miRNAs, WAF1/CIP1 (p21), and hundreds of other downstream genes [135]. However, a mutant p53 will largely reduce the binding to the target DNA and therefore disrupt the signaling of cell division [136,137]. Pan and Nussinov have studied the molecular mechanism of p53-induced DNA-specific interaction by MD simulations [138]. In the simulation, the p53 dimers made slightly conformational change to adopt favorable interactions and packing at the p53 dimer–dimer interface resulting in the DNA bending. The role of cations, Zn^{2+}, in p53 and DNA recognition was investigated by Duan and Nilsson [139]. Their simulations indicated that the presence of Zn^{2+} allows a positive-charged residue Arg248 in the loop L3 of p53 to be inserted into the minor groove for specific contact with the DNA base and prevented the aggregation of p53 [139]. In addition, the possible binding modes of the p53–DNA complex were fully studied with MD simulations by Ma and Levine [140]. It was interesting to find that not all the monomers in p53 tetramer are required to recognize DNA sequence and induce DNA bending. With an arrangement of dimer, p53 can both recognize supercoiled DNA sequence and Holliday junction geometry specifically [140].

The role of water is considered important in DNA and protein binding [141–143]. The release of DNA–protein interfacial water molecules could induce the nonspecific, low-affinity complex to form a specific, high-affinity complex [144,145]. Fuxreiter et al. proposed that interfacial waters can serve as a *hydration fingerprint* of a given DNA sequence when they characterized the structure and energetics of water in the *Bam*HI complex with grand canonical Monte Carlo simulations [146]. They found that variations in water distributions could actually control the release number of waters from a given sequence during the transformation from the loose to the tight complex.

Quantitative measurement of the binding affinity could be of importance to evaluate the interaction between DNA and protein. One of the effective ways is to combine molecular mechanics and Poisson–Boltzmann surface area (MM/PBSA) or generalized Born surface area (MM/GBSA) approach to evaluate the binding free energy [147–150], in which the binding free energy between DNA and protein can be calculated via free-energy component analysis. The MM/GBSA method has been applied to study the cooperative binding of DNA and core binding β (CBFβ) to the Runt domain of the CBFα by the MacKerell group [151]. Contributions to the cooperative binding can be specifically targeted to different regions of the Runt domain protein and individual residues by energetic and structural analysis. A similar MM/PBSA method was utilized to explore structural changes of the TATA-box-binding protein complex and interpret the stereochemical effect that frames the intercalated adduct [152]. It should be noted that both MM/GBSA and MM/PBSA methods could lead to a noticeable large fluctuation in the calculation of free energies, implying current limitations of free-energy calculations for large and highly charged systems (e.g., DNA–protein complexes).

Advanced sampling methods have further been implemented in MD simulations for reconstructing the free-energy landscape in DNA–protein complex systems [153–159]. This so-called metadynamics could overcome the large energy barriers between conformational states by adding positive Gaussian potential to the real energy landscape of the system. The sum of Gaussians is then exploited to reconstruct iteratively an estimator of the free energy and force the system to escape from the local minima. Huang and MacKerell utilized MD-based potential of mean force (PMF) calculations to show that protein M.*Hha*I actively facilitates base flipping via the major groove of the DNA [160,161], where free-energy barriers to flipping are significantly higher in noncognate versus the cognate sequence [162]. Similarly, free-energy calculations in DNA repair proteins with 8-oxoguanine (8-oxoG) complex have shown that both attractive and repulsive interactions are important for the preferential binding of oxoG to the active site [163]. Later, the transition path sampling method was used to study the conformational changes, related energetics, and pathways in DNA polymerase beta with 8-oxoG template and other nucleotides [164].

8.5 INTEGRATING EXPERIMENTAL DATA WITH MOLECULAR DYNAMICS SIMULATIONS

In recent years, computational modeling no longer limits its applications to structures from X-ray crystal or nuclear magnetic resonance (NMR) alone. Novel algorithms were developed to combine computational modeling and low-resolution experimental techniques such as cryo-EM (single-particle cryo-EM) and small-angle x-ray scattering (SAXS). X-ray crystallography and NMR spectroscopy are able to give high-resolution structures at atomic level, but both require purity and stability for protein–nucleic acid systems [165]. Compared with x-ray and NMR, cryo-EM can handle much larger and more flexible assemblies. Cryo-EM provides information of macrobiomolecule assemblies at different functional states but at a low to medium resolution [166]. Another widely used technique for structure characterization is SAXS. SAXS can provide low-resolution (~1–3 nm) information about the shape of a macromolecule [167]. By comparing or fitting atomic structures to low-resolution data, computational methods can help interpret low-resolution data and potentially provide insights into macromolecule functionalities. Because protein–nucleic acid interactions typically involve very high interdomain fluctuations, the hybrid methods can also facilitate the target system to find its functional state.

8.5.1 Combining Electron Microscopy and Molecular Dynamics Simulation

EM single-particle reconstruction has been a central structure determination technique for studying macromolecular assemblies [168]. In recent years, computational methods have been developed to fit atomic structures into EM maps, which allow the further investigation at both atomic and low-resolution levels. Common fitting approaches include rigid-body docking and flexible fitting [169]. Both methods are developed to find the best position and orientation of a given molecule inside the EM electron density, while flexible fitting allows conformational changes of the atomic structure.

Schulten and coworkers resort to molecular dynamics flexible fitting (MDFF) method for 3D EM reconstruction [165]. The MDFF method is an extension of MD simulation technique. In addition to the empirical forces calculated based on force field potential, extra forces proportional to the gradient of the EM density map are applied in MD simulations, driving the atomic structure into the high-density regions and therefore correctly capture the functional state from the experimental data [168]. The empirical force field used in MD simulations ensures the reasonable physical interactions between the particles (Figure 8.4).

In the work of Querol-Audi et al., 3D EM reconstruction was used as an effective tool to study the interaction difference between flap endonuclease 1 (FEN1), DNA and two processivity clamps, proliferating cell nuclear antigen (PCNA), and the checkpoint sliding clamp Rad9/Rad1/Hus1 (9-1-1), respectively [170]. The cluster centroids from MD simulations were used for the 3D EM reconstruction of the ternary complexes at resolution of 18 Å. The combined

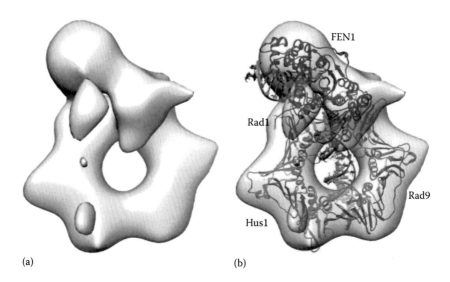

(a)   (b)

Figure 8.4 (a) Electron density map of the 9-1-1/FEN1/DNA complex. (b) MDFF flexible fitting of the 9-1-1/FEN1/DNA complex into a cryo-EM map. (From Dr. Ivanov's lab, Georgia State University, Atlanta, GA.)

analysis based on 2D EM and 3D reconstructed structures showed different interactions between DNA/FEN1 and 9-1-1 than PCNA. In 9-1-1/DNA/FEN1 ternary complex, the DNA and FEN1 were tilted and showed interactions with all three 9-1-1 subunits, which stabilized the DNA passing through the ring for repair purpose. On the contrary, for the PCNA/DNA/FEN1 complex, the more symmetrical PCNA ring showed a biased DNA interaction toward one side of the ring, which was linked to processive and mobile sliding for the DNA replication. For both modules, FEN1 was flexibly tethered to 9-1-1 or PCNA and could swing to the upright position to the 9-1-1 or PCNA to fulfill its function.

8.5.2 Integrating Small-Angle X-Ray Scattering and Molecular Dynamics Simulation

SAXS measures the rotationally averaged scattering intensity of different special frequencies, $I(q)$, which can be converted to a pairwise electron distance distribution function $P(r)$ [171]. Due to the rotational averaging nature, SAXS gives less structural information than x-ray, NMR, and EM. However, SAXS has a wide application range and fast processing time on macromolecule assemblies, allowing high-throughput analyses on a large number of samples [171]. The SAXS data can be used to evaluate the manually built or simulated structures. For MD simulations of large protein–nucleic acid complex, the potential energy landscape could be extremely rugged. However, the SAXS profiles can be computed for intermediate structures via foXS method [172] and compared with experiment, which helps choose the close-to-native structure for further studies. The profile is computed using the Debye formula for spherical scatterers. Moreover, SAXS data can be used for structure prediction through ab initio methods or rigid-body modeling (if the atomic structures of the subunits are known).

An example is the study on the replication protein A (RPA)'s DNA-binding activities, by Brosey et al. [173]. RPA is a modular multidomain protein. Although the crystal structure is available for each subunit, x-ray crystallography on the full-length RPA is very difficult because of the high interdomain flexibility [174]. In this work, SAXS and neutron scattering were combined with all-atom MD simulations to investigate the architecture of RPA's DNA-binding core (RPA-DBC). Molecular models were constructed for DNA-free RPA-DBC and RPA-DBC with DNA at different lengths (10, 20, 30 nt). As an evaluation of the simulations, the computed SAXS profiles were compared with the experiment, and the atomic structures were compared with the ab initio molecular envelope constructed based on the scattering data [175].

The structural analysis suggests that RPA becomes more compact when binding to single-stranded DNA (ssDNA). While DNA-free RPA shows interdomain mobility, RPA tends to be less dynamic on binding ssDNA. Also in this work, only two states (initial and final) were observed for RPA–ssDNA binding, which contrasts the conventionally proposed three transition states (initial, intermediate, and final). Overall, the SAXS/MD and crystallographic studies provide highly complementary information for the exploration of structural details.

8.6 CONCLUSIONS

An increasing number of MD simulations have advanced the understanding of protein–nucleic acid recognition and interaction at atomic details in the past decade. It is more complicated to model the protein–nucleic acid complexes than to model protein alone, because the force field for protein–nucleic acid complex needs to be properly balanced and the electrostatic interaction including solvent and ions should be carefully optimized. One of the possible future directions is to develop force fields with more sophisticated representation of electrostatic interaction (e.g., higher order of moments, explicitly treat the polarization effect). On the other hand, advanced sampling methods such as metadynamics could help overcome the limitation of simulation time and large energy barriers between conformational states. MD simulation combined with free-energy calculation method (e.g., FEP) could further quantitatively measure the binding affinities between protein and nucleic acid. Finally, it is important to bring both experimental and computational techniques together to explore the dynamics of both nucleic acid and protein and address their biological significance in diverse ways.

ACKNOWLEDGMENT

The authors are grateful to the support provided by the IBM Blue Gene Program.

REFERENCES

1. Van Duyne, G. D. and W. Yang. 2008. Protein–nucleic acid complexes: Large, small, old, and new. *Curr Opin Struct Biol* 18:67–69.
2. Phillips, S. E. V. and K. Luger. 2012. Proteins for packaging, partitioning, processing, and proofing of nucleic acids. *Curr Opin Struct Biol* 22:62–64.
3. Berman, H. M., J. Westbrook, Z. Feng, G. Gilliland, T. N. Bhat, H. Weissig, I. N. Shindyalov, and P. E. Bourne. 2000. The Protein Data Bank. *Nucleic Acids Res* 28:235–242.
4. Doherty, A. J., L. C. Serpell, and C. P. Ponting. 1996. The helix-hairpin-helix DNA-binding motif: A structural basis for non-sequence-specific recognition of DNA. *Nucleic Acids Res* 24:2488–2497.
5. Steitz, T. A. 1990. Structural studies of protein nucleic-acid interaction—The sources of sequence-specific binding. *Q Rev Biophys* 23:205–280.
6. Klug, A. and D. Rhodes. 1987. Zinc fingers—A novel protein fold for nucleic-acid recognition. *Cold Spring Harb Symp Quant Biol* 52:473–482.
7. Brennan, R. G. and B. W. Matthews. 1989. The helix-turn-helix DNA-binding motif. *J Biol Chem* 264:1903–1906.
8. Murre, C., G. Bain, M. A. Vandijk, I. Engel, B. A. Furnari, M. E. Massari, J. R. Matthews, M. W. Quong, R. R. Rivera, and M. H. Stuiver. 1994. Structure and function of helix-loop-helix proteins. *BBA-Gene Struct Expression* 1218:129–135.
9. Gajiwala, K. S. and S. K. Burley. 2000. Winged helix proteins. *Curr Opin Struct Biol* 10:110–116.
10. Landschulz, W. H., P. F. Johnson, and S. L. Mcknight. 1988. The leucine zipper—A hypothetical structure common to a new class of DNA-binding proteins. *Science* 240:1759–1764.
11. Garcia-Mayoral, M. F., D. Hollingworth, L. Masino, I. Diaz-Moreno, G. Kelly, R. Gherzi, C. F. Chou, C. Y. Chen, and A. Ramos. 2007. The structure of the C-terminal KH domains of KSRP reveals a noncanonical motif important for mRNA degradation. *Structure* 15:485–498.
12. Bycroft, M., T. J. P. Hubbard, M. Proctor, S. M. V. Freund, and A. G. Murzin. 1997. The solution structure of the S1 RNA binding domain: A member of an ancient nucleic acid-binding fold. *Cell* 88:235–242.
13. Yan, K. S., S. Yan, A. Farooq, A. Han, L. Zeng, and M. M. Zhou. 2003. Structure and conserved RNA binding of the PAZ domain. *Nature* 426:469–474.

14. Zamore, P. D., J. R. Williamson, and R. Lehmann. 1997. The Pumilio protein binds RNA through a conserved domain that defines a new class of RNA-binding proteins. *RNA* 3:1421–1433.
15. Rivas, F. V., N. H. Tolia, J. J. Song, J. P. Aragon, J. D. Liu, G. J. Hannon, and L. Joshua-Tor. 2005. Purified Argonaute2 and an siRNA form recombinant human RISC. *Nat Struct Mol Biol* 12:340–349.
16. Dreyfuss, G., M. S. Swanson, and S. Pinolroma. 1988. Heterogeneous nuclear ribonucleoprotein-particles and the pathway of messenger-RNA formation. *Trends Biochem Sci* 13:86–91.
17. Chambers, J. C., D. Kenan, B. J. Martin, and J. D. Keene. 1988. Genomic structure and amino-acid sequence domains of the human La auto-antigen. *J Biol Chem* 263:18043–18051.
18. Sachs, A. B., R. W. Davis, and R. D. Kornberg. 1987. A single domain of yeast poly(a)-binding protein is necessary and sufficient for RNA-binding and cell viability. *Mol Cell Biol* 7:3268–3276.
19. Dirks, R. M., M. Lin, E. Winfree, and N. A. Pierce. 2004. Paradigms for computational nucleic acid design. *Nucleic Acids Res* 32:1392–1403.
20. MacKerell, A. D. and L. Nilsson. 2008. Molecular dynamics simulations of nucleic acid-protein complexes. *Curr Opin Struct Biol* 18:194–199.
21. Cornell, W. D., P. Cieplak, C. I. Bayly, I. R. Gould, K. M. Merz, D. M. Ferguson, D. C. Spellmeyer, T. Fox, J. W. Caldwell, and P. A. Kollman. 1995. A 2nd generation force-field for the simulation of proteins, nucleic-acids, and organic-molecules. *J Am Chem Soc* 117:5179–5197.
22. Maple, J. R., M. J. Hwang, T. P. Stockfisch, U. Dinur, M. Waldman, C. S. Ewig, and A. T. Hagler. 1994. Derivation of class-II force-fields.1. Methodology and quantum force-field for the alkyl functional-group and alkane molecules. *J Comput Chem* 15:162–182.
23. MacKerell, A. D., D. Bashford, M. Bellott, R. L. Dunbrack, J. D. Evanseck, M. J. Field, S. Fischer et al. 1998. All-atom empirical potential for molecular modeling and dynamics studies of proteins. *J Phys Chem B* 102:3586–3616.
24. Wang, D. Q., F. Freitag, Z. Gattin, H. Haberkern, B. Jaun, M. Siwko, R. Vyas, W. F. van Gunsteren, and J. Dolenc. 2012. Validation of the GROMOS 54A7 force field regarding mixed alpha/beta-peptide molecules. *Helv Chim Acta* 95:2562–2577.
25. Allinger, N. L., Y. H. Yuh, and J.-H. Lii. 1989. Molecular mechanics. The MM3 force field for hydrocarbons. *J Am Chem Soc* 111:8551–8566.
26. Jorgensen, W. L., D. S. Maxwell, and J. TiradoRives. 1996. Development and testing of the OPLS all-atom force field on conformational energetics and properties of organic liquids. *J Am Chem Soc* 118:11225–11236.
27. Winkler, W. C., S. Cohen-Chalamish, and R. R. Breaker. 2002. An mRNA structure that controls gene expression by binding FMN. *Proc Natl Acad Sci USA* 99:15908–15913.
28. Noller, H. F., V. Hoffarth, and L. Zimniak. 1992. Unusual resistance of peptidyl transferase to protein extraction procedures. *Science* 256:1416–1419.
29. Spizzo, R., M. S. Nicoloso, C. M. Croce, and G. A. Calin. 2009. SnapShot: MicroRNAs in cancer. *Cell* 137:586–586 e1.
30. Frohlich, K. S. and J. Vogel. 2009. Activation of gene expression by small RNA. *Curr Opin Microbiol* 12:674–682.

31. Georg, J., B. Voss, I. Scholz, J. Mitschke, A. Wilde, and W. R. Hess. 2009. Evidence for a major role of antisense RNAs in cyanobacterial gene regulation. *Mol Syst Biol* 5:305.

32. Khraiwesh, B., M. A. Arif, G. I. Seumel, S. Ossowski, D. Weigel, R. Reski, and W. Frank. 2010. Transcriptional control of gene expression by microRNAs. *Cell* 140:111–122.

33. Hale, C. R., P. Zhao, S. Olson, M. O. Duff, B. R. Graveley, L. Wells, R. M. Terns, and T. M. P. 2009. RNA-guided RNA cleavage by a CRISPR RNA-cas protein complex. *Cell* 139:945–956.

34. Marraffini, L. A. and E. J. Sontheimer. 2010. CRISPR interference: RNA-directed adaptive immunity in bacteria and archaea. *Nat Rev Genet* 11:181–190.

35. Hamilton, A. J. and D. C. Baulcombe. 1999. A species of small antisense RNA in post-transcriptional gene silencing in plants. *Science* 286:950–952.

36. Lecellier, C. H., P. Dunoyer, K. Arar, J. Lehmann-Che, S. Eyquem, C. Himber, A. Saib, and O. Voinnet. 2005. A cellular microRNA mediates antiviral defense in human cells. *Science* 308:557–560.

37. Buchon, N. and C. Vaury. 2006. RNAi: A defensive RNA-silencing against viruses and transposable elements. *Heredity* 96:195–202.

38. Perez, A., I. Marchan, D. Svozil, J. Sponer, T. E. Cheatham, C. A. Laughton, and M. Orozco. 2007. Refinement of the AMBER force field for nucleic acids: Improving the description of alpha/gamma conformers. *Biophys J* 92:3817–3829.

39. Denning, E. J., U. D. Priyakumar, L. Nilsson, and A. D. Mackerell. 2011. Impact of 2′-hydroxyl sampling on the conformational properties of RNA: Update of the CHARMM all-atom additive force field for RNA. *J Comput Chem* 32:1929–1943.

40. Ban, N., P. Nissen, J. Hansen, P. B. Moore, and T. A. Steitz. 2000. The complete atomic structure of the large ribosomal subunit at 2.4 A resolution. *Science* 289:905–920.

41. Wimberly, B. T., D. E. Brodersen, W. M. Clemons, Jr., R. J. Morgan-Warren, A. P. Carter, C. Vonrhein, T. Hartsch, and V. Ramakrishnan. 2000. Structure of the 30S ribosomal subunit. *Nature* 407:327–339.

42. Brodersen, D. E., W. M. Clemons, Jr., A. P. Carter, B. T. Wimberly, and V. Ramakrishnan. 2002. Crystal structure of the 30 S ribosomal subunit from *Thermus thermophilus*: Structure of the proteins and their interactions with 16 S RNA. *J Mol Biol* 316:725–768.

43. Kazantsev, A. V., A. A. Krivenko, D. J. Harrington, S. R. Holbrook, P. D. Adams, and N. R. Pace. 2005. Crystal structure of a bacterial ribonuclease P RNA. *Proc Natl Acad Sci USA* 102:13392–13397.

44. Torres-Larios, A., K. K. Swinger, A. S. Krasilnikov, T. Pan, and A. Mondragon. 2005. Crystal structure of the RNA component of bacterial ribonuclease P. *Nature* 437:584–587.

45. Serganov, A., L. Huang, and D. J. Patel. 2009. Coenzyme recognition and gene regulation by a flavin mononucleotide riboswitch. *Nature* 458:233–237.

46. Vidovic, I., S. Nottrott, K. Hartmuth, R. Luhrmann, and R. Ficner. 2000. Crystal structure of the spliceosomal 15.5kD protein bound to a U4 snRNA fragment. *Mol Cell* 6:1331–1342.

47. Klein, D. J. and A. R. Ferre-D'Amare. 2006. Structural basis of glmS ribozyme activation by glucosamine-6-phosphate. *Science* 313:1752–1756.

48. Serganov, A., L. Huang, and D. J. Patel. 2008. Structural insights into amino acid binding and gene control by a lysine riboswitch. *Nature* 455:1263–1267.

49. Hainzl, T., S. Huang, and A. E. Sauer-Eriksson. 2002. Structure of the SRP19 RNA complex and implications for signal recognition particle assembly. *Nature* 417:767–771.

50. Reyes, C. M. and P. A. Kollman. 1999. Molecular dynamics studies of U1A-RNA complexes. *RNA* 5:235–244.

51. Tang, Y. and L. Nilsson. 1999. Molecular dynamics simulations of the complex between human U1A protein and hairpin II of U1 small nuclear RNA and of free RNA in solution. *Biophys J* 77:1284–1305.

52. Hermann, T. and E. Westhof. 1999. Simulations of the dynamics at an RNA-protein interface. *Nat Struct Biol* 6:540–544.

53. Zhao, Y., B. L. Kormos, D. L. Beveridge, and A. M. Baranger. 2006. Molecular dynamics simulation studies of a protein–RNA complex with a selectively modified binding interface. *Biopolymers* 81:256–269.

54. Kormos, B. L., A. M. Baranger, and D. L. Beveridge. 2006. Do collective atomic fluctuations account for cooperative effects? Molecular dynamics studies of the U1A-RNA complex. *J Am Chem Soc* 128:8992–8993.

55. Kormos, B. L., A. M. Baranger, and D. L. Beveridge. 2007. A study of collective atomic fluctuations and cooperativity in the U1A-RNA complex based on molecular dynamics simulations. *J Struct Biol* 157:500–513.

56. Showalter, S. A. and K. B. Hall. 2005. Correlated motions in the U1 snRNA stem/loop 2:U1A RBD1 complex. *Biophys J* 89:2046–2058.

57. Showalter, S. A. and K. B. Hall. 2002. A functional role for correlated motion in the N-terminal RNA-binding domain of human U1A protein. *J Mol Biol* 322:533–542.

58. Chen, K. and N. Rajewsky. 2007. The evolution of gene regulation by transcription factors and microRNAs. *Nat Rev Genet* 8:93–103.

59. Kim, V. N., J. Han, and M. C. Siomi. 2009. Biogenesis of small RNAs in animals. *Nat Rev Mol Cell Biol* 10:126–139.

60. Bartel, D. P. 2004. MicroRNAs: Genomics, biogenesis, mechanism, and function. *Cell* 116:281–297.

61. Bartel, D. P. 2009. MicroRNAs: Target recognition and regulatory functions. *Cell* 136:215–233.

62. Reinhart, B. J., F. J. Slack, M. Basson, A. E. Pasquinelli, J. C. Bettinger, A. E. Rougvie, H. R. Horvitz, and G. Ruvkun. 2000. The 21-nucleotide let-7 RNA regulates developmental timing in *Caenorhabditis elegans*. *Nature* 403:901–906.

63. Didiano, D. and O. Hobert. 2008. Molecular architecture of a miRNA-regulated 3′ UTR. *RNA* 14:1297–1317.

64. Hammond, S. M. 2007. MicroRNAs as tumor suppressors. *Nat Genet* 39:582–583.

65. Esquela-Kerscher, A. and F. J. Slack. 2006. Oncomirs—MicroRNAs with a role in cancer. *Nat Rev Cancer* 6:259–269.

66. Poliseno, L., L. Salmena, J. Zhang, B. Carver, W. J. Haveman, and P. P. Pandolfi. 2010. A coding-independent function of gene and pseudogene mRNAs regulates tumour biology. *Nature* 465:1033–1038.

67. Ma, L., J. Teruya-Feldstein, and R. A. Weinberg. 2007. Tumour invasion and metastasis initiated by microRNA-10b in breast cancer. *Nature* 449:682–688.

68. Tay, Y. M., W. L. Tam, Y. S. Ang, P. M. Gaughwin, H. Yang, W. Wang, R. Liu et al. 2008. MicroRNA-134 modulates the differentiation of mouse embryonic stem cells, where it causes post-transcriptional attenuation of Nanog and LRH1. *Stem Cells* 26:17–29.

69. Tay, Y., J. Zhang, A. M. Thomson, B. Lim, and I. Rigoutsos. 2008. MicroRNAs to Nanog, Oct4 and Sox2 coding regions modulate embryonic stem cell differentiation. *Nature* 455:1124–1128.
70. Nelson, P. T., W. X. Wang, and B. W. Rajeev. 2008. MicroRNAs (miRNAs) in neurodegenerative diseases. *Brain Pathol* 18:130–138.
71. Matranga, C., Y. Tomari, C. Shin, D. P. Bartel, and P. D. Zamore. 2005. Passenger-strand cleavage facilitates assembly of siRNA into Ago2-containing RNAi enzyme complexes. *Cell* 123:607–620.
72. Pham, J. W., J. L. Pellino, Y. S. Lee, R. W. Carthew, and E. J. Sontheimer. 2004. A Dicer-2-dependent 80S complex cleaves targeted mRNAs during RNAi in *Drosophila*. *Cell* 117:83–94.
73. Schwarz, D. S., G. Hutvagner, T. Du, Z. S. Xu, N. Aronin, and P. D. Zamore. 2003. Asymmetry in the assembly of the RNAi enzyme complex. *Cell* 115:199–208.
74. Tomari, Y. and P. D. Zamore. 2005. Perspective: Machines for RNAi. *Genes Dev* 19:517–529.
75. Wang, Y., Y. Li, Z. Ma, W. Yang, and C. Ai. 2010. Mechanism of microRNA-target interaction: Molecular dynamics simulations and thermodynamics analysis. *PLoS Comput Biol* 6:e1000866.
76. Filipowicz, W., S. N. Bhattacharyya, and N. Sonenberg. 2008. Mechanisms of post-transcriptional regulation by microRNAs: Are the answers in sight? *Nat Rev Genet* 9:102–114.
77. Chi, S. W., J. B. Zang, A. Mele, and R. B. Darnell. 2009. Argonaute HITS-CLIP decodes microRNA-mRNA interaction maps. *Nature* 460:479–486.
78. Hafner, M., M. Landthaler, L. Burger, M. Khorshid, J. Hausser, P. Berninger, A. Rothballer et al. 2010. Transcriptome-wide identification of RNA-binding protein and microRNA target sites by PAR-CLIP. *Cell* 141:129–141.
79. Thomas, M., J. Lieberman, and A. Lal. 2010. Desperately seeking microRNA targets. *Nat Struct Mol Biol* 17:1169–1174.
80. Zisoulis, D. G., M. T. Lovci, M. L. Wilbert, K. R. Hutt, T. Y. Liang, A. E. Pasquinelli, and G. W. Yeo. 2010. Comprehensive discovery of endogenous Argonaute binding sites in *Caenorhabditis elegans*. *Nat Struct Mol Biol* 17:173–179.
81. Vella, M. C., E. Y. Choi, S. Y. Lin, K. Reinert, and F. J. Slack. 2004. The *C. elegans* microRNA let-7 binds to imperfect let-7 complementary sites from the lin-41 3′ UTR. *Genes Dev* 18:132–137.
82. Fabian, M. R., N. Sonenberg, and W. Filipowicz. 2010. Regulation of mRNA translation and stability by microRNAs. *Annu Rev Biochem* 79:351–379.
83. Selbach, M., B. Schwanhausser, N. Thierfelder, Z. Fang, R. Khanin, and N. Rajewsky. 2008. Widespread changes in protein synthesis induced by microRNAs. *Nature* 455:58–63.
84. Baek, D., J. Villen, C. Shin, F. D. Camargo, S. P. Gygi, and D. P. Bartel. 2008. The impact of microRNAs on protein output. *Nature* 455:64–71.
85. Rodriguez, A., E. Vigorito, S. Clare, M. V. Warren, P. Couttet, D. R. Soond, S. van Dongen et al. 2007. Requirement of bic/microRNA-155 for normal immune function. *Science* 316:608–611.
86. Giraldez, A. J., Y. Mishima, J. Rihel, R. J. Grocock, S. Van Dongen, K. Inoue, A. J. Enright, and A. F. Schier. 2006. Zebrafish MiR-430 promotes deadenylation and clearance of maternal mRNAs. *Science* 312:75–79.

87. Krutzfeldt, J., N. Rajewsky, R. Braich, K. G. Rajeev, T. Tuschl, M. Manoharan, and M. Stoffel. 2005. Silencing of microRNAs in vivo with 'antagomirs'. *Nature* 438:685–689.

88. Xia, Z., P. Clark, T. Huynh, P. Loher, Y. Zhao, H. W. Chen, P. Ren, I. Rigoutsos, and R. Zhou. 2012. Molecular dynamics simulations of Ago silencing complexes reveal a large repertoire of admissible 'seed-less' targets. *Sci Rep* 2:569.

89. Wang, Y., S. Juranek, H. Li, G. Sheng, T. Tuschl, and D. J. Patel. 2008. Structure of an argonaute silencing complex with a seed-containing guide DNA and target RNA duplex. *Nature* 456:921–926.

90. Wang, Y., S. Juranek, H. Li, G. Sheng, G. S. Wardle, T. Tuschl, and D. J. Patel. 2009. Nucleation, propagation and cleavage of target RNAs in Ago silencing complexes. *Nature* 461:754–761.

91. Wang, Y., G. Sheng, S. Juranek, T. Tuschl, and D. J. Patel. 2008. Structure of the guide-strand-containing argonaute silencing complex. *Nature* 456:209–213.

92. Xia, Z., T. Huynh, P. Ren, and R. Zhou. 2013. Large domain motions in Ago protein controlled by the guide DNA-strand seed region determine the Ago-DNA-mRNA complex recognition process. *PLoS One* 8:e54620.

93. Garcia, D. M., D. Baek, C. Shin, G. W. Bell, A. Grimson, and D. P. Bartel. 2011. Weak seed-pairing stability and high target-site abundance decrease the proficiency of lsy-6 and other microRNAs. *Nat Struct Mol Biol* 18:1139–1146.

94. Shin, C., J. W. Nam, K. K. Farh, H. R. Chiang, A. Shkumatava, and D. P. Bartel. 2010. Expanding the microRNA targeting code: Functional sites with centered pairing. *Mol Cell* 38:789–802.

95. Sanbonmatsu, K. Y. 2012. Computational studies of molecular machines: The ribosome. *Curr Opin Struct Biol* 22:168–174.

96. Chacon, P., F. Tama, and W. Wriggers. 2003. Mega-Dalton biomolecular motion captured from electron microscopy reconstructions. *J Mol Biol* 326:485–492.

97. Tama, F., M. Valle, J. Frank, and C. L. Brooks, 3rd. 2003. Dynamic reorganization of the functionally active ribosome explored by normal mode analysis and cryo-electron microscopy. *Proc Natl Acad Sci USA* 100:9319–9323.

98. Wang, Y., A. J. Rader, I. Bahar, and R. L. Jernigan. 2004. Global ribosome motions revealed with elastic network model. *J Struct Biol* 147:302–314.

99. Sanbonmatsu, K. Y., S. Joseph, and C. S. Tung. 2005. Simulating movement of tRNA into the ribosome during decoding. *Proc Natl Acad Sci USA* 102:15854–15859.

100. Baxter-Roshek, J. L., A. N. Petrov, and J. D. Dinman. 2007. Optimization of ribosome structure and function by rRNA base modification. *PLoS One* 2:e174.

101. Meskauskas, A. and J. D. Dinman. 2007. Ribosomal protein L3: Gatekeeper to the A site. *Mol Cell* 25:877–888.

102. Rakauskaite, R. and J. D. Dinman. 2011. Mutations of highly conserved bases in the peptidyltransferase center induce compensatory rearrangements in yeast ribosomes. *RNA* 17:855–864.

103. Burakovsky, D. E., P. V. Sergiev, M. A. Steblyanko, A. V. Kubarenko, A. L. Konevega, A. A. Bogdanov, M. V. Rodnina, and O. A. Dontsova. 2010. Mutations at the accommodation gate of the ribosome impair RF2-dependent translation termination. *RNA* 16:1848–1853.

104. Burakovsky, D. E., P. V. Sergiev, M. A. Steblyanko, A. L. Konevega, A. A. Bogdanov, and O. A. Dontsova. 2011. The structure of helix 89 of 23S rRNA is important for peptidyl transferase function of *Escherichia coli* ribosome. *FEBS Lett* 585:3073–3078.

105. Trabuco, L. G., E. Schreiner, J. Eargle, P. Cornish, T. Ha, Z. Luthey-Schulten, and K. Schulten. 2010. The role of L1 stalk-tRNA interaction in the ribosome elongation cycle. *J Mol Biol* 402:741–760.

106. Trabuco, L. G., C. B. Harrison, E. Schreiner, and K. Schulten. 2010. Recognition of the regulatory nascent chain TnaC by the ribosome. *Structure* 18:627–637.

107. Gumbart, J., L. G. Trabuco, E. Schreiner, E. Villa, and K. Schulten. 2009. Regulation of the protein-conducting channel by a bound ribosome. *Structure* 17:1453–1464.

108. Whitford, P. C., J. N. Onuchic, and K. Y. Sanbonmatsu. 2010. Connecting energy landscapes with experimental rates for aminoacyl-tRNA accommodation in the ribosome. *J Am Chem Soc* 132:13170–13171.

109. Sanbonmatsu, K. Y. and S. Joseph. 2003. Understanding discrimination by the ribosome: Stability testing and groove measurement of codon-anticodon pairs. *J Mol Biol* 328:33–47.

110. Almlof, M., M. Ander, and J. Aqvist. 2007. Energetics of codon-anticodon recognition on the small ribosomal subunit. *Biochemistry* 46:200–209.

111. Wallin, G. and J. Aqvist. 2010. The transition state for peptide bond formation reveals the ribosome as a water trap. *Proc Natl Acad Sci USA* 107:1888–1893.

112. Besseova, I., K. Reblova, N. B. Leontis, and J. Sponer. 2010. Molecular dynamics simulations suggest that RNA three-way junctions can act as flexible RNA structural elements in the ribosome. *Nucleic Acids Res* 38:6247–6264.

113. Reblova, K., F. Razga, W. Li, H. Gao, J. Frank, and J. Sponer. 2010. Dynamics of the base of ribosomal A-site finger revealed by molecular dynamics simulations and Cryo-EM. *Nucleic Acids Res* 38:1325–1340.

114. Vaiana, A. C. and K. Y. Sanbonmatsu. 2009. Stochastic gating and drug-ribosome interactions. *J Mol Biol* 386:648–661.

115. Romanowska, J., P. Setny, and J. Trylska. 2008. Molecular dynamics study of the ribosomal A-site. *J Phys Chem B* 112:15227–15243.

116. Vaiana, A. C., E. Westhof, and P. Auffinger. 2006. A molecular dynamics simulation study of an aminoglycoside/A-site RNA complex: Conformational and hydration patterns. *Biochimie* 88:1061–1073.

117. Ge, X. and B. Roux. 2010. Absolute binding free energy calculations of sparsomycin analogs to the bacterial ribosome. *J Phys Chem B* 114:9525–9539.

118. Xia, Z., Z. Zhu, J. Zhu, and R. Zhou. 2009. Recognition mechanism of siRNA by viral p19 suppressor of RNA silencing: A molecular dynamics study. *Biophys J* 96:1761–1769.

119. Freddolino, P. L., A. S. Arkhipov, S. B. Larson, A. McPherson, and K. Schulten. 2006. Molecular dynamics simulations of the complete satellite tobacco mosaic virus. *Structure* 14:437–449.

120. Shaw, D. E., P. Maragakis, K. Lindorff-Larsen, S. Piana, R. O. Dror, M. P. Eastwood, J. A. Bank et al. 2010. Atomic-level characterization of the structural dynamics of proteins. *Science* 330:341–346.

121. Silhavy, D., A. Molnar, A. Lucioli, G. Szittya, C. Hornyik, M. Tavazza, and J. Burgyan. 2002. A viral protein suppresses RNA silencing and binds silencing-generated, 21- to 25-nucleotide double-stranded RNAs. *EMBO J* 21:3070–3080.

122. Voinnet, O. 2005. Induction and suppression of RNA silencing: Insights from viral infections. *Nat Rev Genet* 6:206–220.

123. Anandalakshmi, R., G. J. Pruss, X. Ge, R. Marathe, A. C. Mallory, T. H. Smith, and V. B. Vance. 1998. A viral suppressor of gene silencing in plants. *Proc Natl Acad Sci USA* 95:13079–13084.

124. Qu, F. and T. J. Morris. 2005. Suppressors of RNA silencing encoded by plant viruses and their role in viral infections. *FEBS Lett* 579:5958–5964.

125. Lakatos, L., G. Szittya, D. Silhavy, and J. Burgyan. 2004. Molecular mechanism of RNA silencing suppression mediated by p19 protein of tombusviruses. *EMBO J* 23:876–884.

126. Dunoyer, P., C. H. Lecellier, E. A. Parizotto, C. Himber, and O. Voinnet. 2004. Probing the microRNA and small interfering RNA pathways with virus-encoded suppressors of RNA silencing. *Plant Cell* 16:1235–1250.

127. Calabrese, J. M. and P. A. Sharp. 2006. Characterization of the short RNAs bound by the P19 suppressor of RNA silencing in mouse embryonic stem cells. *RNA* 12:2092–2102.

128. Vargason, J. M., G. Szittya, J. Burgyan, and T. M. Hall. 2003. Size selective recognition of siRNA by an RNA silencing suppressor. *Cell* 115:799–811.

129. Eriksson, M. A. L., T. Hard, and L. Nilsson. 1995. Molecular-dynamics simulations of the glucocorticoid receptor DNA-binding domain in complex with DNA and free in solution. *Biophys J* 68:402–426.

130. Milner, J. 1997. Structures and functions of the tumour suppressor p53. *Pathol Biol* 45:797–803.

131. Soussi, T. and P. May. 1996. Structural aspects of the p53 protein in relation to gene evolution: A second look. *J Mol Biol* 260:623–637.

132. Wolkowicz, R. and V. Rotter. 1997. The DNA binding regulatory domain of p53: See the C. *Pathol Biol* 45:785–796.

133. Yonish-Rouach, E. 1997. A question of life or death: The p53 tumor suppressor gene. *Pathol Biol* 45:815–823.

134. Friend, S. 1994. P53—A glimpse at the puppet behind the shadow play. *Science* 265:334–335.

135. Mraz, M., K. Malinova, J. Kotaskova, S. Pavlova, B. Tichy, J. Malcikova, K. Stano Kozubik et al. 2009. miR-34a, miR-29c and miR-17-5p are downregulated in CLL patients with TP53 abnormalities. *Leukemia* 23:1159–1163.

136. Bullock, A. N. and A. Fersht. 2001. Rescuing the function of mutant p53. *Nat Rev Cancer* 1:68–76.

137. Hainaut, P. and M. Hollstein. 2000. p53 and human cancer: The first ten thousand mutations. *Adv Cancer Res* 77:81–137.

138. Pan, Y. P. and R. Nussinov. 2007. Structural basis for p53 binding-induced DNA bending. *J Biol Chem* 282:691–699.

139. Duan, J. X. and L. Nilsson. 2006. Effect of Zn^{2+} on DNA recognition and stability of the p53 DNA-binding domain. *Biochemistry* 45:7483–7492.

140. Ma, B. and A. J. Levine. 2007. Probing potential binding modes of the p53 tetramer to DNA based on the symmetries encoded in p53 response elements. *Nucleic Acids Res* 35:7733–7747.

141. Garner, M. M. and D. C. Rau. 1995. Water release associated with specific binding of gal repressor. *EMBO J* 14:1257–1263.

142. Sidorova, N. Y. and D. C. Rau. 1996. Differences in water release for the binding of EcoRI to specific and nonspecific DNA sequences. *Proc Natl Acad Sci USA* 93:12272–12277.

143. Otwinowski, Z., R. W. Schevitz, R. G. Zhang, C. L. Lawson, A. Joachimiak, R. Q. Marmorstein, B. F. Luisi, and P. B. Sigler. 1988. Crystal-structure of Trp repressor operator complex at atomic resolution. *Nature* 335:321–329.

144. Lundback, T. and T. Hard. 1996. Sequence-specific DNA-binding dominated by dehydration. *Proc Natl Acad Sci USA* 93:4754–4759.

145. Robinson, C. R. and S. G. Sligar. 1998. Changes in solvation during DNA binding and cleavage are critical to altered specificity of the EcoRI endonuclease. *Proc Natl Acad Sci USA* 95:2186–2191.

146. Fuxreiter, M., M. Mezei, I. Simon, and R. Osman. 2005. Interfacial water as a "hydration fingerprint" in the noncognate complex of BamHI. *Biophys J* 89:903–911.

147. Cheatham, T. E., 3rd, J. Srinivasan, D. A. Case, and P. A. Kollman. 1998. Molecular dynamics and continuum solvent studies of the stability of polyG-polyC and polyA-polyT DNA duplexes in solution. *J Biomol Struct Dyn* 16:265–280.

148. Kollman, P. A., I. Massova, C. Reyes, B. Kuhn, S. Huo, L. Chong, M. Lee et al. 2000. Calculating structures and free energies of complex molecules: Combining molecular mechanics and continuum models. *Acc Chem Res* 33:889–897.

149. Jayaram, B., D. Sprous, M. A. Young, and D. L. Beveridge. 1998. Free energy analysis of the conformational preferences of A and B forms of DNA in solution. *J Am Chem Soc* 120:10629–10633.

150. Srinivasan, J., T. E. Cheatham, P. Cieplak, P. A. Kollman, and D. A. Case. 1998. Continuum solvent studies of the stability of DNA, RNA, and phosphoramidate—DNA helices. *J Am Chem Soc* 120:9401–9409.

151. Habtemariam, B., V. M. Anisimov, and A. D. MacKerell, Jr. 2005. Cooperative binding of DNA and CBFbeta to the Runt domain of the CBFalpha studied via MD simulations. *Nucleic Acids Res* 33:4212–4222.

152. Zhang, Q. and T. Schlick. 2006. Stereochemistry and position-dependent effects of carcinogens on TATA/TBP binding. *Biophys J* 90:1865–1877.

153. Laio, A. and M. Parrinello. 2002. Escaping free-energy minima. *Proc Natl Acad Sci USA* 99:12562–12566.

154. Huber, T., A. E. Torda, and W. F. van Gunsteren. 1994. Local elevation: A method for improving the searching properties of molecular dynamics simulation. *J Comput Aided Mol Des* 8:695–708.

155. Grubmuller, H. 1995. Predicting slow structural transitions in macromolecular systems: Conformational flooding. *Phys Rev E Stat Phys Plasmas Fluids Relat Interdiscip Topics* 52:2893–2906.

156. Dickson, B. M. 2011. Approaching a parameter-free metadynamics. *Phys Rev E Stat Nonlinear Soft Matter Phys* 84:037701.

157. Zheng, L., M. Chen, and W. Yang. 2008. Random walk in orthogonal space to achieve efficient free-energy simulation of complex systems. *Proc Natl Acad Sci USA* 105:20227–20232.

158. Darve, E. and A. Pohorille. 2001. Calculating free energies using average force. *J Chem Phys* 115:9169–9183.

159. Henin, J. and C. Chipot. 2004. Overcoming free energy barriers using unconstrained molecular dynamics simulations. *J Chem Phys* 121:2904–2914.

160. Huang, N., N. K. Banavali, and A. D. MacKerell, Jr. 2003. Protein-facilitated base flipping in DNA by cytosine-5-methyltransferase. *Proc Natl Acad Sci USA* 100:68–73.

161. Huang, N. and A. D. MacKerell, Jr. 2004. Atomistic view of base flipping in DNA. *Philos Trans A Math Phys Eng Sci* 362:1439–1460.

162. Huang, N. and A. D. MacKerell, Jr. 2005. Specificity in protein–DNA interactions: Energetic recognition by the (cytosine-C5)-methyltransferase from HhaI. *J Mol Biol* 345:265–274.

163. Banerjee, A., W. Yang, M. Karplus, and G. L. Verdine. 2005. Structure of a repair enzyme interrogating undamaged DNA elucidates recognition of damaged DNA. *Nature* 434:612–618.

164. Wang, Y. and T. Schlick. 2007. Distinct energetics and closing pathways for DNA polymerase beta with 8-oxoG template and different incoming nucleotides. *BMC Struct Biol* 7:7.

165. Trabuco, L. G., E. Villa, E. Schreiner, C. B. Harrison, and K. Schulten. 2009. Molecular dynamics flexible fitting: A practical guide to combine cryo-electron microscopy and x-ray crystallography. *Methods* 49:174–180.

166. Mitra, K. and J. Frank. 2006. Ribosome dynamics: Insights from atomic structure modeling into cryo-electron microscopy maps. *Annu Rev Biophys Biomol Struct* 35:299–317.

167. Koch, M. H., P. Vachette, and D. I. Svergun. 2003. Small-angle scattering: A view on the properties, structures and structural changes of biological macromolecules in solution. *Q Rev Biophys* 36:147–227.

168. Trabuco, L. G., E. Villa, K. Mitra, J. Frank, and K. Schulten. 2008. Flexible fitting of atomic structures into electron microscopy maps using molecular dynamics. *Structure* 16:673–683.

169. Wriggers, W. and P. Chacon. 2001. Modeling tricks and fitting techniques for multi-resolution structures. *Structure* 9:779–788.

170. Querol-Audi, J., C. Yan, X. Xu, S. E. Tsutakawa, M. S. Tsai, J. A. Tainer, P. K. Cooper, E. Nogales, and I. Ivanov. 2012. Repair complexes of FEN1 endonuclease, DNA, and Rad9-Hus1-Rad1 are distinguished from their PCNA counterparts by functionally important stability. *Proc Natl Acad Sci USA* 109:8528–8533.

171. Forster, F., B. Webb, K. A. Krukenberg, H. Tsuruta, D. A. Agard, and A. Sali. 2008. Integration of small-angle x-ray scattering data into structural modeling of proteins and their assemblies. *J Mol Biol* 382:1089–1106.

172. Schneidman-Duhovny, D., M. Hammel, and A. Sali. 2010. FoXS: A web server for rapid computation and fitting of SAXS profiles. *Nucleic Acids Res* 38:W540–W544.

173. Brosey, C. A., C. Yan, S. E. Tsutakawa, W. T. Heller, R. P. Rambo, J. A. Tainer, I. Ivanov, and W. J. Chazin. 2013. A new structural framework for integrating replication protein A into DNA processing machinery. *Nucleic Acids Res* 41:2313–2327.

174. Wold, M. S. 1997. Replication protein A: A heterotrimeric, single-stranded DNA-binding protein required for eukaryotic DNA metabolism. *Annu Rev Biochem* 66:61–92.

175. Franke, D. and D. I. Svergun. 2009. DAMMIF, a program for rapid ab-initio shape determination in small-angle scattering. *J Appl Crystallogr* 42:342–346.

Chapter 9

Simulating Membranes and Membrane Proteins

Nicholas Leioatts and Alan Grossfield

CONTENTS

9.1 INTRODUCTION

The past decade has seen a tremendous growth in the size and length of molecular dynamics (MD) simulations, owing to advancements in both hardware (Allen et al., 2001; Shaw et al., 2009; Shirts and Pande, 2000) and software (Bowers et al., 2006; Fitch et al., 2003; Kutzner et al., 2007; Phillips et al., 2005; Van Der Spoel et al., 2005). Molecular dynamics is most useful where experiments are difficult or impossible to perform, and as a result, simulations of integral membrane proteins, particularly G protein-coupled receptors (GPCRs), have been the target of enormous effort (Dror et al., 2011a; Grossfield, 2011). Here, we describe how these simulations are conducted, from initial system construction to final analysis, including assessments of statistical convergence. These sections are written to be a practical didactic guide rather than a systematic review of the literature, and we will point out common traps and pitfalls often encountered and how to avoid them. The content of this chapter is organized around three sections depending loosely on stages of a project: building the system, running the simulation, and analyzing the results.

9.2 CONSIDERATIONS FOR BUILDING MEMBRANE–PROTEIN SYSTEMS

Any resource on membrane simulations would be incomplete if system construction were not addressed; in contrast to soluble proteins and nucleic acids, there are a number of challenges involved in building a liquid crystalline lipid bilayer and embedding a protein in it. Many phospholipids have been parametrized as part of a number of force fields and parameter databases. These databases, such as Lipidbook (http://lipidbook.bioch.ox.ac.uk/package/), are a good reference or starting point when you are designing a system and deciding on membrane composition. However, it is important to note that lipid force fields are still largely a work in progress and are continuously being reparametrized. Presently, three major lipid force field families are in common use (CHARMM, Berger, and GROMOS), and it was not until the recent Lipid11 (Skjevik et al., 2012) release that Amber included well-tested parameters. In this section, we will also discuss applying the canonical (number, volume, and temperature [NVT]), microcanonical (number, volume, and total energy [NVE]), and isothermal–isobaric (number, pressure, and temperature [NPT]) ensembles to membrane protein systems. Finally, various system construction methods are discussed. Several protocols exist for stably inserting the protein into the bilayer efficiently. These are important because this process can be inherently slow—it is possible to collect an entire simulation without realizing that the protein is not properly embedded in the membrane. Correct protein placement is crucial to the validity of your simulation results.

9.2.1   Choosing Membrane Components

Deciding what phospholipids to use in your simulation can be a complex process. You need to consider whether you want to imitate a realistic bilayer (still often limited to 3–4 lipid types), replicate a bilayer used in a specific set of experiments, or extrapolate from a trend shown in the literature. However, having multiple lipid species in a bilayer introduces a new degree of freedom, lateral reorganization of the bilayer, which requires slow motions and thus can greatly slow the statistical convergence of the simulation. For this reason, or because the lipids themselves are not the focus of the particular simulation, a single lipid type is often chosen. For example, 1-palmitoyl-2-oleoyl-phostphatidylcholine (POPC) is a very common choice, particularly for simulations performed with the CHARMM family of force fields; it is highly populated in mammalian membranes, zwitterionic, and well characterized experimentally. By contrast, simulations performed using the GROMACS (Kutzner et al., 2007; Van Der Spoel et al., 2005) suite often use dipalmitoylphosphatidylcholine (DPPC) as the "default" lipid, although it is less representative of natural lipids because it lacks an unsaturated chain. Moreover, DPPC has a high melting temperature and is actually a gel at room temperature, so simulations in a pure DPPC bilayer must be performed at slightly elevated temperatures. While these are good-go-to lipids for simulations of mammalian proteins, they are not always the most appropriate. For example, bacterial outer membranes tend to be enriched in anionic lipids, so it would be critical to include one, such as 1-palmitoyl-2-oleoyl-phosphatidylglycerol (POPG). Similarly, simulations of rhodopsin, found in the rod outer segment disk membranes, should contain lipids with polyunsaturated ω-3 fatty acids if they are to effectively represent physiological conditions.

    It is also critical to determine the size of the simulation cell and how many lipids are needed. This is a case where there are contradictory imperatives. On one hand, it is clearly preferable to keep the simulation cell as small as possible to keep the computational cost manageable. On the other hand, lipid bilayers are mesoscopic structures with long correlation lengths, and perturbations due to the presence of a protein can easily extend 10s of angstroms. Moreover, even neat bilayers can have long-range structure and motions, some of which would be damped out by a small periodic box (Lindahl and Edholm, 2000). This is distinct from soluble protein simulations, where the water has very short correlation times and lengths.

In addition, it is important to include salt at physiological concentrations, not just what is needed to produce a net neutral system; not only is this more realistic, it can make major differences in the way the counterions interact with charged lipids.

9.2.2 Choosing an Ensemble

As simulation ensembles are a general consideration, we will focus here on the caveats that pertain to simulations of membrane proteins. The microcanonical (constant particle NVE) ensemble has the benefit of simplicity: it is the least demanding molecular dynamics algorithm to implement, involving nothing more than Newton's equations of motion. Moreover, it is by far the algorithm most conducive to testing: almost any possible error in implementation of the integrator itself or the force calculation will lead to poor energy conservation and a systematic rise in temperature. Moreover, the fact that no thermostat is applied means that the kinetics are "pure" and thus more likely to be physically correct. However, NVE results are harder to compare to experiments traditionally performed at constant temperature and pressure, and significant tuning is required at the beginning of the simulation: volume is a stiff variable, and even small errors in its initial value (which is hard to estimate, because the partial molar volume of a protein in particular is not well known) lead to significant pressures that may alter the results. Similarly, construction protocols are not perfect, leading to an initial rise in the temperature as the system relaxes and equilibrates. Thus, even simulations that are performed in the NVE ensemble are nearly always first equilibrated using temperature control (Grossfield et al., 2006b).

The canonical ensemble (constant particle NVT) provides a solution to this problem. An "external bath" is set up via temperature coupling to a thermostat. The thermostat will maintain a constant temperature by resampling atom velocities in a predetermined way (Allison and McCammon, 1984; Andersen, 1980; Berendsen et al., 1984; Hoover, 1985; Nosé, 1984). However, this will at least in principle alter the kinetics of the system; the precise degree to which this occurs is not readily known and probably varies with the details of the thermostatting procedure. Moreover, thermostatting makes the system more robust to significant structural changes in the simulation. When changes (or fluctuations) with a large enthalpic component (which would alter the temperature in an NVE ensemble) occur, the heat has a way to dissipate, as is the case in most experimental setups.

Isothermal–isobaric (constant particle NPT) simulations are the ensemble of choice when it comes to data analysis, since most biological experiments are conducted at constant temperature and constant pressure. When using this ensemble, there are a few variations to consider. We will restrict our subsequent discussion to simple "rectangular" boxes; while algorithms such as the

Parrinello–Rahman barostat (Parrinello and Rahman, 1981) can correctly handle changes to the box shape as well as its dimensions, nonorthogonal boxes are rarely used in biological simulations. Other box shapes (e.g., hexagonal in the membrane plane) can be used for membrane simulations, but they are far less common, and as a rule, the discussion later will still apply.

When simulating proteins in solution, the usual choice is to simulate a constant box shape, uniformly scaling all box dimensions; this is intuitive, given that the solvent environment is expected to be isotropic. However, this assumption does not hold for a membrane system, where the bilayer normal (typically chosen to be z) is physically distinct from the membrane plane. Slaving fluctuations along the membrane normal to those in the plane can unphysically alter the membrane's behavior, because both the compressibility and fluctuation timescale for water and lipid differ. Moreover, if uniform scaling is applied during equilibration (when the solvent interaction with the bilayer may drift significantly), the result can be an effective external force that drives the bilayer away from its equilibrium area.

Intuitively, it would seem that the opposite extreme (allowing the three box dimensions to vary independently) would make sense, but in fact, this too can lead to artifacts. If the dimensions in the plane of the membrane are not constrained, there is nothing to prevent the simulation cell from drifting into an extreme aspect ratio. This would present two problems: in the shorter dimension, there could be significant periodicity artifacts, inducing unphysical correlations and increasing the likelihood of unphysical interactions between a membrane protein and its periodic images. Moreover, if the long dimension grows significantly larger than its original value, the grid used for long-range electrostatics calculations may become inappropriate; if the same number of grid points are used to span a dramatically larger box, the accuracy of the calculated forces could be diminished.

For this reason, the most common choice is to fix the ratio of the two box dimensions in the plane of the membrane (x and y), while allowing the z dimension to fluctuate independently. However, there are still a number of choices to be made. Ideally, one would simply apply uniform pressure to each dimension. However, the area per lipid is determined by a very delicate balance of interactions, and it is not unusual for force fields to have trouble reproducing it. For example, the CHARMM27 lipid force field was notorious for forming gel-like phases at ambient temperatures (Feller and Pastor, 1996); the recent update to the CHARMM36 force field has reduced but not eliminated these problems (Klauda et al., 2010). The Berger lipid force field (Berger et al., 1997), commonly used with the GROMACS simulation package, performs better in this respect. However, this was achieved in part by using extremely large partial charges on the carbonyl groups attaching the fatty acids to the lipid glycerol backbone, which can induce artifacts in the interactions between lipids and salt, particularly sodium.

There are two commonly used strategies for handling this problem. One is to fix the *x* and *y* box dimensions while allowing the *z* dimension to vary; this ensemble, called NPAT (the "A" stands for "area"), removes the worst potential artifacts of constant volume calculations, but the risk remains that certain fluctuations, those that would change the membrane area (e.g., large-scale protein rearrangement), could be suppressed. This is particularly important if one wishes to simulate the process of something binding to the membrane, as, for instance, when one simulates antimicrobial peptides' interactions with bilayers. In this case, it is absolutely essential that the membrane area be able to fluctuate. For these cases, the most common solution is to simulate under a laterally applied tension (the so-called NPγT ensemble); the tension is used to keep the area per lipid in reasonable agreement with experiment. However, the required tension is a fairly sensitive parameter that can vary with system size and composition; the best practice is to tune it prior to running a protein simulation using an equivalent-sized neat bilayer of the same composition.

As discussed earlier, there are two reasons to apply thermostats and barostats. One is to give the system a means to relax during the equilibration phase, because accurately determining the initial volume and energy is far more challenging than determining a target pressure and temperature. The second is to generate correct fluctuations, such that the system can be treated using classical statistical mechanics, generally the NVT or NPT ensemble.

There are a large number of choices available to implement both thermostats and barostats, many of which satisfy both of these goals. For example, temperature control mechanisms can be loosely grouped as either velocity replacement or velocity scaling approaches. Velocity replacement, or the Andersen style, algorithms act by periodically replacing a subset of particle velocities with new ones generated randomly from the Maxwell–Boltzmann distribution (Andersen, 1980). The original implementation defined a "collision frequency" that determined the odds that any given particle's velocity would be altered on a given time step (Andersen, 1980), but alternatively, one could periodically resample all particle velocities at once, the so-called "catastrophic Andersen" algorithm (Fitch et al., 2003; Hurst et al., 2010; Romo et al., 2010). Although no analytic proof exists to show that this approach generates the canonical ensemble (Frenkel and Smit, 2002), it has passed numerical tests on small systems, strongly suggesting its validity. One downside to this approach is that by altering the velocities, one can clearly alter the system's kinetics.

Alternatively, one can use stochastic dynamics methods such as the Langevin dynamics (Allen and Tildesley, 1987; Allison and McCammon, 1984), where interaction with a temperature bath is mimicked by adding two terms to the integration: a friction force opposing the instantaneous velocity and a random force intended to replicate collisions with the external bath.

This method is very robust and computationally inexpensive to implement, but does not produce reproducible trajectories unless the random number seed is retained.

Velocity scaling algorithms function by smoothly scaling the existing velocities (rather than replacing them) and are thus perceived to be less disruptive to the system's kinetics. The most obvious approach, rescaling the velocities on each step such that the instantaneous temperature matches the target temperature, generates the isokinetic ensemble, which does not possess the same fluctuations as the canonical ensemble; for a finite-sized system, the temperature should fluctuate. For this reason, Berendsen and coworkers created a new thermostat that scales the velocities partway back toward the target temperature, thus allowing the system temperature to fluctuate about the target (Berendsen et al., 1984). However, the Berendsen thermostats do not generate fluctuations in any known thermodynamic ensemble. A more correct approach is to use one of the extended system methods, where the instantaneous temperature is treated as an independent degree of freedom, with dynamics driven by its deviation from the target (Hoover, 1985; Martyna et al., 1992; Nosé, 1984). These methods generate continuous, reversible dynamics in the correct ensemble. Additionally, extended system methods have a conserved quantity (analogous to total energy in an NVE simulation) that can be monitored to ensure integration is proceeding accurately.

Similarly, there are a number of pressure-control options available. One can periodically interrupt the dynamics to perform a Monte Carlo–like step in volume space (Frenkel and Smit, 2002), provided one uses the correct acceptance criterion. This method generates the correct NPT fluctuations and does not require the pressure calculation via the virial during dynamics. Continuously varying methods include the Berendsen algorithm, which (in direct analogy to the thermostat) scales the volume such that the pressure is moved partway back toward the target pressure; as with the thermostat, the Berendsen barostat does not reproduce the fluctuations of any known thermodynamic ensemble. Alternatively, one could use the extended system barostat derived by Andersen, which can be shown to reproduce the correct fluctuations (Andersen, 1980); this algorithm can be visualized as monitoring the dynamics of a piston that modulates the system volume. The tendency for this piston to "ring" or undergo long-timescale-correlated motion can be reduced either by periodically resampling the piston's velocity (as recommended in the original derivation [Andersen, 1980]) or by subjecting it to stochastic dynamics, as is done in the Langevin piston algorithm (Feller et al., 1995).

It is true that the degree to which an "incorrect" thermostat alters the system's fluctuations decreases with system size, so it is not totally clear what effect using these algorithms has on the quality of the output of a large simulation, such as a lipid bilayer in water. However, virtually, all of the major simulation packages implement at least one algorithm that generates correct fluctuations, and these algorithms are generally stable and well behaved and impose no significant computational cost above that for the less correct methods. For this reason, we believe there is no good reason to use the methods that do not generate correct statistics (e.g., the Berendsen barostat and thermostat), except perhaps as part of an initial construction protocol.

9.2.3   Building a Starting Structure

One often overlooked aspect of simulating membrane systems is the procedure used for initial construction. In the context of a macromolecule in water, this is generally quite simple: download the protein structure from the protein data bank (PDB), add hydrogens and perform any other initial cleanup, and then embed the protein in water, usually by superposing the protein with a pre-equilibrated box of water and removing those molecules that overlap the protein. Then add the necessary salt ions (usually by replacing water molecules at random). Frequently, a brief period of dynamics is performed where the protein is restrained before the production dynamics begins. As a rule, the results are not sensitive to the precise protocol, because the dynamics of water are significantly faster than those of the protein, such that the hydration is "self-healing."

This is less true when simulating a lipid bilayer, especially with a protein embedded in it. The obvious analog of the aqueous procedure would be to build the protein, then insert it into an existing bilayer by removing overlapping lipids and water. The problem is that lipids are relatively large and move slowly compared to water. As a result, deleting "overlapping" lipids could lead to a large, jagged hole in the membrane that is not completely filled by the protein. Since water diffuses rapidly, this can cause formation of hydration pockets on the protein surface in the hydrophobic core of the membrane, and while the dynamics should cause this water to leave over time, the process may occur slowly, wasting much of the simulation time simply to recover from a physically unlikely starting structure.

The first protocol used to handle this problem was devised by Woolf and Roux (1996). They started with a library of lipid conformations from a previous neat membrane simulation and used an automated procedure to pack these lipids around the protein. Basically, they started by randomly distributing van der Waals' spheres to represent the lipid headgroups and subjecting them to a short stochastic simulation to spread them out. They then used those spheres to place lipids from the library, performing systematic rotations and translations to obtain effective packing without clashes. Other groups have applied analogous protocols, replacing the rotation and translation steps with extensive potential energy minimization while gradually "turning on" intermolecular interactions (Grossfield et al., 2006b). This is essentially the strategy applied by the CHARMM-GUI membrane builder, a user-friendly web-based package for building simulations (Jo et al., 2009).

Alternatively, one can use the "INFLATEGRO" strategy, implemented as part of GROMACS (Kandt et al., 2007; Schmidt and Kandt, 2012). Here, one begins with a pre-equilibrated membrane patch. The lipid centers of mass are scaled outward in the plane of the membrane (after removing an appropriate number of lipids to account for the protein volume). After placing the protein, the lipids are subjected to repeated cycles of energy minimization interspersed with scaling of their centers of mass, again resulting in a well-packed bilayer with minimal clashes.

The membrane builder distributed with lightweight object-oriented structure (LOOS) (an analysis suite designed to be simulation-package agnostic) uses an algorithm that combines these two strategies (Romo and Grossfield, 2009, 2012). It uses a library of lipid structures and does not require a preexisting bilayer, but uses a scaling/minimization strategy analogous to that of INFLATEGRO. One of its major advantages is its ability to construct bilayers with arbitrarily complex compositions (including asymmetric leaflets), with or without a protein.

9.3 CONSIDERATIONS FOR PRODUCTION SIMULATIONS

9.3.1 Designing a Simulation

The first question considered in designing a simulation should always be as follows: what is the scientific question? Without knowing the question, it is difficult to know which (if any) simulation method is appropriate. As a rule, just picking a protein and watching it wiggle in an MD simulation is not a recipe for successful science. Instead, a better approach is to carefully consider the scientific question first, in order to decide which approaches are likely to work given the available resources.

For example, one must carefully choose the system conditions: the number and type of lipids, the amount of water, salt concentrations, and other factors like the temperature and pH (usually represented only implicitly by choice of charge state for ionizable moieties). In addition to system conditions, one must also choose the resolution of the model. This can only be done after consideration of which degrees of freedom are likely to be essential. Although it may seem counterintuitive, it is often the case that the most accurate or detailed model is *not* the one that will be most helpful in addressing a scientific problem. However, increasing the level of detail almost always increases the cost of performing the calculation, leaving one with a choice: do I use the more detailed model and accept the fact that I may not be able to get good statistics or use a simpler model and risk not capturing the relevant phenomena? The optimal choice *varies* with the problem. Although all-atom (or united atom) models are still the default choice for many applications, coarse-grained MD calculations have become very popular in the membrane simulation community (Ayton and Voth, 2009), particularly those based on the MARTINI force field (Marrink et al., 2007; Monticelli et al., 2008; Seo et al., 2012). MARTINI allows for rapid sampling and has been successfully applied to a number of interesting membrane phenomena, such as antimicrobial peptide binding (Horn et al., 2012; Mátyus et al., 2007; Polyansky et al., 2010; Rzepiela et al., 2010) and lipid–protein interactions (Horn et al., 2013; MacCallum and Tieleman, 2011; MacCallum et al., 2008; Periole et al., 2007; Provasi et al., 2011). However, the model's limitations—potentially oversimplified electrostatics and poor protein backbone energetics—mean that it cannot be applied to a broad swath of problems, including protein dynamics and ion channels. By contrast, there are phenomena, such as protonation/deprotonation events, that are not readily captured even by all-atom models, due to the classical nature of the associated force fields.

Furthermore, the various choices are often interrelated: one might decide that ideally the protein should be diluted in the membrane, at no more than 1000:1 lipid–protein ratio, because that is how the experiments were done. However, is that matching worth the additional computational cost? The answers could be very different if one is performing all-atom or coarse-grained simulations, given the enormous differences in computational expense. Depending on what one is studying—for instance, internal protein dynamics versus the structure of nearby lipids—it may or may not be worthwhile to reduce the system size by simulating fewer lipids.

9.3.2 Why Simulate Membrane Systems?

Membrane proteins are notoriously difficult to work with experimentally. Many are expressed in low quantities, and they are often not stable without hydrophobic contacts (typically made by protein–membrane interactions), so detergents or membranes are needed to keep them in their native form. This makes them difficult to purify and manipulate, in contrast to many soluble proteins. Moreover, many structural biology tools, such as solution NMR and crystallography, are far more challenging to apply to membrane proteins.

The relative paucity of high-resolution experimental data makes molecular simulations correspondingly more valuable. Although MD simulations of membrane systems can be challenging to construct and perform (see Section 9.2), they can provide information with unrivaled resolution in time and space. They can aid in the interpretation of experimental data and can suggest new hypotheses for future experimental testing. Given the rapid improvements seen in computational performance, it seems likely this capability will only increase in the future.

9.3.2.1 Neat Membrane Simulations

The first MD simulations of membranes were unsurprisingly of neat membranes (Venable et al., 1993). Since that time, these simulations have dramatically improved in their ability to reproduce experimental results, such as x-ray and neutron scattering or solid-state NMR (see Section 9.4). More importantly, these calculations help put a physical image onto the results, as, for instance, in the case of nuclear Overhauser effect spectroscopy (NOESY) cross-relaxation rate (Feller et al., 1999). Often, they supply additional information that is not readily available experimentally. For example, much of our understanding of the highly fluid structure of ω-3 polyunsaturated lipids comes from work combining experiment and molecular dynamics simulations (Carrillo-Tripp and Feller, 2005; Eldho et al., 2003; Feller, 2008; Feller and Gawrisch, 2005; Feller et al., 2002; Grossfield et al., 2006a).

9.3.2.2 Membrane Protein Simulations

Membrane protein simulations have already yielded insights that are uniquely valuable. Here, we highlight as an example a few interesting studies of GPCRs. We refer interested readers to recent reviews of GPCRs (Grossfield, 2011; Johnston and Filizola, 2011) and channel proteins (Lindahl and Sansom, 2008; Vargas et al., 2012).

GPCRs in particular have been the subject of some of the longest all-atom simulations reported to date. They have been used to probe potential ligand-binding pathways (Dror et al., 2011b; Hurst et al., 2010). Diverse paths have been found, from the ligand first partitioning into the lipid bilayer and then diffusing to the protein to binding first to a more exposed domain on the extracellular face, before inserting into the binding pocket—a transition that required the efflux of a large amount of water. Further studies of ligands in their receptor have revealed a degree of conformational flexibility (Grossfield et al., 2008; Lau et al., 2007; Martínez-Mayorga et al., 2006; Shan et al., 2012). In particular, Lyman and coworkers showed that adenosine receptor ligands are very dynamic, sampling a number of conformations in the binding pocket of the inactive protein (Lee and Lyman, 2012).

Much work has been done to show how bilayer components interact with and influence GPCR dynamics (Grossfield et al., 2006a, b; Horn et al., 2013; Khelashvili et al., 2009). These simulations helped reveal the physics underlying the modulation of rhodopsin function by polyunsaturated lipid chains, cholesterol, and different headgroups.

Several labs have also studied long-timescale events such as the activation mechanism (Dror et al., 2011a; Grossfield et al., 2008; Martínez-Mayorga et al., 2006). In the work by Dror et al. (2011a), the protein spontaneously moved from an active-like conformation to an inactive-like conformation in the absence of an activating ligand. They were able to qualitatively analyze this process, identifying a "connector region" that loosely coupled the ligand-binding site to the active site.

Finally, simulations have made significant contributions to evaluating the importance of internal waters to GPCR function (Grossfield et al., 2008; Hurst et al., 2010; Jardón-Valadez et al., 2009, 2010); although their structural role was revealed by previous crystal structures (Okada et al., 2002), the simulations tracked dramatic changes in hydration that occur during the photocycle, a result since confirmed by solid-state NMR (Grossfield et al., 2008).

9.3.3   Ru(i)nning Simulations

In a sense, running the simulation is the easiest part of MD: the system contents and external conditions are chosen, and it is just a matter of making the sure the simulation continues running and does not crash. However, it is not wise to wait until the simulation has completed to begin analysis, because by then it may be too late to correct flaws in the setup, etc. To this end, there are a number of "diagnostic analyses" one should do on the fly, simple calculations that can reveal critical flaws. For example, when simulating a molecule with a native state (e.g., a folded protein), it is a good idea to monitor the root mean square deviation (RMSD) to your starting structure. Overall, RMSD is not a particularly good measure of structural similarity: it is size dependent, and even for moderate-sized fluctuations, different structures can have the same RMSD. However, it is very inexpensive to calculate, and a continuous upward drift in RMSD suggests that the system is not well equilibrated and perhaps that it is falling apart.

Similarly, for membrane systems, it is crucial to monitor the area per lipid. While a stable area per lipid does not necessarily indicate that all is well with the system, one that continuously drifts (or that assumes unphysically large or small values) is a clear signal that there may be a problem. As discussed earlier in Section 9.2.2, area per lipid is a tricky quantity to get right in a force field, and the surface tension required to maintain appropriate values can be sensitive to the temperature, force field, lipid composition, and size of the bilayer patch. Moreover, the fluctuation in the area per lipid is one of the slower modes of a lipid bilayer sampled by all-atom molecular dynamics, so visual inspection of this time series can suggest the timescale needed for adequate sampling; at the very least, if the trajectory only sees a single fluctuation period (which can last on the order of 100 ns or longer), then it is almost certainly too short to say much about global bilayer structure. Somewhat along the same lines, it is a good idea to monitor the distribution of various components along the membrane normal (see Section 9.4.2.1 and Figure 9.1), as this can reveal disruptions in the membrane structure.

The purpose of these analyses is not so much to produce publishable interpretations of the simulation, but rather simply to prevent (or at least detect)

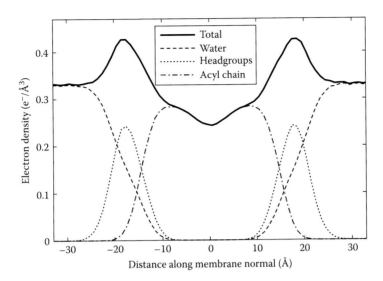

Figure 9.1 Electron density profile along the membrane normal. The solid curve shows the total electron density for a neat POPC lipid membrane, while the other curves show the density distributions for the water, lipid headgroups, and acyl chains. (Data provided by Tod Romo, unpublished.)

catastrophic failures. As such, they should be used to complement, but not replace, visual inspection of the trajectory. A trajectory viewer (such as VMD [Humphrey et al., 1996]) is a crucial tool, as you can often spot a problem "by eye" without knowing the best observable associated with it. Indeed, passing all of these visual and numerical tests does not in itself indicate that the trajectory is "correct," merely that its flaws are not yet known. Still, the best practice is to perform all of these tests and monitor the results carefully. Perhaps, the best way to do so is to automate them as part of the production process. For example, a routine protocol would be to bring down the latest production data each day, merge it into a single large trajectory (while centering either the protein or the membrane itself at the origin), run a set of automated analyses such as the ones described earlier, and then view the trajectory in VMD.

9.3.4   How Long Is Long Enough?

Although all-atom MD is clearly a powerful technique for understanding membranes and membrane proteins, it has one primary weakness: the computational cost of performing the simulations is extraordinarily high, so much so that obtaining good statistics can be nearly impossible. At first blush, worrying about this may seem excessive, given the self-evident value of being able to actually view molecules moving. However, our ability to draw conclusions from what we see is based entirely on the statistics: has the system reached equilibrium? Is there enough data to state that the observed phenomena are not due simply to luck?

In many cases, statistical evaluation of uncertainties and calculating error bars are not particularly challenging. Given sampling of a quantity A, the uncertainty in the calculated average is simply the standard error (SE):

$$SE(A) = \frac{\sigma(A)}{\sqrt{N}} \qquad (9.1)$$

where
   σ is the standard deviation
   N is the number of data points

However, this equation applies *only* to independent and ideal samples, while MD produces correlated data, meaning that N samples do not contain a full N points worth of independent information. Indeed, all practitioners of MD are aware of this: we typically record coordinates only intermittently, something like every 5 or even 100 ps. If reducing statistical error was simply a matter of recording more points, we would simply write out coordinates more frequently. Choosing otherwise is an admission that we know there is little extra information in sampling smaller time intervals.

Thus, the real question is often as follows: given that we have fewer than N points of information in N samples, how much less do we have? Or, to ask another way, what is the correlation time? To a good approximation, we can define

$$N_{eff} \approx \frac{N}{\tau_{corr}} \qquad (9.2)$$

where
   N_{eff} is the effective number of data points
   τ_{corr} is the correlation time

The problem, however, is that in complex system such as a lipid bilayer or a protein, there is not one correlation time but many, and it is not always obvious which is the relevant one. It seems intuitive that one would calculate the correlation time directly from the time series of the observable of interest. However, this can be deceptive when the quantity fluctuates rapidly (suggesting a short τ) but is coupled to other slower motions. For example, the motions of an individual lipid chain are relatively quick, but are modulated by the slower fluctuations of the bilayer as a whole. This example will be discussed further in Section 9.4.2.2.

The best method for assessing the SE of a scalar quantity from a single time series is block averaging (Flyvbjerg and Petersen, 1989; Grossfield and Zuckerman, 2009). This approach uses the variance found in subsets of the time series to draw conclusions about the overall uncertainty, as demonstrated in Figure 9.2. One starts with a time series (or the equivalent from a quasi-dynamical method such as Monte Carlo) for a quantity A containing N data points. The trajectory is divided into M segments, each containing N/M points (see Figure 9.2a). For each of the M blocks, one takes the average value of A, then uses them to compute the SE of the block averages:

$$\text{BSE}(A, M) = \frac{\sigma_M(A)}{\sqrt{M}}. \tag{9.3}$$

This quantity can be viewed as an estimate of the true SE (and thus the statistical uncertainty). However, when the blocks are short and M is large, the averages in consecutive blocks are correlated with each other, with the result that the blocked standard error (BSE) systematically underestimates the true SE. However, Figure 9.2b shows that as the number of points per block $n = N/m$ becomes large, the blocks become uncorrelated and the BSE approaches the true SE. Thus, the best approach is to plot the BSE as a function of n. The BSE increases monotonically (except for statistical noise) with n and will eventually reach a plateau, where it becomes independent of n. At this point, the blocks are large enough to be statistically independent, and the plateau value of the BSE can be used to approximate the true SE.

One challenge in applying the block averaging approach is that the plateau region of the curve can be noisy, because large n implies a small number of blocks. For this reason, one often needs to visually inspect the curve and decide where the plateau region begins. In practice, it is also prudent to use an average over a range of block sizes in the plateau region rather than read a single value off the curve. As a result, there is no easy way to fully automate the calculation. Despite this inconvenience, we suggest that the block averaging approach be considered the gold standard for calculating the uncertainty in a single quantity.

In the case of biomolecules such as proteins, which have a native state that dominates the configuration, one can attempt to extract the longest correlation times (the macrofluctuations of the protein) and use them as a proxy for the equilibration time. Several labs have pursued this goal, most notably those of Zuckerman and Grossfield (Grossfield and Zuckerman, 2009). Zuckerman and coworkers have implemented a number of methods based on structural histograms: by grouping the protein structures in some way, one can assess either the rate at which the groups' populations converge (Grossfield et al., 2007; Lyman and Zuckerman, 2006, 2007) or the rate of interconversion between states (Zhang et al., 2010). Alternatively, Grossfield and coworkers have focused

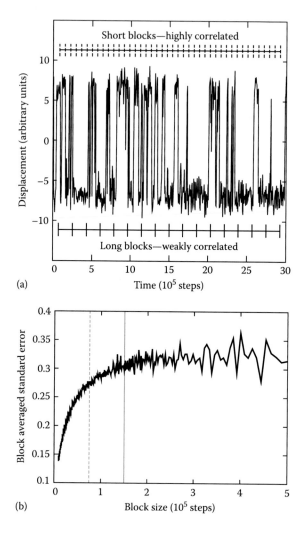

(a)

(b)

Figure 9.2   Evaluating statistical error via block averaging. (a) A time series generated for a simple double-well potential. (b) The standard deviation of the block averages as a function of the block size; the plateau value is the best estimate of the true SE. Two example block lengths are superimposed on the plot in (a). The shorter-size blocks are marked in (b) with a dashed line (before the plateau), and the longer-size blocks are marked with a solid line (after the plateau).

on methods based in the convergence of the principal components computed from the trajectory (Grossfield et al., 2007; Romo and Grossfield, 2011a,b). The results from the two approaches are largely consistent, although some differences in interpretation remain.

All of these methods focus on how to extract statistical uncertainties from a single trajectory and neglect the best strategy (universally applied in experimental work), which is simply to repeat the experiment. If one runs 20 (or 2 million) independent trajectories, one can treat the average from each as a single measurement and (provided the systems are *truly* independent) use simple statistics via Equation 9.1. However, even this strategy is not necessarily simple to apply: each trajectory must be sufficiently long that it conducts some sampling and is not simply determined by the details of the construction procedure. In effect, there is a characteristic event time that sets the minimum simulation time to be performed. In many simple systems, such as butane, the event time (in this case, the time required for a transition of the central torsion angle, *once the transition has actually started*) is quite short, and the vast majority of the simulation time is spent waiting for the transitions to occur. By contrast, the transitions seen in large systems such as a GPCR can require 10s or even 100s of nanoseconds to occur (Hurst et al., 2010; Romo et al., 2010).

This brings us to our final caution: there is no ready method to know what has *not* been seen in a simulation. If we have a single transition from the starting state to a new one, we know that the statistics are not great, but until that transition occurred, we did not even know the second state existed. This sort of unknown (Rumsfeld, 2002) remains the greatest challenge in assessing simulation results.

9.4 ANALYZING THE SIMULATION AND COMPARING TO EXPERIMENT

Pulling the interesting parts of a simulation together into a coherent story is often a daunting challenge in computational biology. One reason is the embarrassment of riches presented by a simulation: once the trajectory is complete, one can calculate almost any property, as long as it is a function of the atomic coordinates. This presents both a challenge and an opportunity: membrane protein systems are $O(10^5-10^7)$ atoms or more, but only a few degrees of freedom are really functionally interesting. Luckily, in most cases, there are ways to define these degrees of freedom. In practice, you will have some preconceived notion about what the system will do and what you hoped to learn from the simulation (or else you never would have run it in the first place), and this can guide your choices in how to analyze the data. In this respect, literature review is especially important to a computational biologist. In order for a simulation to make an impact, it is critical that it addresses a question the experimental community finds interesting, and the best way to know this is to immerse yourself in the literature.

We must warn investigators, young and old, that specific analyses can be misleading, particularly without careful statistics: if you go looking for a pattern in the data, you will usually find it, whether or not it is actually there (this type of occurrence is "apophenia," a word every simulator should know). For this reason, the best approach is to calculate a large number of related properties and in general consider the system from as many directions as possible, to make sure you are seeing the whole phenomenon and not just a convenient slice. To do so, it is critical to have the means to quickly create code to perform novel analyses, which can be demanding if one is limited to using the tools distributed with the common simulation packages. For this reason, it is a good investment to learn one or more analysis suites such as MDAnalysis (Michaud-Agrawal et al., 2011) or (our particular favorite) LOOS (Romo and Grossfield, 2012). One of the real strengths of simulation is its ability to produce whatever analysis one wants without the restrictions found experimentally, and it would be foolish to give that up based on the challenges of developing code within the framework of a particular MD package.

That said, it is often crucial to focus analysis on those quantities that can be measured experimentally. The purpose of this is validation: while simulations have continually gotten better, the fact is that one needs to provide evidence that the simulation is correctly capturing natural phenomena before proceeding to compute things that are not accessible experimentally.

This distinction lets us split data analysis into two categories: comparison to the interpretation of experimental results and comparison to the experimental observable itself. The former is the more common approach; where experiment is by definition limited to what can be measured, simulations let you calculate pretty much any quantity, and as a result, it is often satisfying to go directly to the quantity of interest.

However, this can often lead to erroneous disagreement between simulations and experimental results, because there are often several steps and numerous built-in assumptions between the raw experimental results and the resulting interpretations. While these assumptions are often reasonable—for example, fluctuations are normally distributed—they are not necessarily perfectly accurate and are not needed when simulations directly measure the underlying distributions. As a result, when attempting to validate a simulation via comparison to experiment, the best choice is to compare the simulation to the underlying raw data (e.g., NMR spectrum or x-ray scattering structure factors) rather than their interpretation (orientation of a moiety or electron density profile of a membrane).

On a technical note, we must briefly address the implementation of tools for analyzing data. This topic itself has filled many manuscripts, and we refer readers to the excellent books by Frenkel and Smit (2002), Rapaport (2004), and Press et al. (2007) to gain a further knowledge on implementing analysis tools. Many prebuilt tools are also offered in simulation software packages. However, reliance solely on the tools implemented by others is a pitfall that will limit your capability. Therefore, a good working knowledge of a scripting (if not a compiled) language will be of great benefit.

Many additional analysis suites exist (Michaud-Agrawal et al., 2011; Miller et al., 2008; Romo and Grossfield, 2009, 2012), with the goal of expanding the type of questions computational biologists can address. LOOS (analysis library) (Romo and Grossfield, 2012), the package maintained by the Grossfield lab, is different from most analysis suites in that our main goal is a core library to facilitate easy tool development. While LOOS is distributed with roughly 100 stand-alone applications, it is in essence a tool for making tools. The goal of LOOS is to trivialize the overhead of writing analysis software: reading and writing files from a number of simulation packages (we currently support the native file formats of Amber, CHARMM, GROMACS, NAMD, and Tinker) as well as performing the microtasks that make up other analysis, such as selecting atoms, measuring distances and angles, superposing structures, and computing principle axes. Turning all of these tasks into one-line operations allows the user to focus on the analysis she wishes to perform rather than the sometimes messy numerics behind the scene. The core of LOOS is written in C++, but one can develop applications in either C++ or python, with minimal differences in performance. This suite is freely available from http://loos.sourceforge.net/.

9.4.1 Common Analyses

One common analysis is to calculate the number density of one group of atoms around another, in order to identify preferential interactions and colocations. For instance, you may be interested in how many lipids of type b are within a given distance of another lipid, a. In this case, we would call a the probe and b the target.

The typical 1D form of this analysis (with some normalization applied) is called the radial distribution function (RDF). This result is plotted as a function of distance, yielding the number of targets that are within a sphere of some radius. Formally, the RDF is defined as

$$g_{ab}(r) = \frac{n_{ab}(r)/V(r)}{\rho_{0,b}} \qquad (9.4)$$

where
$n_{ab}(r)$ is the number of b molecules at distance r away from a molecules
$V(r)$ is the volume of the slice of space at distance r
$\rho_{0,b}$ is the average density of b in the system as a whole

Of course, when computing an RDF using real data, one uses finite bins rather than continuous space. The functional form of $V(r)$ depends on the choice of dimensionality of the calculation. For example, in solution, one would typically define r to be the standard 3D distance, so the volume would be

$$V(r) = \frac{4\pi}{3}\left(\left(r + \frac{\delta}{2}\right)^3 - \left(r - \frac{\delta}{2}\right)^3\right) \tag{9.5}$$

where δ is the bin width. With membranes, we are often more interested in the distribution of objects in the plane of the membrane, so we use a 2D distance $r = \sqrt{\Delta x^2 + \Delta y^2}$ and volume

$$V(r) = \pi\left(\left(r + \frac{\delta}{2}\right)^2 - \left(r - \frac{\delta}{2}\right)^2\right). \tag{9.6}$$

Physically, the RDF can be thought of as a normalized probability density, where a value of 1 is the result seen for an ideal gas.

As an example application, we can consider the packing of lipids around a membrane protein. This can be interesting, for instance, when examining the distribution of polyunsaturated lipids around rhodopsin (Feller et al., 2003; Grossfield et al., 2006a,b; Horn et al., 2013) or when looking at the origins of headgroup effects on protein function (Soubias and Gawrisch, 2011; Soubias et al., 2010). In Figure 9.3a, we show lipid molecules distributed around rhodopsin, while Figure 9.3b shows the 2D lateral RDF for two headgroup species (PE and PC) about the protein, as calculated from a series of coarse-grained MD simulations (Horn et al., 2013).

While the RDF is a powerful quantity, as its name implies, it considers only radial information, ignoring rotational degrees of freedom. This amounts to considering a and b as spheres (for a conventional 3D distribution) or cylinders (for a 2D lateral distribution). This can be a good approximation (for instance, when examining the distribution of one lipid species around another), but for others, it can be badly flawed. For example, to treat a GPCR—a large protein with seven transmembrane helices, each with its own distinct sequence—as a featureless cylinder would discard much of what makes the molecule interesting. In this case, one could instead compute the 2D distribution about the protein, retaining these degrees of freedom. However, the calculation is somewhat more involved than a simple RDF, because one must first remove the protein's translational and rotational degrees of freedom by aligning each frame to the starting structure. Moreover, because the data are spread

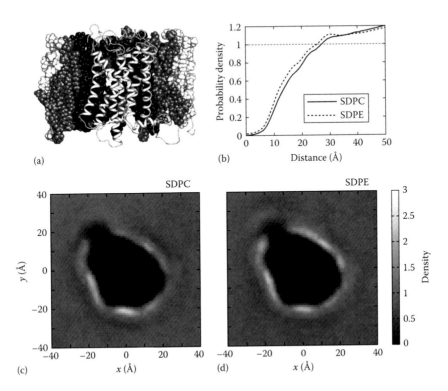

Figure 9.3 Distributions of lipid headgroups around a protein. (a) Rhodopsin (white cartoon), with lipids in the first, second, and third "solvation shells" around it shown as black, gray, and white spheres, respectively. (b) The lateral RDFs for PC and PE headgroups about the protein. (c and d) The 2D distribution of PC and PE about rhodopsin. (Data previously published as Horn, J.N., Kao, T.-C., and Grossfield, A., *Adv. Exp. Med. Biol.*, 796, 75, 2014.)

out over two dimensions, it is intrinsically more noisy than the equivalent 1D data. Figure 9.3c and d shows the distribution of PC and PE lipids about the protein without rotational averaging; it is clear that there are sites on the surface where headgroups pack more strongly than others, information that cannot be gleaned from the 1D RDF.

For that matter, one can also examine the full 3D distribution about the protein, using much the same procedure; after aligning the protein, the lipids are distributed into voxels and the average density computed. The same caveats apply as in the 2D case (most notably, the lack of spatial averaging means the data tends to be noisier), but the amount of information available dramatically increases. This sort of analysis has been used to explore both the distribution of lipids around cholesterol (Pitman et al., 2004) and proteins (Horn et al., 2013) and hydration inside them (Romo et al., 2010).

9.4.2   Comparison to Experimental Observables

One of the key tests of a simulation is its ability to reproduce experimental results. Part of this is simple validation: we need some evidence that the simulation is representative of realistic conditions before we are willing to trust it to predict new phenomena. However, simulations have the advantage that they produce not just the final result but also the underlying distribution, in much the same way that single-molecule techniques do when compared to bulk spectroscopic measurements. This means that many of the assumptions often used in interpreting experimental data—turning the raw measurements into physically intuitive results—can be directly tested using simulation data. For this reason, we recommend comparing simulation results to the rawest available form of the data, rather than to the interpretation of that data.

9.4.2.1   Density Profiles

Because lipid membranes are in a liquid crystalline state under physiological conditions, there is often a paucity of high-resolution information about them. The information that does exist often describes the distribution of chemical moieties along the membrane normal (the unique axis of the crystal). For example, this information can be extracted from scattering experiments on ordered samples. X-ray scattering is determined by the electron density distribution, while neutron scattering depends on the distribution of nuclei, although the scattering is not simply related to nuclear mass. White and coworkers combined these techniques to "solve" the structure of liquid crystalline bilayers by decomposing the membrane into specific components (e.g., water, acyl chain) (Wiener and White, 1991a,b, 1992; Wiener et al., 1991).

This kind of information is available directly from a molecular dynamics trajectory; at each time point, the membrane center of mass is aligned at $z = 0$, and the distributions of different atom types can be computed by simple histogramming. Computing the mass density distribution is nearly as easy, simply requiring reweighting the number distributions by the mass of each nuclei. Obtaining the electron density is also relatively simple, using the weighting $e_{tot,i} = N_i e^0 - q_i$, where e^0 is the charge on an electron, N_i is the atomic number, and q_i is the partial charge for that atom, taken from the force field. The results of this kind of calculation are shown for a POPC membrane in Figure 9.1.

Working with a quantity such as electron density is very convenient, in that its meaning is physically intuitive and connects directly to what we want to know, the distribution of different membrane components. However, there are complications when comparing to electron density distributions from experiment. The two primary methods for extracting electron density from experimental structure factors, direct Fourier reconstruction and functional modeling, both require external information that is not available experimentally (Sachs et al., 2003). As a result of these assumptions, the density profiles calculated from an MD trajectory and an experiment could disagree not because the simulation or experiment was wrong, but because of flaws in those external assumptions.

For this reason, several groups have argued that it is better to work in the other direction and compute the bilayer structure factors (which can be measured unambiguously) from the simulation (Benz et al., 2006, 2005; Castro-Román et al., 2006; Kucerka et al., 2008; Sachs et al., 2003). To do so, one begins by computing the electron density profile as described earlier and then performs a Fourier transform:

$$F(q) = \int_{-D/2}^{D/2} (\rho(z) - \rho_W)\cos(qz)dz \tag{9.7}$$

where
  $F(q)$ are the structure factors
  D is average box length along the membrane normal z
  $\rho(z)$ is the simulated electron density
  ρ_W is the electron density of bulk water

At this point, most of the all-atom simulations performed are small enough that bilayer undulations are damped out by the periodic boundary conditions. However, as computers get faster, larger simulations where these undulations are present are becoming more common (Lindahl and Edholm, 2000). These undulations complicate the calculation described previously, effectively "blurring" the electron density (Brandt et al., 2011; Braun et al., 2011). This happens because the direction of the bilayer normal can vary locally, rather than remaining fixed along the z axis, such that electron-dense moieties are systematically spread in z. Sachs and coworkers solved this problem by identifying the motions associated with the "undulating reference frame" (Brandt et al., 2011), estimating their contribution to the structure factor spectrum, and filtering them out to produce the spectrum for a "flat" bilayer (Braun et al., 2011).

The electrostatic potential along the membrane normal can also be computed from a density distribution, although this quantity is removed from the raw experimental data by an additional step. To make this calculation, one first computes the charge density profile, with each atom weighted by its partial charge q_i. From this, the electrostatic potential can be computed as (Sachs et al., 2004)

$$\varphi(z) = -\frac{1}{\varepsilon_0}\int_0^z (z-u)\rho(u)du - \frac{z}{\varepsilon_0 D}\int_0^D (D-u)\rho(u)du \qquad (9.8)$$

where
  ρ is the charge density
  D is the box dimension along the membrane normal
  ε_0 is the permittivity of free space

The equation was solved by arbitrarily setting $\varphi(0) = 0$; the second integral results from assuming that the potential is continuous at the periodic boundary. More recently, other groups have argued that the second term is unnecessary (Gurtovenko and Vattulainen, 2008), but its absence can lead to unphysical consequences, including a dependence of the shape of the potential profile on the choice of origin and the presence of discontinuities in the potential in regions where the charge density is zero.

9.4.2.2 Order Parameters from ²H NMR

One of the most common experiments to characterize membrane structure is measuring the lipid order parameters, summarized in Figure 9.4. The experiment proceeds as follows: lipid acyl chains are partially or fully deuterated in a system with no other deuteriums and the spectrum is measured, with oriented samples. This is typically done with the magnetic field applied along the membrane normal, although the experiment can also be done with vesicles and using de-Pake-ing techniques (Sternin et al., 1983). If we consider a single deuterated site (e.g., carbon 7 of the saturated chain of POPC), the spectrum will be a classical Pake doublet, and the frequency difference between the two peaks will be

$$\Delta v = \frac{3}{4}\frac{e^2 qQ}{h}\langle 3\cos^2 \Theta_{C-D} - 1 \rangle \qquad (9.9)$$

where
  $e^2 qQ/h$ is the quadrupolar coupling constant
  Θ_{C-D} is the angle between the carbon–deuterium bond vector and the applied magnetic field (see Figure 9.4a)
  the angle brackets denote the thermodynamic average

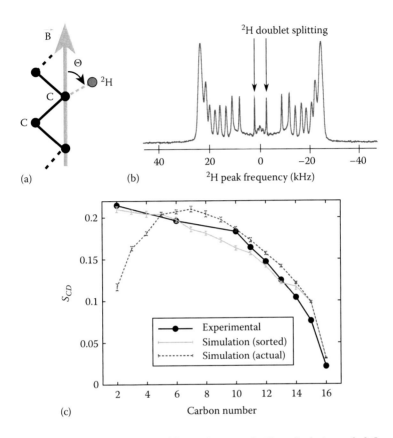

Figure 9.4 ^{2}H order parameters. (a) A schematic lipid acyl chain and defines Θ. The external magnetic field ($\vec{B}$) is indicated by the gray arrow. (b) An example experimental spectrum, for a perdeuterated lipid system. (Data provided by Denise Greathouse, unpublished; Data previously published as Romo, T.D., *Biochim. Biophys. Acta*, 1808(8), 2019, 2011.) (c) Example order parameters extracted from an experiment and the equivalent data from a molecular dynamics simulation. Note that the experimental data have been sorted to decrease monotonically.

As a rule (particularly when comparing to simulation), the geometric component is extracted as the order parameter

$$S_{CD} = \frac{1}{2}|\langle 3\cos^2\Theta_{C-D} - 1\rangle|. \tag{9.10}$$

This quantity is easily computed from an all-atom molecular dynamics simulation, assuming the magnetic field is applied along the membrane normal. If the force field does not include explicit acyl hydrogens, their position can be estimated by taking the average of the carbon's bond vectors to those preceding and following it in the chain and assuming ideal methylene geometry.

The comparison between simulation and experiment is generally somewhat more complicated, because one rarely has a sample with only one site deuterated. More commonly, the sample is perdeuterated on a single chain, which gives a set of overlapping doublets, centered at the same chemical shift (see Figure 9.4b). This means that (1) there is no experimental way to know which pair of peaks is due to which carbon position and (2) the location of any given peak may be shifted somewhat by the spectral signal from other deuterons. The former problem is generally handled in experimental papers by assuming that the amplitude of the splitting decreases monotonically as one moves down the chain, away from the lipid headgroup. However, Figure 9.4c shows typical simulation data, which contradicts this assumption. The reason is there are actually two ways to produce a small splitting. The first comes from a bond vector that samples a broad range of orientations; this is the assumption made when sorting carbons by their peak splitting. However, a chain that is relatively ordered (has a single well-defined orientation) but has a systematic tilt can produce a smaller splitting than one that is less ordered but untilted, and the chemical geometry of the headgroup–glycerol–acyl chain linkage can produce such a tilt. As a result, it is not unusual for simulation data to show lower values for the first carbon or two in the chain (Figure 9.4c); in that case, the only fair way to compare the simulation and experiment is to sort the simulation data (where we can identify each carbon individually) is descending order as well.

A second question that is crucial to comparing the simulation to experiment (or comparing two simulations) is how to estimate the error bars on S_{CD} values computed from a simulation. Again focusing on a single deuterated position, the question comes down to how many independent samples one has obtained. For example, one could treat the value of each C–^{2}H bond at each time point as independent and compute the SE in the mean. However, this produces implausibly small error bars, which makes sense: at the very least, the two ^{2}H's bound to a given carbon are very strongly correlated with each other. One could instead lump these two bonds together and consider each lipid independently. This leads to larger but still implausibly small error bars, because the correlation between neighboring lipids is neglected. The question then becomes how long a length scale does the correlation persist, and how can we estimate it? In a simulation where the box size is allowed to fluctuate, the answer is probably "all the way," because the fluctuations in the lateral area directly modulate all of the lipids' chain orientations simultaneously. Furthermore, this quantity is one of the slower fluctuations available to the system, on the timescale of 10s to 100s of nanoseconds, which means that there is significant correlation in time as well.

Arguably, the best way to handle this problem is to run multiple independent trajectories and treat the average S_{CD} from each simulation as an independent measurement. However, this can be prohibitively costly in many cases, and so it would be valuable to have a good estimator for the case where only one trajectory per system is produced.

One approach, which to our knowledge has only been implemented in the package LOOS (Romo and Grossfield, 2009, 2012), is to use block averaging on the average instantaneous SCD for the whole system. That is, for any given position in the chain, one computes the value for each $C-{}^2H$ bond in a given configuration and averages the values for the whole system. Next, take the time series of system averages, and apply block averaging (Flyvbjerg and Petersen, 1989) to estimate the BSE (see Section 9.3.4). To use the algorithm effectively (at least as it is implemented in LOOS), one must run the order parameter calculation twice. The first time, one computes the actual order parameter values and requests a block averaging calculation; in addition to outputting the order parameters themselves, the BSE is returned for each carbon position as a function of the length of the blocks. These plots must be visually inspected to determine where the plateau region begins and thus what range of block sizes should be averaged when computing the full BSE. With this information, one reruns the order parameter calculation, giving this range of block sizes as an input parameter. This yields the best estimate of the statistical uncertainty for each carbon order parameter.

9.4.2.3 Computing ${}^2H$ NMR Spectra

Although order parameters of the kind described earlier are by far the most common means of comparing simulation and ${}^2H$ NMR, they are arguably not the best one. As with electron density and x-ray scattering (described in Section 9.4.2.1), it can be more reliable to compare the simulation to the original results, in this case, the ${}^2H$ spectrum.

The theory describing the spectral lineshape due to quadrupolar splitting was worked out by Brown and coworkers (Nevzorov et al., 1999), including the dependence on bond orientation and the disorder in the alignment of an ensemble of samples (the mosaic spread). This theory was later used to examine the orientation of the retinal ligand in the binding pocket of rhodopsin (Salgado et al., 2004). They first chemically synthesized three distinct retinal compounds, each with a single methyl group deuterated. (Performing NMR on rhodopsin with one of these retinals bound gives a simple spectrum with a single doublet peak.) They reconstituted rhodopsin with oriented bilayers and measured the spectra while varying the tilt in the bilayer with respect to the magnetic field. They then determined the orientation of the retinal ionone ring by comparing the spectra for the three methyl groups with those computed using the lineshape theory (Nevzorov et al., 1999) and selecting the single structure that best matched the experiment.

While powerful and successful in its own right, this approach did require the assumption that there was only a single conformation for retinal's ionone ring in the dark state of rhodopsin. However, extensive molecular dynamics simulations showed that not one but two distinct orientations were sampled (Lau et al., 2007). To reconcile these apparently contradictory results, they combined the lineshape theory and the molecular dynamics simulations. They computed the probability distribution for the bond orientations for each of the methyl groups and used it to compute a weighted average spectrum (in a sense, the spectrum predicted by the simulation). The results showed that the weighted combination of the two states produced a spectrum that matched the experiment significantly better than either state alone did, emphasizing the importance of comparing simulations as directly as possible to the experiments themselves. The same approach was also used to determine the orientation of retinal in the Meta-I state, settling a controversy in the literature regarding the role of a pair of glutamate residues in the binding pocket (Martínez-Mayorga et al., 2006).

9.5 CONCLUSION

Membrane proteins are notoriously hard to work with experimentally, and as such their study continues to be greatly augmented by molecular simulations. The current state of hardware and algorithms has put us on the timescale of microsecond-length simulations (for all-atom systems), allowing us to ask more interesting questions and answer them with better statistical confidence. Moreover, new analytical tools are increasing our ability to extract the maximum information from the simulations we perform. At the same time, experimental methods are improving as well, giving the simulators more data to validate their calculations and also more starting structures for new simulations. As a result, the potential for collaborative advancement combining simulation and experiment has never been higher.

REFERENCES

Allen, F., G. Almasi, W. Andreoni, D. Beece, B. J. Berne, A. Bright, J. Brunheroto, C. et al. (2001). Blue Gene: A vision for protein science using a petaflop supercomputer. *IBM Syst J 40*, 310.

Allen, M. P. and D. J. Tildesley (1987). *Computer Simulations of Liquids*. Oxford, U.K.: Clarendon Press.

Allison, S. A. and J. A. McCammon (1984, February). Multistep brownian dynamics: Application to short wormlike chains. *Biopolymers 23*(2), 363–375.

Andersen, H. C. (1980). Molecular-dynamics simulations at constant pressure and/or temperature. *J Chem Phys 72*, 2384–2393.

Ayton, G. S. and G. A. Voth (2009, April). Systematic multiscale simulation of membrane protein systems. *Curr Opin Struct Biol 19*(2), 138–144.

Benz, R. W., F. Castro-Román, D. J. Tobias, and S. H. White (2005, February). Experimental validation of molecular dynamics simulations of lipid bilayers: A new approach. *Biophys J 88*(2), 805–817.

Benz, R. W., H. Nanda, F. Castro-Román, S. H. White, and D. J. Tobias (2006, November). Diffraction-based density restraints for membrane and membrane-peptide molecular dynamics simulations. *Biophys J 91*(10), 3617–3629.

Berendsen, H. J. C., J. P. M. Postma, W. F. van Gunsteren, A. DiNola, and J. R. Haak (1984). Molecular dynamics with coupling to an external bath. *J Chem Phys 81*, 3684–3690.

Berger, O., O. Edholm, and F. Jähnig (1997, May). Molecular dynamics simulations of a fluid bilayer of dipalmitoylphosphatidylcholine at full hydration, constant pressure, and constant temperature. *Biophys J 72*(5), 2002–2013.

Bowers, K. J., E. Chow, H. Xu, R. O. Dror, M. P. Eastwood, B. A. Gregersen, J. L. Klepeis et al. (2006, November). Scalable algorithms for molecular dynamics simulations on commodity clusters. *Proceedings of the ACM/IEEE Conference on Supercomputing (SC06)*, Tampa, FL.

Brandt, E. G., A. R. Braun, J. N. Sachs, J. F. Nagle, and O. Edholm (2011, May). Interpretation of fluctuation spectra in lipid bilayer simulations. *Biophys J 100*(9), 2104–2111.

Braun, A. R., E. G. Brandt, O. Edholm, J. F. Nagle, and J. N. Sachs (2011, May). Determination of electron density profiles and area from simulations of undulating membranes. *Biophys J 100*(9), 2112–2120.

Carrillo-Tripp, M. and S. E. Feller (2005, August). Evidence for a mechanism by which omega-3 polyunsaturated lipids may affect membrane protein function. *Biochemistry 44*(30), 10164–10169.

Castro-Román, F., R. W. Benz, S. H. White, and D. J. Tobias (2006, November). Investigation of finite system-size effects in molecular dynamics simulations of lipid bilayers. *J Phys Chem B 110*(47), 24157–24164.

Dror, R. O., D. H. Arlow, P. Maragakis, T. J. Mildorf, A. C. Pan, H. Xu, D. W. Borhani, and D. E. Shaw (2011a, November). Activation mechanism of the Î²2-adrenergic receptor. *Proc Natl Acad Sci USA 108*(46), 18684–18689.

Dror, R. O., A. C. Pan, D. H. Arlow, D. W. Borhani, P. Maragakis, Y. Shan, H. Xu, and D. E. Shaw (2011b, August). Pathway and mechanism of drug binding to g-protein-coupled receptors. *Proc Natl Acad Sci USA 108*(32), 13118–13123.

Eldho, N. V., S. E. Feller, S. Tristram-Nagle, I. V. Polozov, and K. Gawrisch (2003, May). Polyunsaturated docosahexaenoic vs docosapentaenoic acid-differences in lipid matrix properties from the loss of one double bond. *J Am Chem Soc 125*(21), 6409–6421.

Feller, S. E. (2008, May). Acyl chain conformations in phospholipid bilayers: A comparative study of docosahexaenoic acid and saturated fatty acids. *Chem Phys Lipids 153*(1), 76–80.

Feller, S. E. and K. Gawrisch (2005, August). Properties of docosahexaenoic-acid-containing lipids and their influence on the function of rhodopsin. *Curr Opin Struct Biol 15*(4), 416–422.

Feller, S. E., K. Gawrisch, and A. D. MacKerell (2002, January). Polyunsaturated fatty acids in lipid bilayers: Intrinsic and environmental contributions to their unique physical properties. *J Am Chem Soc 124*(2), 318–326.

Feller, S. E., K. Gawrisch, and T. B. Woolf (2003, April). Rhodopsin exhibits a preference for solvation by polyunsaturated docosohexaenoic acid. *J Am Chem Soc 125*(15), 4434–4435.

Feller, S. E., D. Huster, and K. Gawrisch (1999). Interpretation of NOESY cross-relaxation rates from molecular dynamics simulation of a lipid bilayer. *J Am Chem Soc 121*, 8963–8964.

Feller, S. E. and R. W. Pastor (1996, September). On simulating lipid bilayers with an applied surface tension: Periodic boundary conditions and undulations. *Biophys J 71*(3), 1350–1355.

Feller, S. E., Y. Zhang, R. W. Pastor, and B. R. Brooks (1995). Constant pressure molecular dynamics simulation: The Langevin piston method. *J Chem Phys 103*, 4613–4621.

Fitch, B. G., R. S. Germain, M. Mendell, J. Pitera, M. Pitman, A. Rayshubskiy, Y. Sham et al. (2003). Blue Matter, an application framework for molecular simulation on Blue Gene. *J Parallel Distrib Comput 63*, 759–773.

Flyvbjerg, H. and H. G. Petersen (1989). Error estimates on averages of correlated data. *J Chem Phys 91*, 461–466.

Frenkel, D. and B. Smit (2002). *Understanding Molecular Simulation* (2nd edn.). San Diego, CA: Academic Press.

Grossfield, A. (2011, July). Recent progress in the study of g protein-coupled receptors with molecular dynamics computer simulations. *Biochim Biophys Acta 1808*(7), 1868–1878.

Grossfield, A., S. E. Feller, and M. C. Pitman (2006a, May). Contribution of omega-3 fatty acids to the thermodynamics of membrane protein solvation. *J Phys Chem B 110*(18), 8907–8909.

Grossfield, A., S. E. Feller, and M. C. Pitman (2006b, March). A role for direct interactions in the modulation of rhodopsin by omega-3 polyunsaturated lipids. *Proc Natl Acad Sci USA 103*(13), 4888–4893.

Grossfield, A., S. E. Feller, and M. C. Pitman (2007, April). Convergence of molecular dynamics simulations of membrane proteins. *Proteins: Struct Funct Bioinf 67*(1), 31–40.

Grossfield, A., M. C. Pitman, S. E. Feller, O. Soubias, and K. Gawrisch (2008, August). Internal hydration increases during activation of the G-protein-coupled receptor rhodopsin. *J Mol Biol 381*(2), 478–486.

Grossfield, A. and D. M. Zuckerman (2009, January). Quantifying uncertainty and sampling quality in biomolecular simulations. *Annu Rep Comput Chem 5*, 23–48.

Gurtovenko, A. A. and I. Vattulainen (2008, April). Membrane potential and electrostatics of phospholipid bilayers with asymmetric transmembrane distribution of anionic lipids. *J Phys Chem B 112*(15), 4629–4634.

Hoover, W. G. (1985). Canonical dynamics: Equilibrium phase-space distributions. *Phys Rev A 31*, 1695–1697.

Horn, J. N., T.-C. Kao, and A. Grossfield (2014). Coarse-grained molecular dynamics provides insight into the interactions of lipids and cholesterol with rhodopsin. *Adv Exp Med Biol 796*, 75–94.

Horn, J. N., J. D. Sengillo, D. Lin, T. D. Romo, and A. Grossfield (2012, February). Characterization of a potent antimicrobial lipopeptide via coarse-grained molecular dynamics. *Biochim Biophys Acta 1818*(2), 212–218.

Humphrey, W., A. Dalke, and K. Schulten (1996, February). VMD: Visual molecular dynamics. *J Mol Graph 14*(1), 33–38, 27–28.

Hurst, D. P., A. Grossfield, D. L. Lynch, S. Feller, T. D. Romo, K. Gawrisch, M. C. Pitman, and P. H. Reggio (2010, June). A lipid pathway for ligand binding is necessary for a cannabinoid G protein-coupled receptor. *J Biol Chem 285*(23), 17954–17964.

Jardón-Valadez, E., A.-N. Bondar, and D. J. Tobias (2009, April). Dynamics of the internal water molecules in squid rhodopsin. *Biophys J 96*(7), 2572–2576.

Jardón-Valadez, E., A.-N. Bondar, and D. J. Tobias (2010, October). Coupling of retinal, protein, and water dynamics in squid rhodopsin. *Biophys J 99*(7), 2200–2207.

Jo, S., J. B. Lim, J. B. Klauda, and W. Im (2009, July). Charmm-gui membrane builder for mixed bilayers and its application to yeast membranes. *Biophys J 97*(1), 50–58.

Johnston, J. M. and M. Filizola (2011, August). Showcasing modern molecular dynamics simulations of membrane proteins through g protein-coupled receptors. *Curr Opin Struct Biol 21*(4), 552–558.

Kandt, C., W. L. Ash, and D. P. Tieleman (2007, April). Setting up and running molecular dynamics simulations of membrane proteins. *Methods 41*(4), 475–488.

Khelashvili, G., A. Grossfield, S. E. Feller, M. C. Pitman, and H. Weinstein (2009, August). Structural and dynamic effects of cholesterol at preferred sites of interaction with rhodopsin identified from microsecond length molecular dynamics simulations. *Proteins 76*(2), 403–417.

Klauda, J. B., R. M. Venable, J. A. Freites, J. W. O'Connor, D. J. Tobias, C. Mondragon-Ramirez, I. Vorobyov, A. D. MacKerell, and R. W. Pastor (2010, June). Update of the charmm all-atom additive force field for lipids: Validation on six lipid types. *J Phys Chem B 114*(23), 7830–7843.

Kucerka, N., J. F. Nagle, J. N. Sachs, S. E. Feller, J. Pencer, A. Jackson, and J. Katsaras (2008, September). Lipid bilayer structure determined by the simultaneous analysis of neutron and x-ray scattering data. *Biophys J 95*(5), 2356–2367.

Kutzner, C., D. van der Spoel, M. Fechner, E. Lindahl, U. W. Schmitt, B. L. de Groot, and H. Grubmüller (2007, September). Speeding up parallel gromacs on high-latency networks. *J Comput Chem 28*(12), 2075–2084.

Lau, P.-W., A. Grossfield, S. E. Feller, M. C. Pitman, and M. F. Brown (2007, September). Dynamic structure of retinylidene ligand of rhodopsin probed by molecular simulations. *J Mol Biol 372*(4), 906–917.

Lee, J. Y. and E. Lyman (2012). Agonist dynamics and conformational selection during microsecond simulations of the a2a adenosine receptor. *Biophys J 102*, 2114–2120.

Lindahl, E. and O. Edholm (2000). Mesoscopic undulations and thickness fluctuations in lipid bilayers from molecular dynamics simulations. *Biophys J 79*, 426–433.

Lindahl, E. and M. S. P. Sansom (2008, August). Membrane proteins: Molecular dynamics simulations. *Curr Opin Struct Biol 18*(4), 425–431.

Lyman, E. and D. M. Zuckerman (2006, July). Ensemble-based convergence analysis of biomolecular trajectories. *Biophys J 91*(1), 164–172.

Lyman, E. and D. M. Zuckerman (2007, November). On the structural convergence of biomolecular simulations by determination of the effective sample size. *J Phys Chem B 111*(44), 12876–12882.

MacCallum, J. L., W. F. D. Bennett, and D. P. Tieleman (2008, May). Distribution of amino acids in a lipid bilayer from computer simulations. *Biophys J 94*(9), 3393–3404.

MacCallum, J. L. and D. P. Tieleman (2011, December). Hydrophobicity scales: A thermodynamic looking glass into lipid–protein interactions. *Trends Biochem Sci 36*(12), 653–662.

Marrink, S. J., H. J. Risselada, S. Yefimov, D. P. Tieleman, and A. H. de Vries (2007, July). The martini force field: Coarse grained model for biomolecular simulations. *J Phys Chem B 111*(27), 7812–7824.

Martínez-Mayorga, K., M. C. Pitman, A. Grossfield, S. E. Feller, and M. F. Brown (2006, December). Retinal counterion switch mechanism in vision evaluated by molecular simulations. *J Am Chem Soc 128*(51), 16502–16503.

Martyna, G. J., M. L. Klein, and M. E. Tuckerman (1992). Nosé-hoover chains: The canonical ensemble via continuous dynamics. *J Chem Phys 97*, 2635–2645.

Mátyus, E., C. Kandt, and D. P. Tieleman (2007). Computer simulation of antimicrobial peptides. *Curr Med Chem 14*(26), 2789–2798.

Michaud-Agrawal, N., E. J. Denning, T. B. Woolf, and O. Beckstein (2011, April). MDAnalysis: A toolkit for the analysis of molecular dynamics simulations. *J Comput Chem 32*, 2319–2327.

Miller, B. T., R. P. Singh, J. B. Klauda, M. Hodoscek, B. R. Brooks, and H. L. Woodcock, 3rd (2008, September). CHARMMing: A new, flexible web portal for CHARMM. *J Chem Inf Model 48*(9), 1920–1929.

Monticelli, L., S. Kandasamy, X. Periole, R. Larson, D. Tieleman, and S. Marrink (2008). The martini coarse grained forcefield: Extension to proteins. *J Chem Theory Comput 4*, 819–839.

Nevzorov, A. A., S. Moltke, M. P. Heyn, and M. F. Brown (1999). Solid-state NMR line shapes of uniaxially oriented immobile systems. *J Am Chem Soc 121*(33), 7636–7643.

Nosé, S. (1984). A unified formulation of the constant temperature molecular dynamics method. *J Chem Phys 81*, 511–519.

Okada, T., Y. Fujiyoshi, M. Silow, J. Navarro, E. M. Landau, and Y. Shichida (2002, April). Functional role of internal water molecules in rhodopsin revealed by X-ray crystallography. *Proc Natl Acad Sci USA 99*(9), 5982–5987.

Parrinello, M. and A. Rahman (1981). Polymorphic transitions in single crystals: A new molecular dynamics method. *J Appl Phys 52*(12), 7182–7190.

Periole, X., T. Huber, S.-J. Marrink, and T. P. Sakmar (2007, August). G protein-coupled receptors self-assemble in dynamics simulations of model bilayers. *J Am Chem Soc 129*(33), 10126–10132.

Phillips, J. C., R. Braun, W. Wang, J. Gumbart, E. Tajkhorshid, E. Villa, C. Chipot, R. D. Skeel, L. Kalé, and K. Schulten (2005, December). Scalable molecular dynamics with NAMD. *J Comput Chem 26*(16), 1781–802.

Pitman, M., F. Suits, A. D. MacKerell Jr., and S. E. Feller (2004). Molecular-level organization saturated and polyunsaturated fatty acids in a phosphatidylcholine bilayer containing cholesterol. *Biochemistry 43*(49), 15318–15328.

Polyansky, A. A., R. Ramaswamy, P. E. Volynsky, I. F. Sbalzarini, S. J. Marrink, and R. G. Efremov (2010). Antimicrobial peptides induce growth of phosphatidylglycerol domains in a model bacterial membrane. *J Phys Chem Lett 1*(20), 3108–3111.

Press, W. H., S. A. Teukolsky, W. T. Vetterling, and B. P. Flannery (2007). *Numerical Recipes: The Art of Scientific Computing.* New York: Cambridge University Press.

Provasi, D., M. C. Artacho, A. Negri, J. C. Mobarec, and M. Filizola (2011, October). Ligand-induced modulation of the free-energy landscape of g protein-coupled receptors explored by adaptive biasing techniques. *PLoS Comput Biol 7*(10), e1002193.

Rapaport, D. C. (2004). *The Art of Molecular Dynamics Simulation* (2nd edn.). Cambridge, U.K.: Cambridge University Press, ISBN: 9780521825689.

Romo, T. D., L. A. Bradney, D. V. Greathouse, and A. Grossfield (2011, August). Membrane binding of an acyl-lactoferricin b antimicrobial peptide from solid-state NMR experiments and molecular dynamics simulations. *Biochim Biophys Acta 1808*(8), 2019–2030.

Romo, T. D. and A. Grossfield (2009). LOOS: An extensible platform for the structural analysis of simulations. *Conf Proc IEEE Eng Med Biol Soc 2009*, 2332–2335.

Romo, T. D. and A. Grossfield (2011a). Block covariance overlap method and convergence in molecular dynamics simulation. *J Chem Theory Comput 7*(0), 2464–2472.

Romo, T. D. and A. Grossfield (2011b, January). Validating and improving elastic network models with molecular dynamics simulations. *Proteins: Struct Funct Bioinf 79*(1), 23–34.

Romo, T. D. and A. Grossfield (2012). LOOS: Lightweight object oriented structure analysis. http://loos.sourceforge.net.

Romo, T. D., A. Grossfield, and M. C. Pitman (2010, January). Concerted interconversion between ionic lock substates of the beta(2) adrenergic receptor revealed by microsecond timescale molecular dynamics. *Biophys J 98*(1), 76–84.

Rumsfeld, D. (2002, February). News transcript: DoD news briefing—Secretary rumsfeld and gen. Myers, United States Department of Defense, Washington, DC. http://www.defense.gov/transcripts/transcript.aspx?transcriptid=2636, accessed April 11, 2014.

Rzepiela, A. J., D. Sengupta, N. Goga, and S. J. Marrink (2010). Membrane poration by antimicrobial peptides combining atomistic and coarse-grained descriptions. *Faraday Discuss 144*, 431–443; discussion 445–481.

Sachs, J. N., P. S. Crozier, and T. B. Woolf (2004, December). Atomistic simulations of biologically realistic transmembrane potential gradients. *J Chem Phys 121*(22), 10847–10851.

Sachs, J. N., H. I. Petrache, and T. B. Woolf (2003, December). Interpretation of small angle x-ray measurements guided by molecular dynamics simulations of lipid bilayers. *Chem Phys Lipids 126*(2), 211–223.

Salgado, G. F. J., A. V. Struts, K. Tanaka, N. Fujioka, K. Nakanishi, and M. F. Brown (2004, October). Deuterium NMR structure of retinal in the ground state of rhodopsin. *Biochemistry 43*(40), 12819–12828.

Schmidt, T. H. and C. Kandt (2012, October). Lambada and inflategro2: Efficient membrane alignment and insertion of membrane proteins for molecular dynamics simulations. *J Chem Inf Model 52*(10), 2657–2669.

Seo, M., S. Rauscher, R. Pomés, and D. P. Tieleman (2012, May). Improving internal peptide dynamics in the coarse-grained martini model: Toward large-scale simulations of amyloid- and elastin-like peptides. *J Chem Theory Comput 8*(5), 1774–1785.

Shan, J., G. Khelashvili, S. Mondal, E. L. Mehler, and H. Weinstein (2012). Ligand-dependent conformations and dynamics of the serotonin 5-ht(2a) receptor determine its activation and membrane-driven oligomerization properties. *PLoS Comput Biol 8*(4), e1002473.

Shaw, D. E., R. O. Dror, J. K. Salmon, J. P. Grossman, K. M. Mackenzie, J. A. Bank, C. Young et al. (2009). Millisecond-scale molecular dynamics simulations on anton. *Proceedings of the Conference on High Performance Computing Networking, Storage and Analysis, SC'09*, ACM, New York, pp. 39:1–39:11.

Shirts, M. and V. S. Pande (2000, December). Computing: Screen savers of the world unite! *Science 290*(5498), 1903–1904.

Skjevik, Å. A., B. D. Madej, R. C. Walker, and K. Teigen (2012, September). LIPID11: A modular framework for lipid simulations using amber. *J Phys Chem B 116*(36), 11124–11136.

Soubias, O. and K. Gawrisch (2011, September). The role of the lipid matrix for structure and function of the gpcr rhodopsin. *Biochim Biophys Acta 1818*, 234–240.

Soubias, O., W. E. Teague, K. G. Hines, D. C. Mitchell, and K. Gawrisch (2010, August). Contribution of membrane elastic energy to rhodopsin function. *Biophys J 99*(3), 817–824.

Sternin, E., M. Bloom, and A. L. Mackay (1983). De-pake-ing of NMR-spectra. *J Magn Reson 55*, 274–282.

Van Der Spoel, D., E. Lindahl, B. Hess, G. Groenhof, A. E. Mark, and H. J. C. Berendsen (2005, December). Gromacs: Fast, flexible, and free. *J Comput Chem 26*(16), 1701–1718.

Vargas, E., V. Yarov-Yarovoy, F. Khalili-Araghi, W. A. Catterall, M. L. Klein, M. Tarek, E. Lindahl et al. (2012, December). An emerging consensus on voltage-dependent gating from computational modeling and molecular dynamics simulations. *J Gen Physiol 140*(6), 587–594.

Venable, R. M., Y. Zhang, B. J. Hardy, and R. W. Pastor (1993). Molecular dynamics simulations of a lipid bilayer and of hexadecane: An investigation of membrane fluidity. *Science 262*, 223–225.

Wiener, M. C., G. I. King, and S. H. White (1991, September). Structure of a fluid dioleoylphosphatidylcholine bilayer determined by joint refinement of x-ray and neutron diffraction data. I. Scaling of neutron data and the distributions of double bonds and water. *Biophys J 60*(3), 568–576.

Wiener, M. C. and S. H. White (1991a, January). Fluid bilayer structure determination by the combined use of x-ray and neutron diffraction. II. "Composition-space" refinement method. *Biophys J 59*(1), 174–185.

Wiener, M. C. and S. H. White (1991b, July). Transbilayer distribution of bromine in fluid bilayers containing a specifically brominated analogue of dioleoylphosphatidylcholine. *Biochemistry 30*(28), 6997–7008.

Wiener, M. C. and S. H. White (1992, February). Structure of a fluid dioleoylphosphatidylcholine bilayer determined by joint refinement of x-ray and neutron diffraction data. III. Complete structure. *Biophys J 61*(2), 434–447.

Woolf, T. B. and B. Roux (1996). Structure, energetics and dynamics of lipid-protein interactions: A molecular dynamics study of the gramicidin A channel in a DMPC bilayer. *Prot Struct Funct Gen 24*, 92–114.

Zhang, X., D. Bhatt, and D. M. Zuckerman (2010, September). Automated sampling assessment for molecular simulations using the effective sample size. *J Chem Theory Comput 6*(10), 3048–3057.

Chapter 10

Protein and Nanoparticle Interactions

Perspectives of Nanomedicine and Nanotoxicity

Seung-gu Kang and Ruhong Zhou

CONTENTS

10.1   INTRODUCTION

Nanotechnology has emerged as a revolutionary new science and technology with great potential to improve many aspects of our daily lives. Although it is still in its early stage of development with some limitations that require more fundamental research, nanotechnology has enabled and facilitated novel research in many areas, such as nanomaterials and nanomedicine [1–3]. The intrinsic structural and functional flexibilities of nanoparticles make them an ideal candidate for resolving shortcomings that existed in conventional medicines such as low target selectivity and high dosage needed [4–7]. A variety of material platforms are being actively developed for these purposes, including polymers, lipids, and inorganic solid or carbon-based nanoparticles [8–11]. Due to the large surface-to-volume ratio of these nanomaterials, sophisticated nanosystems can be developed with various functional modules for applications in sensing environmental changes, targeting specific cell receptors, and maximizing drug payload [5,12,13].

Despite the fact that nanoscale medicine has made significant breakthroughs in the laboratory and advanced rapidly in clinical trials, the potential adverse health effects caused by "nanotoxicity" remain as an important issue that needs serious attention [14–16]. Questions on how nanoparticles interact with biomolecules [17] and how it would affect their biological function at the molecular and system levels are fundamentally important for understanding nanotoxicity associated with these nanomaterials [18,19]. Molecular level understanding is crucial not only for toxicological impact as a potential pathogen to the host biology but also for pharmacokinetic efficacy of trial nanomedicines, thus serving as basis for either reducing harmful effect or designing better therapeutics.

Although it is relatively easy to establish the cause and effect with experiments, it is very difficult to reveal the underlying molecular mechanism due to the low temporal and spatial resolutions even with the most sophisticated experimental techniques available in a highly inhomogeneous environment. As an alternative, computational approaches can complement existing experimental approaches to investigate the nano–bio interface with atomic resolution [20]. In this chapter, we review recent advances of in silico modeling for the interactions between proteins and nanoparticles from two very different perspectives, namely, nanomedicine and nanotoxicity. Using results from previous studies of systems containing the metallofullerenol $Gd@C_{82}(OH)_{22}$ with proteins such as matrix metalloproteinase (MMP)-9, collagen complex, WW domain, and Src homology 3 (SH3) domain, we describe how molecular dynamics simulations reveal specific interactions with the target proteins and how they could provide pharmacokinetic activity on one hand and toxic potential on the other hand.

We start the next section with the force field development for metallofullere-nol Gd@C$_{82}$(OH)$_{22}$, followed by the validation of the empirical force field. Then, we introduce nanomedicinal applications of Gd@C$_{82}$(OH)$_{22}$ such as inhibiting the MMP-9 [21] and stabilizing the collagen complex [22] in the cancer media. There, we mainly focus on whether Gd@C$_{82}$(OH)$_{22}$ would have specific bindings on the target proteins and what are the driving forces for the binding and their impact on cancer prohibition. Finally, we discuss the potential nanotoxicity of Gd@C$_{82}$(OH)$_{22}$ and how it interacts with the WW [23] and SH3 [24] domains, two representative protein domains involved in signal transduction pathways.

10.2 COMPUTATIONAL MODELING FOR PROTEIN–NANOPARTICLE INTERACTION

10.2.1 First-Principal Calculation for Force Parameterization of Endohedral Metallofullerenol Gd@C$_{82}$(OH)$_{22}$

Classical force fields of fullerene derivatives were parameterized all based on first-principal density functional theory (DFT) with the spin-polarized general-ized gradient approximation (GGA) using the Perdew–Burke–Ernzerhof (PBE) exchange-correlation functional [25]. The Slater basis sets were constructed at the level of triple-ζ plus one polarization function. The heavy atoms were treated with the frozen-core approximation (i.e., for [1S^2] of O and C and for [1S^2–4d^{10}] of Gd, respectively) [26]. The zero-order-regular approximation (ZORA) was applied for the scalar relativistic effect [27]. All calculations were performed with the Amsterdam Density Functional (ADF 2010.01) program [28,29]. Geometry optimization was first performed for Gd^{3+}@[C$_{82}$(OH)$_{22}$]$^{3-}$ and C$_{82}$(OH)$_{22}$ with septet and singlet spin multiplicities for their respective electronic ground states. The geometries were confirmed to be in the energy minima by calculating the vibrational frequencies. The atomic point charges were finally derived by simulating multiple moments (e.g., monopole, dipole, and quadrupole) of corresponding systems in order to avoid strong basis set dependence as found in Mullikan analysis. Dispersion terms for fullerenol and Gd were adapted from corresponding atom types in chemistry at Harvard macromolecular mechanics (CHARMM) force field and Ref. 30.

10.2.2 Validation of Gd@C$_{82}$(OH)$_{22}$ Force Parameters

Although some of the experimental results are available for electronic proper-ties of Gd@C$_{82}$ [31,32] or Gd@C$_{82}$(OH)$_{22}$ [33,34] in solid state, no experimental data on multiple moments, such as dipole and quadrupole, of metallofullerenol Gd@C$_{82}$(OH)$_{22}$ have been determined. This would be largely due to the difficul-ties related to aggregation in gas phase or aqueous solution phase. In case of

Gd@C$_{82}$, however, recent studies have shown that DFT level quantum mechanical calculations agree with experimental observations such as the ground-state spin multiplicity and optimized geometry of Gd atom in the fullerene cage [35]. For example, Gd atom has been observed in the vicinity of a C–C double bond on the C$_2$ molecular axis of C$_{82}$, in contrast to the other Group 3 metallofullerenes M@C$_{82}$ (i.e., M = La, Sc, etc.), where La and Sc atoms are located near a six-membered ring in the opposite side of the C–C double bond [32]. This geometry has been successfully predicted in the DFT calculations [35]. This made it possible to accurately predict the ground-state spin multiplicity (i.e., M = 7) as determined from magnetic studies [36,37]. Interestingly, in our DFT calculations, we also found that Gd atom is stabilized near the C–C double bond rather than the opposite six-membered ring.

In addition to the multiple-moment prediction, the atomic partial charges were further evaluated in order to see if the empirical parameters provide accurate energetics for a short-range interaction. Similar to the protocol for the CHARMM force field [38,39], the interaction energy profiles between Gd@C$_{82}$(OH)$_{22}$ and one probe water molecule were computed as a function of intermolecular distance using both quantum mechanics (QM) and molecular mechanics (MM). Figure 10.1 shows one representative energy profile for the probe water molecule approaching toward Gd@C$_{82}$(OH)$_{22}$ along an arbitrary direction (Gd@C$_{82}$(OH)$_{22}$ shows a roughly isotropic spherical structure with

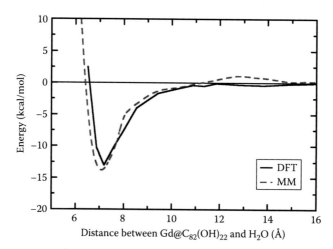

Figure 10.1   Potential energy profiles between QM (DFT) and MM for Gd@C$_{82}$(OH)$_{22}$ and a probe water (TIP3P) as a function of intermolecular separation. Our developed empirical force field parameters (i.e., atomic partial charges) for Gd@C$_{82}$(OH)$_{22}$ reasonably reproduce the QM DFT-based energy profile, in particular, the minimum energy value (about –14 kcal/mol) and its position (about 7.1 Å).

the 22 OH groups distributed uniformly on the sphere). Our comparison demonstrates that the MM partial charges for the metallofullerenol $Gd@C_{82}(OH)_{22}$ do seem to reasonably reproduce the QM energy profile, in particular, the minimum energy value (about –14 kcal/mol) and its position (about 7.1 Å). Profiles along other directions show similar behavior. We attribute this good agreement in QM and MM energy profiles to the exceptionally well-delocalized electron density over the fullerenol cage $C_{82}(OH)_{22}$.

The delocalization of electron density on the fullerenol cage was further validated by reparameterizing the atomic partial charges with a "wrong" Gd position. As mentioned earlier, the other Group 3 metallofullerenes, $La@C_{82}$ or $Sc@C_{82}$, have their La/Sc atoms located near a six-membered ring, that is, in the opposite side of the C–C double bond where Gd sits [32]. We thus located the Gd atom in the opposite of its correct position to be at La/Sc's location and then redid the charge fitting and compared how the nearby carbon atom charges change with the Gd position. Table 10.1 summarizes the partial charges of the six nearest carbon atoms at either the C–C double bond or the six-membered ring site with Gd at both positions. A careful comparison between atomic charges at the same site with two different charge positions (i.e., one with Gd at the C–C double bond and the other with Gd at the six-membered ring side) shows that the average charge differences (Δq) are only –0.011e and –0.003e at the C–C double bond side and six-membered ring site, respectively. This implies that the absorbed electrons from the endohedral Gd are fairly well distributed (or delocalized) over the fullerenol cage $C_{82}(OH)_{22}$, which fortunately makes the

TABLE 10.1 ATOMIC CHARGE COMPARISON BETWEEN CHARGE FITS PARAMETERIZED AT TWO DIFFERENT GD POSITIONS OF $Gd@C_{82}(OH)_{22}$

Gd at C–C Double Bond (C=C)				Gd at Six-Membered Ring (C_6)				Charge Difference	
C=C Side		C_6 Side		C=C Side		C_6 Side		C=C Side	C_6 Side
Atom	$q_{C=C}$	Atom	q_{C_6}	Atom	$q_{C=C}$	Atom	q_{C_6}	$\Delta q_{C=C}$	Δq_{C_6}
C_{37}	–0.193	C_{29}	–0.036	C_{37}	–0.149	C_{29}	–0.098	–0.044	0.062
C_{46}	–0.134	C_{30}	–0.035	C_{46}	–0.082	C_{30}	–0.092	–0.051	0.057
C_{55}	0.367	C_{53}	–0.113	C_{55}	0.350	C_{53}	–0.137	0.017	0.025
C_{56}	0.361	C_{54}	–0.085	C_{56}	0.342	C_{54}	–0.137	0.020	0.052
C_{27}	0.345	C_{40}	0.331	C_{27}	0.348	C_{40}	0.396	–0.003	–0.065
C_{28}	0.352	C_{41}	0.250	C_{28}	0.354	C_{41}	0.400	–0.002	–0.150
Average	0.183		0.052		0.194		0.055	–0.011	–0.003

partial charges (force field parameters) relatively insensitive (or robust) to the motion of the encaged Gd atom as well as external molecules such as waters.

Overall, the MM energy profile seems to reproduce the QM energy profiles reasonably well, which implies additional scaling factors might not be necessary for our current empirical atomic partial charges. Subsequent further tests on charge delocalization confirm the reliability of the empirical force field in a slightly perturbed environment such as internal motion of Gd or external approaching water molecules.

10.2.3 Molecular Dynamics Simulation Protocols

All systems exampled here were based on atomistic models described with the CHARMM22 protein force field in explicit TIP3P [40] water solvent. The bonding and nonbonding interactions between atoms were enumerated by the CHARMM22 protein force field [38]. The long-ranged electrostatic interactions were enumerated by the particle mesh Ewald (PME) method [41]. The short-ranged dispersion interaction was treated with a cutoff of 12 Å. Periodic boundary conditions were applied for all three directions in order to avoid surface abnormality. All molecular dynamics simulations were carried out with the NAMD2 software package (nanoscale molecular dynamics program) [42] optimized on an IBM Blue Gene supercomputer [43].

10.2.3.1 MMP-9 and Gd@C_{82}(OH)$_{22}$

A structure of MMP-9 was obtained from the coordinates determined by x-ray crystallography for the 159-residue catalytic domain (residue 110–215 and 391–443; PDB code, 2OVZ) after mutating Q402 back to the native residue Glu [44]. MMP-9 along with four Gd@C_{82}(OH)$_{22}$ molecules was prepared in a cube of 90 Å × 90 Å × 90 Å water box, where Gd@C_{82}(OH)$_{22}$ had no initial contact with the protein. Multiple Gd@C_{82}(OH)$_{22}$ molecules were employed to expedite the sampling of various binding modes between Gd@C_{82}(OH)$_{22}$ and MMP-9, as well as to understand the clustered behavior among the Gd@C_{82}(OH)$_{22}$ molecules themselves. Then, the composite system was ionized with 100 mM NaCl solution in order to mimic the physiological condition as much as possible, which generated a system of ~60,000 atoms. Before the production runs, the system was first subjected to 20,000 steps of energy minimization, followed by a 250 ps equilibration with a 0.5 fs time step. In isobaric isothermal (NPT) ensemble (i.e., 1 atm and 310 K), five independent simulations were performed with a 2 fs time step, generating trajectories up to 500 ns. Each run was independently configured with different intermolecular orientations in order to avoid a sampling bias. Similarly, control systems were run for normal fullerenols C_{82}(OH)$_{22}$ and C_{60}(OH)$_{22}$, where the former is intended to understand the Gd(III) ion effect and the latter is for comparison with the experimental results.

10.2.3.2   Collagen Complex and Gd@C$_{82}$(OH)$_{22}$

The collagen triplex was based on a triad of polypeptide chains of x-ray crystal structure (PDB code, 1WZB). Each peptide is composed of 29 residues with a sequence of X-Y-Gly repeated to make an 8.7 nm long chain, where X was replaced with Pro instead of the original hydroxyproline (Hyp), while Y was kept unchanged to the original Hyp. In our simulation, we devised two different collagen systems in order to understand the Gd@C$_{82}$(OH)$_{22}$ impact on collagen assembly depending on structural hierarchy. For structural stability of the collagen triplex, we configured a single collagen triplex with 14 Gd@C$_{82}$(OH)$_{22}$, each was at least 15 Å away from the complex. Next, we expand our system with four collagen complexes with 20 Gd@C$_{82}$(OH)$_{22}$, where the collagen bundle has been configured to have quasihexagonal packing of collagen assembly [45] with each triplex parallelly separated from the rest by at least one water layer. Afterward, each system was prepared with TIP3P water solvent with 100 mM NaCl, resulting in a water box of 78 Å × 72 Å × 146 Å for the single collagen triplex with ~81,000 atoms and a water box of 100 Å × 80 Å × 126 Å for the collagen bundle with ~100,000 atoms.

10.2.3.3   WW Domain and Gd@C$_{82}$(OH)$_{22}$

The human Yes-associated protein WW domain (hYAP65, L30K) was utilized as our protein system with the cocrystallized 10-residue-long proline-rich motif (PRM) (PDB code, 1JMQ) [46]. Only the functional unit from residues between L13 and P42 was selected for majority of our investigation as in previous studies [47–49], due to their seminal role in biological function and structural stability of the WW domain [50,51]. As described in the next section, we prepared two different configurations for (1) the intrinsic binding mode and (2) the inhibitory dynamics of Gd@C$_{82}$(OH)$_{22}$. For the former, the Gd@C$_{82}$(OH)$_{22}$ molecules have been set apart from the WW domain and each other as similar to the case of MMP-9. For the latter, the PRM ligand was also included with one Gd@C$_{82}$(OH)$_{22}$ around the WW domain, where both PRM and Gd@C$_{82}$(OH)$_{22}$ were positioned at least 30 Å away from the WW domain to avoid initial contacts among them. Following the similar protocol used in MMP-9, each system was simulated with 5 runs; each run was at least 200 ns long and started with an initial configuration of different intermolecular orientation and random velocity.

10.2.3.4   SH3 Domain and Gd@C$_{82}$(OH)$_{22}$

The N-terminal SH3 domain of c-Crk was used in our simulation along with its cocrystallized PRM (PDB code, 1CKB) [52], where the PRM Sos (PPPVPPRR) was selected from two documented PRMs (i.e., C3G and Sos) due to its relatively broad reactivity for various SH3 domains (e.g., c-Crk, v-Crk, and Grb2). Three different systems were devised with (1) a binary system with Gd@C$_{82}$(OH)$_{22}$

and SH3 domain, (2) a ternary system with $Gd@C_{82}(OH)_{22}$, PRM, and SH3 domain, and (3) another binary system with PRM and SH3 domain. Similar to WW domain, the first two systems are for understanding the intrinsic binding dynamics and possible inhibitory effect of $Gd@C_{82}(OH)_{22}$ toward SH3 domain. The positive control system (3) is added to assure the $Gd@C_{82}(OH)_{22}$ effect on the native binding behavior of PRM. For each system, five independent simulations were performed to generate at least 200 ns long trajectories, each with a different starting configuration. The rest of the simulation conditions are basically same as those used in the WW domain study.

10.3 MODELING NANOPHARMACOLOGY FOR PROTEIN–NANOPARTICLE INTERACTIONS

10.3.1 Antitumoral Efficacy of Metallofullerenol $Gd@C_{82}(OH)_{22}$

Fullerene derivatives are one of the mostly investigated carbon allotropes for biomedical applications [53,54]. Their large carbonaceous surfaces are readily modified with various chemicals, which enable the fullerenes as effective nanocarriers for diagnostics and therapeutics [55,56]. In particular, endohedral metallofullerenes have attracted special attentions for their potential applications as bioimaging agents, where the hollow fullerene cages serve as a rigid template for magnetosensitive heavy metal(s), leaving the exterior surface for further chemical modification. Among others, gadolinium (Gd) has been extensively recruited as a contrasting agent for magnetic resonance imaging (MRI) due to its strong paramagnetism at room temperature [57], as intravenously injected with organic chelation to avoid cytotoxicity (e.g., Magnevist). Recent study showed that Gd^{3+} ion encaged in fullerene with multiple hydroxyl groups (i.e., $Gd@C_{82}(OH)_x$) could remarkably enhance the proton relaxivity by more than 20 times better than the commercial organometallic Gd complexes [58].

Besides the high contrasting capacity for MRI, more interestingly, Zhao and his coworkers showed that gadolinium–fullerenol $Gd@C_{82}(OH)_{22}$ has a specific antitumoral efficacy from in vivo tumor-bearing mouse models for both hematoma (H_{22}) [59] and breast cancer cells [60]. $Gd@C_{82}(OH)_{22}$ clearly showed equivalent or even better performance in restricting cancer growth and migration even with much smaller dose than that treated with representative conventional drugs like cyclophosphamide (CTX) and cisplatin (CDDP) [59]. Regarding its anticancer mechanism, subsequent studies showed that $Gd@C_{82}(OH)_{22}$ is not directly involved in killing the cancer, but rather gives an indirect effect on the cancer environment. For example, $Gd@C_{82}(OH)_{22}$ can enhance cellular immune response by stimulating the T cells and macrophages [61]. In cotreatment with CDDP, $Gd@C_{82}(OH)_{22}$ is shown to reactivate CDDP-resistant cells so that they reinternalize drug molecules into cancer cells. Inherited from the intrinsic property of fullerene, $Gd@C_{82}(OH)_{22}$ can also function as an effective scavenger for reactive oxygen species (ROS), protecting normal cells from ROS and lipid peroxidation [62].

More specifically, the microvessel density (MVD) on the tumor surface was observed to be reduced over 40% with $Gd@C_{82}(OH)_{22}$ treatment, implying the antiangiogenic activity of $Gd@C_{82}(OH)_{22}$ [63]. The microvessel formation is important for cancer prognosis since it controls the oxygen and nutrient supply for the cancer cells to grow and eventually migrate to other tissues. More directly, $Gd@C_{82}(OH)_{22}$ was evidenced to downregulate more than 10 different proangiogenic factors [63], which further supports the indirect role of $Gd@C_{82}(OH)_{22}$ in anticancer mechanism that the metallofullerenol is able to effectively defoliate cancer tissue by restricting angiogenesis and ultimately isolating the tumor in a local domain enclosed by the extracellular matrix (ECM).

10.3.2   Inhibitory Mechanism of $Gd@C_{82}(OH)_{22}$ on MMP-9

More recently with in vivo human pancreatic cancer model xenografted in mice, we found that $Gd@C_{82}(OH)_{22}$ effectively restricts tumor growth by over ~50.1% compared with the saline control, again accompanied with clear decrease in MVD on the tumor surface [21]. The antiangiogenic activity of $Gd@C_{82}(OH)_{22}$ was thus investigated in more detail with in vivo cellular and in vitro biochemical assays for representative proangiogenic enzymes. In fact, we recognized that $Gd@C_{82}(OH)_{22}$ can prominently downregulate MMP-2 and MMP-9 in both mRNA expression and enzyme activity levels with more dramatic changes for MMP-9, while the control with $C_{60}(OH)_{22}$ has no noticeable effect on the enzyme activity as with the saline solution.

In normal physiological condition, MMPs, tightly regulated, are playing a seminal role such as in tissue remodeling and wound healing [64]. Once recruited in the tumorigenic environment, however, MMPs tend to be overexpressed and draw mass of blood microvessels, often resulting in rough surface morphology by irregularly and massively formed vessels under loose control [65]. Along the cancer prognosis, MMPs are also involved in degrading ECM layer, which is a critical step for cancer cells to escape from their local domain to migrate to other tissues. Although our findings clearly indicated that $Gd@C_{82}(OH)_{22}$ does affect the MMPs, it was unclear, in molecular level, how $Gd@C_{82}(OH)_{22}$ interacts with the protein and inhibits its function.

To gain more detailed understanding, we used atomistic molecular dynamics simulations for MMP-9 with $Gd@C_{82}(OH)_{22}$, $C_{82}(OH)_{22}$ (control 1), or $C_{60}(OH)_{22}$ (control 2) as described in Section 10.2.3.

10.3.3 Nondestructive interaction of $Gd@C_{82}(OH)_{22}$ on MMP-9

Our molecular dynamics simulations revealed general features on interaction among the $Gd@C_{82}(OH)_{22}$'s and between $Gd@C_{82}(OH)_{22}$ and MMP-9. First, we found that $Gd@C_{82}(OH)_{22}$ molecules are probable to cluster together before and/or after their binding on MMP-9. It was consistent with experiments, where $Gd@C_{82}(OH)_{22}$ particles were aggregated with an ~22 nm diameter from the synchrotron radiation small angle x-ray scattering (SR-SAXS) and atomic force microscopy (AFM) [59]. This indicates that the solubility has been highly enhanced by the massive hydroxylation (i.e., 22 OH groups), but still $Gd@C_{82}(OH)_{22}$ is hydrophobic enough to form nanoclusters in aqueous media. One representative trajectory demonstrated that $Gd@C_{82}(OH)_{22}$ is clustered for the first ~40 ns, before they interact with MMP-9 at the loops in the lower rim at 61.0 ns (Figure 10.2). The "particulated" interaction of $Gd@C_{82}(OH)_{22}$ was also proposed as a potential mechanism for the simultaneous targeting on multiple proangiogenic factors, which is in contrast with the conventional small molecular base drugs aiming for a specific single target [63].

With regard to the interaction with protein, $Gd@C_{82}(OH)_{22}$ was found to have no noticeable impact on the global structure of MMP-9. The root-mean-square deviation (rmsd) of MMP-9 fluctuated between 1 and 3 Å regardless of the types of nanoparticles (i.e., $Gd@C_{82}(OH)_{22}$, $C_{82}(OH)_{22}$, and $C_{60}(OH)_{22}$). Nevertheless, the residue-specific root-mean-square fluctuation (rmsf) revealed that solvent-exposed residues, especially loops L_{34}, L_{45}, S1', and SC, more sensitively respond, as accommodating nanoparticles. With no explicit structural impact like global or at least local deformations, we discarded the possibility of an inhibitory mechanism through a protein structural deformation that often occurred between protein and hydrophobic nanoparticles like carbon nanotube (CNT) and graphene [47,66,67]. In our studies, pristine CNT or graphene was found to be able to seriously destroy the native fold of proteins via strong $\pi-\pi$ and/or hydrophobic interactions with the aromatic and/or hydrophobic residues, respectively. However, the lack of hydrophobicity of $Gd@C_{82}(OH)_{22}$ molecules might be deficient to induce a large structural change on MMP-9. Therefore, this proposes an alternative possibility for $Gd@C_{82}(OH)_{22}$ to inhibit MMP-9 through a specific binding on certain sites of the target protein like conventional drugs.

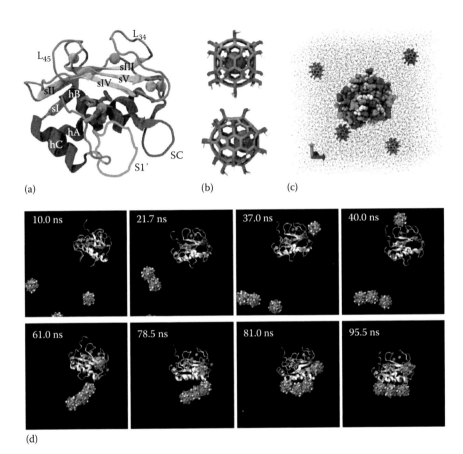

Figure 10.2 Molecular dynamics simulations for MMP-9 and Gd@C$_{82}$(OH)$_{22}$. (a) X-ray crystal structure of the catalytic domain of MMP-9: with Zn^{2+} and Ca^{2+} shown in van der Waals balls. (b) Endohedral metallofullerenol Gd@C$_{82}$(OH)$_{22}$, where Gd atom located inside of the C$_{82}$(OH)$_{22}$ cage. (c) Molecular dynamics setup, where MMP-9, surrounded by four Gd@C$_{82}$(OH)$_{22}$ molecules, is solvated with about 22,000 water molecules. (d) Temporal snapshots of Gd@C$_{82}$(OH)$_{22}$ binding onto MMP-9. The metallofullerenol Gd@C$_{82}$(OH)$_{22}$ is clustered before binding to MMP-9 on the hydrophobic patch near the ligand specificity loop S1′.

10.3.3.1 Indirect Inhibition via an Allosteric Modulation on the Ligand Specificity Loop

Discarding the direct structural deformation pathway, we investigated in more detail if there exist preferred binding modes of Gd@C$_{82}$(OH)$_{22}$ on MMP-9; how different they are, if any, from the control fullerenols (C$_{82}$(OH)$_{22}$ and C$_{60}$(OH)$_{22}$); and how they would affect the proteolytic function of MMP-9. First, we calculated the residue-specific contact probability for each fullerene derivative by an equation defined in the following:

$$p(r) = \frac{\sum_{l=1}^{n} \sum_{t=1}^{N_l} \sum_{m=1}^{4} \delta_t^l(m,r)}{\sum_{l=1}^{n} \sum_{t=1}^{N_l} \sum_{m=1}^{4}}. \tag{10.1}$$

The contacting probability $p(r)$ of a residue r is obtained by counting the contacts $\delta_t^l(m,r)$ between the residue r and each nanoparticle m over all available simulation frames (N_l) and trajectories (n), where $\delta_t^l(m,r)$ is 1 when any heavy atom pairs between residue r and nanoparticle m are less than or equal to 5 Å at time t and trajectory l, otherwise 0. Figure 10.3 shows the residue-specific contact profiles for metallofullerenol Gd@C$_{82}$(OH)$_{22}$ and both fullerenols C$_{82}$(OH)$_{22}$ and C$_{60}$(OH)$_{22}$ with respective protein surfaces gray scaled according to contact probabilities.

The profiles provide a steady-state picture of nanoparticle interactions on MMP-9 surface. The distributions and frequencies of contacting residues are significantly different among fullerene derivatives. Gd@C$_{82}$(OH)$_{22}$ is clearly distinguished from normal fullerenols, as intensively interacting with the lower rim of MMP-9 especially at S1′ loop. In contrast, fullerenols C$_{82}$(OH)$_{22}$ and C$_{60}$(OH)$_{22}$ tend to contact with MMP-9 in more broad area including both upper and lower rims of MMP-9. In all cases, however, no explicit evidence has been found for that nanoparticles directly interact with the catalytic site residues of MMP-9. Despite difficulties in completely excluding the possibility, our simulations indicate that it is less likely for Gd@C$_{82}$(OH)$_{22}$ to inhibit MMP-9 via the direct binding on the Zn^{2+}-coordinated active site.

On the other hand, the favorable interaction on the S1′ loop proposed us an alternative route for MMP-9 inhibition of Gd@C$_{82}$(OH)$_{22}$. The S1′ loop, known as the ligand specificity loop, has long been recognized as a characteristic region in the catalytic domain due to their functional role in ligand recognition, which is largely attributed to its unique amino acid sequence and length among all available MMPs. Thus, Gd@C$_{82}$(OH)$_{22}$ could indirectly inhibit MMP-9 by allosterically modulating the ligand recognition at the ligand specificity S1′ loop instead of directly blocking the enzymatic site of MMP-9. In fact, the direct targeting on the zinc-active site has raised a long-known target specificity issue in developing small molecule-based MMP inhibitors (MMPi's) [68]. Due to the

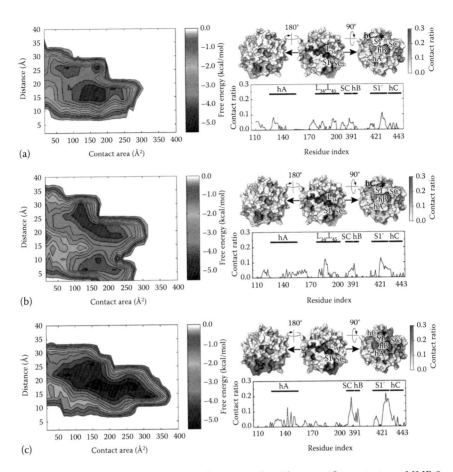

Figure 10.3 Binding free energy landscapes and residue-specific contacts on MMP-9 (a, b). Binding free energy surfaces for fullerenol $C_{60}(OH)_{22}$ (a) and $C_{82}(OH)_{22}$ (b) on MMP-9 shows a nonspecific binding mode (left), and almost all surface residues of MMP-9 contribute in contact with fullerenol derivatives (right). (c) Metallofullerenol $Gd@C_{82}(OH)_{22}$ has a specific binding mode with MMP-9 (left) and interacts with the ligand specificity S1′ loop and SC loop (right). Residue was assigned to be in a contact when any atom in the residue is within 5.0 Å of any atom of fullerene derivatives.

highly homologous active site structures among MMPs, organic inhibitors often failed to bind a specific target even with a high binding affinity, possibly causing serious side effects, which indeed hampered MMPi to be approved in FDA even though MMPs have been recognized as critical targets for antimetastatic therapeutics [64,69]. In this regard, several recent studies demonstrated that the allosteric control by exosite interaction could be an effective alternative for inhibiting MMPs [70].

To gain more insight into the binding thermodynamics, we calculated potential of mean force (PMF) for nanoparticle binding on MMP-9 (Figure 10.3). The binding free energy surface was constructed by histogram method [47,71] as defined in the following:

$$W(x_{area}, y_{dist}) = -RT \ln p(x_{area}, y_{dist}). \qquad (10.2)$$

The probability $p(x_{area}, y_{dist})$ was obtained by counting events in a bin (x_{area}, y_{dist}) along with two reaction coordinates, where x_{area} and y_{dist} represent minimum distances between nanoparticle and the catalytic Zn^{2+} ion of MMP-9 and contact area between nanoparticle and MMP-9, respectively. The binding free energy surface clarifies the characteristics of the binding modes found in the contact analysis. While the distributed binding of $C_{82}(OH)_{22}$ and $C_{60}(OH)_{22}$ is evidenced in the multiple binding modes in their binding free energy surfaces, the intensive contact of $Gd@C_{82}(OH)_{22}$ near the ligand specificity S1' loop is shown in the large single energy basin with ~0.4 and ~0.3 kcal/mol stronger binding affinity than those of $C_{82}(OH)_{22}$ and $C_{60}(OH)_{22}$, respectively. Thus, PMF together with site-specific contacts confirms that $Gd@C_{82}(OH)_{22}$ can form a specific and thermodynamically favorable binding at the ligand specificity S1' loop of MMP-9, emphasizing the indirect inhibition pathway of $Gd@C_{82}(OH)_{22}$ on MMP-9.

10.3.3.2 Potential Inhibitory Pathway and Driving Forces

Understanding the binding mechanism and relevant driving forces is critical in assessing drug pharmacokinetics as well as designing medicines of better efficacy. With regard to the binding on the ligand specificity S1' loop, $Gd@C_{82}(OH)_{22}$ in one trajectory has been demonstrated that interacts with MMP-9 along with a specified binding route on the surface of MMP-9. There, $Gd@C_{82}(OH)_{22}$ binding could be characterized with three distinctive phases depending on the major interaction types with MMP-9: (1) nonspecific long-range electrostatic interaction, (2) nonspecific hydrophobic interaction, and (3) a more specific combined electrostatic and hydrophobic interaction. In the first phase, a long-range electrostatic interaction with surface-exposed charged residues plays an important role, which guides the nanoparticle to the first touch on the protein surface. Interestingly, isoelectric surface shows that a strong negative field is formed around the Zn-coordinated catalytic site of MMP-9, whereas neutral or positive fields are stretched out near the S1' loop. Given the negatively charged fullerenol cage $[C_{82}(OH)_{22}]^{3-}$ induced by the encaged Gd^{3+}, $Gd@C_{82}(OH)_{22}$, in a long distance, tends to interact with positively charged surface residues (e.g., K433 and R440) rather than to approach the active site of the strong negative electrostatic shield. Once $Gd@C_{82}(OH)_{22}$ is driven near the surface, however, it readily moves to nearby hydrophobic patch by way of short-ranged nonspecific hydrophobic interactions, albeit it is transient. Considering that $Gd@C_{82}(OH)_{22}$ is still hydrophobic to some extent, interaction with hydrophobic residues would be energetically beneficial due to relatively low-desolvation free energy. Crossing the hydrophobic patch, eventually, $Gd@C_{82}(OH)_{22}$ ends up in between the S1′ and SC loops with a large contact area. There, $Gd@C_{82}(OH)_{22}$ stabilization seems to be more ideal and well balanced among hydrophobic interactions (e.g., L212, F396, L397, P429, and P430), π–cation interactions (e.g., R143 and K214), and hydrogen bonds (e.g., T426, E427, and backbone amide and carbonyl atoms) (Figure 10.4).

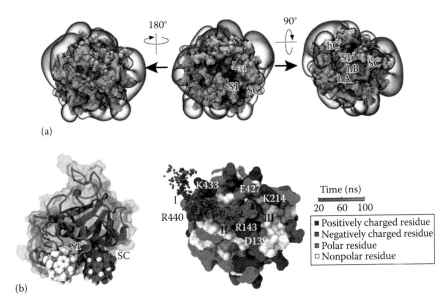

(a)

(b)

Figure 10.4 (See color insert.) Electrostatic potential map and representative inhibitory pathway. (a) While a large negative potential prevails around the active site of MMP-9 (center), a hydrophobic patch with positive hot spots is stretched out over the area encompassing the ligand specificity S1′ and SC loops (right). The electrostatic potential was computed by using APBS 1.3 tool (Adaptive Poisson-Boltzmann Solver) [72]. Blue and red lobes represent isoelectric potential surfaces of +0.5 and –0.5 kT, respectively. (b, left) A representative binding mode shows that $Gd@C_{82}(OH)_{22}$ (a solid ball) binds between the S1′ ligand specificity loop (green ribbon) and the SC loop (purple ribbon) leading to the ligand binding groove. Alternatively, $Gd@C_{82}(OH)_{22}$ (a gray ball) can bind at the back entrance of the S1′ cavity leading into the active site (ball and stick for active sites and orange ball for the catalytic Zn^{2+}). (b, right) A possible binding pathway—depending on major driving forces, $Gd@C_{82}(OH)_{22}$ binds on MMP-9 along with three different phases: phase I, a diffusion-controlled nonspecific electrostatic interaction; phase II, a transient nonspecific hydrophobic interaction; and phase III, a specific hydrophobic and hydrogen-bonded stable binding.

10.3.4 Mechanical Impact of Gd@C$_{82}$(OH)$_{22}$ on Collagen Complexes

In the previous section, in silico study clarified how metallofullerenol Gd@C$_{82}$(OH)$_{22}$ interacts with MMP-9 and furthermore how it inhibits the enzyme activity. Turning our focus to the ECM, we investigated how the nanoparticle could affect the structural integrity of collagen complex, a major constituent of the ECM [73,74] as well as the proteolytic target substrate for MMP-2/MMP-9. In vivo test with mice showed that Gd@C$_{82}$(OH)$_{22}$ treatment thickens the fibrous ECM layer encaging the tumor tissue [59,60]. The collagen content is known to decrease as the tumor advances [75]. Hence, the phenomenal antimetastatic activity with Gd@C$_{82}$(OH)$_{22}$ is believed to be closely related with the enhanced resistance to ECM degradation. Although the ECM degradation is mainly lowered by Gd@C$_{82}$(OH)$_{22}$ inhibition to MMPs, recent study also showed that Gd@C$_{82}$(OH)$_{22}$ can directly interact with collagen [60], where the mechanical properties (e.g., density and stiffness) are remarkably changed. Thus, it may provide another aspect of understanding on how Gd@C$_{82}$(OH)$_{22}$ plays as an antimetastatic nanomedicine in cancer media. As described in the earlier section, Gd@C$_{82}$(OH)$_{22}$ interaction with collagen complex is approached with two levels of structural hierarchy: (1) a single collagen triplex and (2) a tetramer of collagen triplexes.

With regard to a single collagen triplex, the simulation was initiated with 14 Gd@C$_{82}$(OH)$_{22}$ molecules distributed over water solvent with no direct contact with the collagen. Overall, Gd@C$_{82}$(OH)$_{22}$ was frequently interacting with the collagen triplex at various different sites, as accompanied with intermolecular aggregation among the Gd@C$_{82}$(OH)$_{22}$ molecules (Figure 10.5). A more detailed analysis revealed that Gd@C$_{82}$(OH)$_{22}$ has preferential bindings from N-terminus to the middle, followed by C-terminus (i.e., on average up to 4 and 2 in the N-terminus and the middle, respectively, while only 1 in the C-terminus). The nanoparticle adsorption contributed structural integrity of collagen triplex. Compared to the control without any nanoparticle, the residue-specific rmsd and rmsf show that (1) Gd@C$_{82}$(OH)$_{22}$ can sustain the triplex closer to its crystal structure especially at the N- and C-termini and (2) also suppresses thermal fluctuations in both ends, respectively, with more prominent effect on the C-terminus.

The site preference of adsorption may be explained by multiple different energy sources between the collagen triplex and Gd@C$_{82}$(OH)$_{22}$. The long-range electrostatic interaction may play an important role in determining the initial contact rate along the axis of collagen triplex. That is, the N-terminal contact can be intensified by favorable long-range electrostatic interaction between the positively charged N-termini from three chains and the negative fullerenol cage of Gd@C$_{82}$(OH)$_{22}$ while giving a relatively low chance to the middle

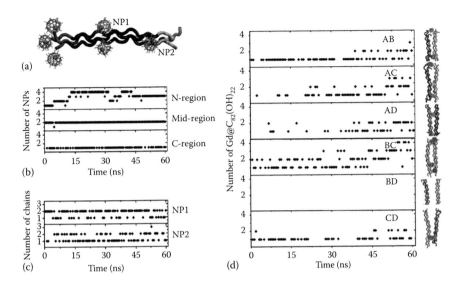

Figure 10.5 $Gd@C_{82}(OH)_{22}$ binding dynamics on collagen complex. (a) Representative snapshot of collagen triplex bound by $Gd@C_{82}(OH)_{22}$ ($t = 50$ ns). (b) Number of adsorbed $Gd@C_{82}(OH)_{22}$ molecules in each region of a collagen triplex. (c) Number of collagen peptide chains with which the representative $Gd@C_{82}(OH)_{22}$ molecule interacts. The adsorbed $Gd@C_{82}(OH)_{22}$ can simultaneously interact with two peptide chains, acting as the "nanoparticle-mediated bridge" between two chains. (d) The number of $Gd@C_{82}(OH)_{22}$ simultaneously contacts with collagen triplex pairs: AB, AC, AD, BC, BD, and CD, together with corresponding snapshots at $t = 60$ ns (on the right side).

and even lower one in the C-terminal region by the repulsive interaction with the negative charges at the C-terminal ends. However, once $Gd@C_{82}(OH)_{22}$ begins to contact with surface residues in close distance, $Gd@C_{82}(OH)_{22}$ can afford to utilize diverse short-ranged interactions such as hydrophobic interaction and hydrogen bonds. As such, even though a relatively small number of $Gd@C_{82}(OH)_{22}$ are adsorbed at the C-terminus, $Gd@C_{82}(OH)_{22}$, once attached, may act as an effective structural stabilizer through stable short-range interactions.

Structurally, $Gd@C_{82}(OH)_{22}$ was shown to enhance the interchain hydrogen bonding probabilities, especially at both terminal regions. For example, while on average, 0.8 and 1.4 hydrogen bonds exist at the N- and C-terminal regions of the collagen triplex, respectively, in the absence of $Gd@C_{82}(OH)_{22}$. The interchain hydrogen bonds become increased to 0.9 and 1.4, respectively, in the presence of $Gd@C_{82}(OH)_{22}$. This implies that the reduced thermal fluctuations with $Gd@C_{82}(OH)_{22}$ might facilitate the hydrogen bond formations or vice versa. In addition, simulations often showed that Gd@C$_{82}$(OH)$_{22}$ interacts not only with each individual chain of the triplex but also with multiple chains at the same time (Figure 10.5). The interchain hydrogen bonds are a well-known element in retaining the collagen triplex structure [76–81]. In this regard, the so-called nanoparticle-mediated "bridge" was proposed as a plausible mechanism to explain how Gd@C$_{82}$(OH)$_{22}$ contributes to the structural stability of the collagen complex. Indeed, Gd@C$_{82}$(OH)$_{22}$ medicates the interchain interactions of collagen triplex by binding in between the collagen monomer chains and holding them tightly through hydrogen bonds.

Furthermore, Gd@C$_{82}$(OH)$_{22}$ can interact with a bundle of collagen triplexes (Figure 10.5). In simulation with a complex of four collagen triplexes, Gd@C$_{82}$(OH)$_{22}$ molecules are likely to bind on at least one of the collagen triplexes with high probability, where 19 out of 20 Ga@C$_{82}$(OH)$_{22}$ molecules were found to be adsorbed. The nanoparticle adsorption highly enhanced the bundle stability (i.e., rmsd $\approx$ 4 Å), compared to the control simulations with no nanoparticles (i.e., rmsd $\approx$ 12 Å). Similar to the case of a single triplex, Gd@C$_{82}$(OH)$_{22}$ was shown to effectively mediate the interactions between the collagen triplexes themselves. Up to four Gd@C$_{82}$(OH)$_{22}$ molecules were found in between pairs of collagen triplexes, forming stable hydrogen bonds. This contributes not only to the conformational stabilization of individual triplex but also to the organized orientation among the collagen triplexes. Interestingly, the ordered arrangement has been validated in circular dichroism (CD) spectra of the collagen solution. The triple helical structures are characterized with a positive and negative peak at 220 and 197 nm, respectively [82], where two peaks increase with Gd@C$_{82}$(OH)$_{22}$. Although the result clearly shows that Gd@C$_{82}$(OH)$_{22}$ enhances a short-ranged oligomer or microfibril formation, it requires further studies on how the nanoparticle insertion would affect the native collagen fibrils and whether it would disturb or enhance the long-ranged order.

10.4   MODELING NANOTOXICOLOGY OF GD@C$_{82}$(OH)$_{22}$

In the previous section, we introduced how metallofullerenol Gd@C$_{82}$(OH)$_{22}$ interacts with a cancer-promoting enzyme MMP-9 and how the nanoparticle can modulate the collagen structure, a major constituent of ECM. Using a large-scale atomistic molecular dynamics simulation, we demonstrated how Gd@C$_{82}$(OH)$_{22}$ can function as an antimetastatic cancer therapeutics. Although this sheds light on its potential application as a direct "nanodrug" of a specific inhibitory and regulatory function, here, we introduce a small different aspect of Gd@C$_{82}$(OH)$_{22}$ and protein interactions from the perspective of nanotoxicity. For our discussion, we took two representative proteins WW domain and SH3 domain, which are ubiquitous and important in signal transduction pathways for mediating protein–protein interactions [83–85]. Even though the following discussions are solely based on theoretical prediction, it might give some insight into how the nanoparticle, if anticipated, could disturb the signaling and related regulatory mechanism in our body.

10.4.1   Direct Inhibition of Gd@C$_{82}$(OH)$_{22}$ on WW Domain

First, we investigated how Gd@C$_{82}$(OH)$_{22}$ interacts with the WW domain. The WW domain ubiquitously exists in regulatory pathways as mediating signal transduction by recognizing the PRMs. Due to its involvement in the epithelial sodium channel function, it has been speculated that the WW domain could be involved in several human diseases like Alzheimer's disease [86,87]. In our simulation, we configured two different systems between Gd@C$_{82}$(OH)$_{22}$ and the WW domain: (1) a binary system of the WW domain surrounded by four Gd@C$_{82}$(OH)$_{22}$ with no initial touch and (2) a ternary system of the WW domain surrounded by one Gd@C$_{82}$(OH)$_{22}$ and its native PRM ligand. The former was prepared in order to understand the intrinsic binding dynamics of Gd@C$_{82}$(OH)$_{22}$ on the WW domain, and the latter was for more deeper understanding on a potential inhibitory function of Gd@C$_{82}$(OH)$_{22}$ in the presence of the native ligand of the WW domain.

To begin with the binary system, our simulation showed that the WW domain is highly liable to the Gd@C$_{82}$(OH)$_{22}$ molecules. Metallofullerenol Gd@C$_{82}$(OH)$_{22}$ is not only interacting with the WW domain but also directly blocking the putative binding site for PRM. Similar to the case of MMP-9, Gd@C$_{82}$(OH)$_{22}$ hardly affects the overall fold of the WW domain (i.e., rmsd $\approx$ 1–3 Å). Due to the surface OH groups, Gd@C$_{82}$(OH)$_{22}$ is not hydrophobic enough to induce global or local structural deformation of the WW domain. Instead, Gd@C$_{82}$(OH)$_{22}$ was clearly shown to directly interact with various critical residues for the PRM recognition, including Y28, K30, H32, Q35, T37, and W39 [46,50,88]. For example, two prolines of the PPxY motif of PRM are packed into the deep groove made of the aromatic Y28 and W39, while the tyrosine of the PPxY motif is stacked on the shallow groove formed by residues K30, H32, and Q35. Besides, the PRM backbone is also stabilized by the hydrogen bond with T37. Moreover, the PMF indicated that the binding is quite stable (i.e., ΔG = −5.44 kcal/mol), where the PMF was similarly measured with Equation 10.1 for two reaction coordinates: the minimum distance between Gd@C$_{82}$(OH)$_{22}$ and two signature residues Y28 and W39 and the contact area between Gd@C$_{82}$(OH)$_{22}$ and the WW domain. In Figure 10.6, representative binding modes demonstrate that the Gd@C$_{82}$(OH)$_{22}$ molecules are clustered on the WW domain, directly blocking both deep and shallow PRM binding grooves. As discussed in the previous examples, Gd@C$_{82}$(OH)$_{22}$ seems to favorably occupy the PRM binding site of the WW domain via multiple different interactions between Gd@C$_{82}$(OH)$_{22}$ and the binding site residues, including π–π interaction with the aromatic residues

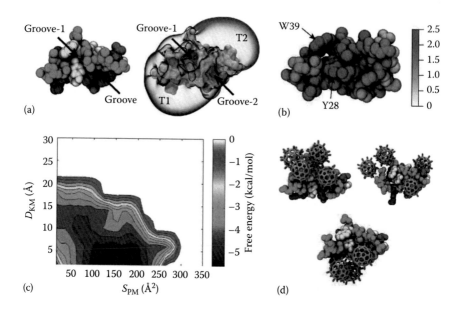

Figure 10.6   **(See color insert.)** Residue-specific contacts and binding free energy surface from the binary system of $Gd@C_{82}(OH)_{22}$ and the WW domain. (a) The WW domain is colored according to the residue types with Y28 and W38 highlighted in yellow: negatively charged, positively charged, and polar and nonpolar residues are colored in red, blue, green, and white, respectively. (b) Site-specific contact ratio of the WW domain, where residues in PRM binding site have high contact ratio to $Gd@C_{82}(OH)_{22}$. (c) The binding free energy surface, where D_{KM} is the minimum distance between $Gd@C_{82}(OH)_{22}$ and the signature residues (Y28 and W39) of the WW domain and S_{PM} is the contacting surface area. (d) Representative binding modes found in the global minimum. Yellow, key residues; white, hydrophobic; green, noncharged polar; red, negatively charged; and blue, positively charged residues.

Y28 and W39, π–cation interaction between the positively charged K30 and aromatic carbon cage of fullerenol $C_{82}(OH)_{22}$, and hydrogen bonds through the multiple hydroxyl groups of $Gd@C_{82}(OH)_{22}$.

Thus, the specific binding of $Gd@C_{82}(OH)_{22}$ may interfere with the native function of the WW domain that mediates the protein–protein interaction through the PRM recognition. In order to better understand the potential disturbance, we then studied the ternary system of the WW domain with both $Gd@C_{82}(OH)_{22}$ and PRM ligand (i.e., $G_1TPPPPYTVG_{10}$). The simulation shows that $Gd@C_{82}(OH)_{22}$ generally wins over the native ligand in almost all highly contacting sites of the WW domain. Even the PRM itself, especially around the central $(P_3)PPPY_7$ motif, shows higher preference to $Gd@C_{82}(OH)_{22}$ rather than the WW domain, which might be attributed to the relative low entropy cost of the rigid polyproline II (PPII) helix formation and hydrophobic interaction with Gd@C$_{82}$(OH)$_{22}$. This implies that the native interaction between the WW domain and PRM is very likely disturbed in the presence of Gd@C$_{82}$(OH)$_{22}$.

In more detail, the PMF analysis provides additional evidence that Gd@C$_{82}$(OH)$_{22}$ still prefers to directly bind on the PRM binding site of the WW domain (i.e., ΔG = −4.78 kcal/mol), even though an off-binding mode begins to be populated in the presence of PRM (i.e., ΔG = −4.57 kcal/mol). The representative structures display that Gd@C$_{82}$(OH)$_{22}$ is occupying the ligand binding sites, effectively interacting with the key aromatic residues Y28 and W39, while distracting PRM from its putative binding site (Figure 10.7b.i). Even in the off-binding mode (Figure 10.7b.ii), PRM is seriously distracted by Gd@C$_{82}$(OH)$_{22}$. On the other aspect, the PMF of PRM reveals that the global energy minimum has been populated for the off-site binding of PRM distracted by Gd@C$_{82}$(OH)$_{22}$ (i.e., indicated by i with ΔG = −4.67 kcal/mol). Even though the on-site binding was found as indicated in iii with ΔG = −4.06 kcal/mol at the PMF surface (Figure 10.7c), Gd@C$_{82}$(OH)$_{22}$ is still more competitive in binding at the ligand binding site of the WW domain by about $\Delta\Delta G$ = −0.72 kcal/mol. Nonetheless, the PRM still remains to be distracted by preoccupying Gd@C$_{82}$(OH)$_{22}$ near the PRM binding site (Figure 10.7d). Our results indicate that the signal transduction pathway could be seriously inhibited by Gd@C$_{82}$(OH)$_{22}$ that directly blocks the PRM binding site of the WW domain.

10.4.2 Indirect Disturbance of PRM Binding on the SH3 Domain

Next, we explored Gd@C$_{82}$(OH)$_{22}$ interaction with the SH3 domain, another highly promiscuous protein–protein interaction mediator involved in signaling and regulatory pathways. Similar to the WW domain, we investigated the intrinsic binding property and PRM inhibitory action of Gd@C$_{82}$(OH)$_{22}$ by configuring a binary system of the SH3 domain and Gd@C$_{82}$(OH)$_{22}$ and a ternary complex of the SH3 domain with both Gd@C$_{82}$(OH)$_{22}$ and a native PRM ligand (P$_1$PPVPPRR$_8$), respectively.

In the binary system, our simulation displays that Gd@C$_{82}$(OH)$_{22}$ prefers to interact with the SH3 domain especially around the PRM binding site. Although this is similar to the WW domain, much larger area is involved in contact with Gd@C$_{82}$(OH)$_{22}$, including the PRM binding site as well as the RT and n-Src loops. The binding free energy surface reflects the fact that two binding modes (i.e., on-site and off-site bindings) are almost equally populated in the SH3 domain, although the off-site mode is slightly more stable than the on-site one by $\Delta\Delta G \approx$ −0.04 kcal/mol. The PRM binding site of the SH3 domain is characterized by linear arrangement of the aromatic and hydrophobic residues

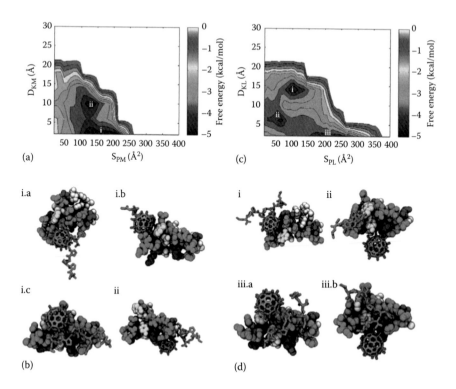

Figure 10.7 Binding free energy surface and representative binding modes from the ternary system of the WW domain with $Gd@C_{82}(OH)_{22}$ and PRM ligand. (a) Binding free energy landscape between the WW domain and $Gd@C_{82}(OH)_{22}$ and (b) representative structures found in local energy minima. (c) Binding free energy landscape between the WW domain and PRM and (d) representative structures found in local energy minima. D and S of two axes in the free energy diagrams indicate distance and contact area, respectively. The subscripts P, K, L, and M represent protein, key residues, PRM ligand, and $Gd@C_{82}(OH)_{22}$, respectively. The local energy minima are rank ordered starting from the global minimum in (a, c).

(i.e., F141, W169, P183, and Y186) for the PxxP motif of the PRM in the PPII helix. Moreover, the RT loop, composed of a cluster of negatively charged residues D147, E148, E149, and D150, is known to recognize PRM and guide its binding orientation through long-range electrostatic interaction with the C-terminal arginine of PRM [52].

The representative binding modes found in the two energy basins demonstrate either how $Gd@C_{82}(OH)_{22}$ directly blocks the PRM binding site or interacts with the RT loop. Compared to the WW domain, the PRM binding site of the SH3 domain is being shielded with a large negative electrostatic field generated by the negative residues of the RT and n-Src loops, which makes Gd@C$_{82}$(OH)$_{22}$ binding more diffused over the area including the binding site and surrounding regions. It should be emphasized that even though it seems more likely for Gd@C$_{82}$(OH)$_{22}$ to be attracted by positively charged residues in long distance, nanoparticular interaction becomes more diversified in short distance depending on various surface residue types. The metallofullerenol can get along with hydrophobic, hydrophilic, and charged residues by utilizing the hydrophobic and aromatic fullerene carbons, hydrogen bonding OH groups, and charged Gd^{3+} ion. Similar to the WW domain, our result implies that Gd@C$_{82}$(OH)$_{22}$ may also interfere with the protein–protein mediation by the SH3 domain, even though it could be less serious than that with the WW domain, as shown in the dual and diffused binding modes (Figure 10.8).

With the ternary system, next, we explored in more detail how Gd@C$_{82}$(OH)$_{22}$ would affect the native function of SH3 domain that recognizes the PRM

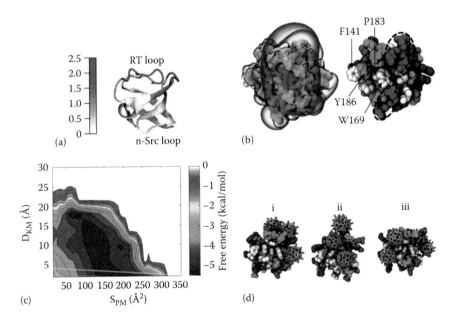

Figure 10.8 **(See color insert.)** Residue-specific contacts and binding free energy surface from the binary system of Gd@C$_{82}$(OH)$_{22}$ and the WW domain. (a) Residue-specific contacts of Gd@C$_{82}$(OH)$_{22}$ on the SH3 domain. (b) Electrostatic potential surface of the SH3 domain, where a large negative electrostatic field is formed around the PRM binding site by negative residue clusters at the n-Src and RT loops. (c) Binding free energy landscape in the binary system and (d) representative structures in the low-energy basin, where residues are colored with residue types as described in Figure 10.2. Key residues are marked in yellow.

ligand, which is then compared with the PRM binding on the SH3 domain in the absence of $Gd@C_{82}(OH)_{22}$. Strikingly, the $Gd@C_{82}(OH)_{22}$ binding, equilibrated between the on- and off-site bindings in the binary system, has been dramatically shifted toward the off-site mode in the presence of PRM, where the binding free energy surface revealed two off-site modes, while only a marginal population was found for the direct binding (Figure 10.9a). Even with the low probability for the direct blockage of $Gd@C_{82}(OH)_{22}$, our simulation still indicates that the PRM could be highly distracted from its putative binding on the SH3 domain by $Gd@C_{82}(OH)_{22}$ adsorbed at the off-sites. The representative binding mode found in the energy minima of $Gd@C_{82}(OH)_{22}$ and SH3 domain (i.e., Figure 10.9b) shows that PRM is always being distracted from the putative binding site. This is also reflected in the PMF surface between the PRM and SH3 domain in the presence of $Gd@C_{82}(OH)_{22}$ (Figure 10.9c and d), where two energy minima are almost equally populated far away from the native PRM binding site of the SH3 domain.

In contrast, the native binding mode was highly favored between the PRM and SH3 domain in the absence of $Gd@C_{82}(OH)_{22}$ (Figure 10.9e). As shown in Figure 10.9f, one trajectory demonstrated that PRM not only recognizes the putative binding site of the SH3 domain but also completes the binding with almost identical native contacts as observed in the crystal structure. Thus, this emphasizes that the binding dynamics of $Gd@C_{82}(OH)_{22}$ (or PRM) on the SH3 domain is sensitive to the existence of the PRM ligand (or $Gd@C_{82}(OH)_{22}$). Indeed, while both PRM and $Gd@C_{82}(OH)_{22}$ favor the on-site binding when each individually interacts with the SH3 domain, the off-site binding has become more favored if both PRM and $Gd@C_{82}(OH)_{22}$ interact together with the protein. This may be understood with a preferential adsorption between PRM and $Gd@C_{82}(OH)_{22}$. Our contact analysis reveals that PRM, especially its two C-terminal Arg's (R_7R_8), frequently contacts with $Gd@C_{82}(OH)_{22}$ via the stable $\pi-\pi$ and $\pi-$cation interactions. This reduces the available hydrophobic surface of $Gd@C_{82}(OH)_{22}$ that is critical for the interaction with the hydrophobic residues (i.e., F141, W169, P183, and Y186) in the native binding site of the SH3 domain. Instead, it possibly facilitates $Gd@C_{82}(OH)_{22}$ to interact with the hydrophilic residues at the off-sites. Moreover, the arginine interception may lead to the distraction of the whole PRM ligand from its putative binding mode, since arginine plays a critical role in initiating the PRM binding by electrostatic interaction with negative residues in the RT loop. Therefore, $Gd@C_{82}(OH)_{22}$ may interfere with the SH3 domain by more indirectly distracting PRM at the off-sites, rather than directly blocking the native binding site as shown in the WW domain. However, both cases implicate that the protein–protein interaction medicated by the WW and SH3 domains could be impaired by the exposure of $Gd@C_{82}(OH)_{22}$.

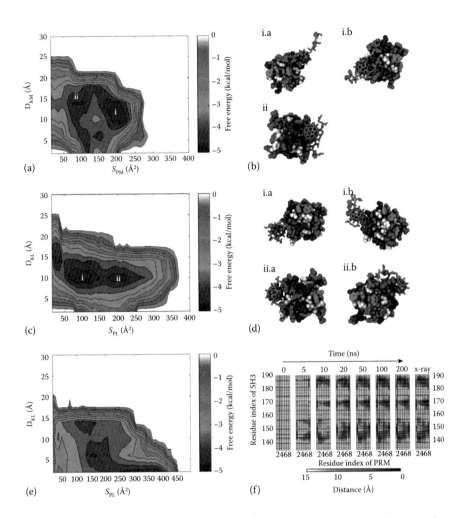

Figure 10.9   Binding free energy surfaces and representative structures between the SH3 domain and PRM with and without Gd@C$_{82}$(OH)$_{22}$. (a) Binding free energy surface between the SH3 domain and Gd@C$_{82}$(OH)$_{22}$ and (b) representative structures found in the energy minima. (c) Binding free energy surface between the SH3 domain and PRM in the presence of Gd@C$_{82}$(OH)$_{22}$ and (d) representative structure found in the energy minima. (e) Binding free energy surface between the SH3 domain and PRM in the absence of Gd@C$_{82}$(OH)$_{22}$ and (f) time evolution of detailed native contact map of one representative trajectory.

10.5   CONCLUSION AND FUTURE PERSPECTIVES

In this chapter, we discussed the interaction between nanoparticles and proteins from the perspectives of both nanomedicine and nanotoxicology. Using the results from our studies of endohedral metallofullerenol $Gd@C_{82}(OH)_{22}$, we first showed how $Gd@C_{82}(OH)_{22}$ can be used in the antimetastasis of cancer cells. In order to reveal the atomic details of the interactions between the nanoparticle $Gd@C_{82}(OH)_{22}$ and target proteins, it is necessary to carry out large-scale atomistic molecular dynamics simulations. With MMP-9, a promoter and inhibitor for ECM degradation and angiogenesis, $Gd@C_{82}(OH)_{22}$ was clearly shown to have a specific binding at its ligand specificity S1′ loop implicating the important role of $Gd@C_{82}(OH)_{22}$, not only in the binding mechanism (and relevant driving forces) of nanoparticle–protein interaction but also in its inhibition mechanism of the enzyme activity as shown in our in vitro biochemical assay. In addition, $Gd@C_{82}(OH)_{22}$ was shown to influence the mechanical properties of the collagen complex, a major constituent of ECM layer, where $Gd@C_{82}(OH)_{22}$ stabilized a single collagen triplex as well as a bundle of triplexes, effectively mediating their interchain interactions.

From another perspective, we also introduced theoretical predictions on how $Gd@C_{82}(OH)_{22}$ interacts with the WW and SH3 domains, two ubiquitous proteins found in signal transduction via protein–protein mediation. Even with differences in detailed mechanisms for the two domains, $Gd@C_{82}(OH)_{22}$ appeared to effectively inhibit the protein–protein interactions mediated by these domains. Toward the WW domain, $Gd@C_{82}(OH)_{22}$ could seriously interfere with the PRM binding through direct blockage on the PRM binding site, whereas for the SH3 domain, it competes with the native ligand PRM for the same binding pocket.

Our findings provide new ideas about the future direction and open up potential opportunities in therapeutic nanomaterials as a direct "nanodrug," in contrast with the trend where nanomedicines are used mainly as a nanocarrier for conventional small molecular drugs. While the current nanomedicines are highly focused on selectivity enhancement at the tissue and/or cell levels, the nanodrugs would target subcellular compartments or even deeper levels of specific proteins. This means more sophisticated nanosystems can be designed, which requires more stereotypic understanding on pharmacokinetics following the route of the central dogma and related regulatory systems. The interaction and its inhibitory mechanism could be more complicated and less intuitive compared to the conventional small molecule-based drug developments, as shown in the exosite interaction of $Gd@C_{82}(OH)_{22}$ at the S1′ loop of MMP-9. Furthermore, whether nanodrugs may cause undesired side effects is still unclear at the moment. As exemplified with the WW and SH3 domains, this may require a long-term assessment on the potential toxicity.

As such, large- and multiscale computational efforts would be inevitable for de novo design and assessment of nanodrugs in multicomponent heterogeneous environment, including an accurate force parameterization for the nanomolecule and appropriate modeling approaches encompassing essential features between the nanoparticle and the surrounding rest.

REFERENCES

1. Doane, T. L. and C. Burda. 2012. The unique role of nanoparticles in nanomedicine: Imaging, drug delivery and therapy. *Chemical Society Reviews* 41:2885–2911.
2. Schroeder, A., D. A. Heller, M. M. Winslow, J. E. Dahlman, G. W. Pratt, R. Langer, T. Jacks, and D. G. Anderson. 2012. Treating metastatic cancer with nanotechnology. *Nature Reviews Cancer* 12:39–50.
3. Lee, P. Y. and K. K. Y. Wong. 2011. Nanomedicine: A new frontier in cancer therapeutics. *Current Drug Delivery* 8:245–253.
4. Zhang, X. Q., X. Xu, N. Bertrand, E. Pridgen, A. Swami, and O. C. Farokhzad. 2012. Interactions of nanomaterials and biological systems: Implications to personalized nanomedicine. *Advanced Drug Delivery Reviews* 64:1363–1384.
5. Euliss, L. E., J. A. DuPont, S. Gratton, and J. M. DeSimone. 2006. Imparting size, shape, and composition control of materials for nanomedicine. *Chemical Society Reviews* 35:1095–1104.
6. Shapira, A., Y. D. Livney, H. J. Broxterman, and Y. G. Assaraf. 2011. Nanomedicine for targeted cancer therapy: Towards the overcoming of drug resistance. *Drug Resistance Update* 14:150–163.
7. Nie, S. M. 2010. Understanding and overcoming major barriers in cancer nanomedicine. *Nanomedicine* 5:523–528.
8. Cabral, H., N. Nishiyama, and K. Kataoka. 2011. Supramolecular nanodevices: From design validation to theranostic nanomedicine. *Accounts of Chemical Research* 44:999–1008.
9. Al-Jamal, W. T. and K. Kostarelos. 2011. Liposomes: From a clinically established drug delivery system to a nanoparticle platform for theranostic nanomedicine. *Accounts of Chemical Research* 44:1094–1104.
10. Tan, A., L. Yildirimer, J. Rajadas, H. De La Pena, G. Pastorin, and A. Seifalian. 2011. Quantum dots and carbon nanotubes in oncology: A review on emerging theranostic applications in nanomedicine. *Nanomedicine* 6:1101–1114.
11. Cha, C., S. R. Shin, N. Annabi, M. R. Dokmeci, and A. Khademhosseini. 2013. Carbon-based nanomaterials: Multifunctional materials for biomedical engineering. *ACS Nano* 7:2891–2897.
12. Mi, Y., Y. J. Guo, and S. S. Feng. 2012. Nanomedicine for multimodality treatment of cancer. *Nanomedicine* 7:1791–1794.
13. Mout, R., D. F. Moyano, S. Rana, and V. M. Rotello. 2012. Surface functionalization of nanoparticles for nanomedicine. *Chemical Society Reviews* 41:2539–2544.
14. Nel, A. E., L. Madler, D. Velegol, T. Xia, E. M. Hoek, P. Somasundaran, F. Klaessig, V. Castranova, and M. Thompson. 2009. Understanding biophysicochemical interactions at the nano-bio interface. *Nature Materials* 8:543–557.

15. Nel, A., T. Xia, L. Madler, and N. Li. 2006. Toxic potential of materials at the nano-level. *Science* 311:622–627.
16. Xia, T., N. Li, and A. E. Nel. 2009. Potential health impact of nanoparticles. *Annual Review of Public Health* 30:137–150.
17. Mahmoudi, M., I. Lynch, M. R. Ejtehadi, M. P. Monopoli, F. B. Bombelli, and S. Laurent. 2011. Protein–nanoparticle interactions: Opportunities and challenges. *Chemical Reviews* 111:5610–5637.
18. Kumari, A. and S. K. Yadav. 2011. Cellular interactions of therapeutically delivered nanoparticles. *Expert Opinion on Drug Delivery* 8:141–151.
19. Mahmoudi, M., K. Azadmanesh, M. A. Shokrgozar, W. S. Journeay, and S. Laurent. 2011. Effect of nanoparticles on the cell life cycle. *Chemical Reviews* 111:3407–3432.
20. Makarucha, A. J., N. Todorova, and I. Yarovsky. 2011. Nanomaterials in biological environment: A review of computer modelling studies. *European Biophysics Journal—Biophysics Letters* 40:103–115.
21. Kang, S. G., G. Zhou, P. Yang, Y. Liu, B. Sun, T. Huynh, H. Meng et al. 2012. Molecular mechanism of pancreatic tumor metastasis inhibition by $Gd@C_{82}(OH)_{22}$ and its implication for de novo design of nanomedicine. *Proceedings of the National Academy of Sciences of the United States of America* 109:15431–15436.
22. Yin, X., L. Zhao, S. G. Kang, J. Pan, Y. Song, M. Zhang, G. Xing et al. 2013. Impacts of fullerene derivatives on regulating the structure and assembly of collagen molecules. *Nanoscale* 5:7341–7348.
23. Kang, S. G., T. Huynh, and R. Zhou. 2012. Non-destructive inhibition of metallofullerenol Gd@C(82)(OH)(22) on WW domain: Implication on signal transduction pathway. *Scientific Reports* 2:957.
24. Kang, S. G., T. Huynh, and R. Zhou. 2013. Metallofullerenol $Gd@C_{82}(OH)_{22}$ distracts the proline-rich-motif from putative binding on the SH3 domain. *Nanoscale* 5:2703–2712.
25. Perdew, J. P., K. Burke, and M. Ernzerhof. 1996. Generalized gradient approximation made simple. *Physical Review Letters* 77:3865–3868.
26. Van Lenthe, E. and E. J. Baerends. 2003. Optimized slater-type basis sets for the elements 1–118. *Journal of Computational Chemistry* 24:1142–1156.
27. Vanlenthe, E., E. J. Baerends, and J. G. Snijders. 1993. Relativistic regular 2-component hamiltonians. *Journal of Chemical Physics* 99:4597–4610.
28. Velde, G. T., F. M. Bickelhaupt, E. J. Baerends, C. F. Guerra, S. J. A. Van Gisbergen, J. G. Snijders, and T. Ziegler. 2001. Chemistry with ADF. *Journal of Computational Chemistry* 22:931–967.
29. Guerra, C. F., J. G. Snijders, G. te Velde, and E. J. Baerends. 1998. Towards an order-N DFT method. *Theoretical Chemistry Accounts* 99:391–403.
30. Clavaguera, C., F. Calvo, and J. P. Dognon. 2006. Theoretical study of the hydrated Gd3+ ion: Structure, dynamics, and charge transfer. *Journal of Chemical Physics* 124:074505.
31. Shinohara, H. 2000. Endohedral metallofullerenes. *Reports on Progress in Physics* 63:843–892.
32. Nishibori, E., K. Iwata, M. Sakata, M. Takata, H. Tanaka, H. Kato, and H. Shinohara. 2004. Anomalous endohedral structure of Gd@C-82 metallofullerenes. *Physical Review B* 69:113412.

33. Tang, J., G. M. Xing, H. Yuan, W. B. Cao, L. Jing, X. F. Gao, L. Qu et al. 2005. Tuning electronic properties of metallic atom in bondage to a nanospace. *Journal of Physical Chemistry B* 109:8779–8785.

34. Tang, J., G. M. Xing, Y. L. Zhao, L. Jing, X. F. Gao, Y. Cheng, H. Yuan et al. 2006. Periodical variation of electronic properties in polyhydroxylated metallofullerene materials. *Advanced Materials* 18:1458–1462.

35. Senapati, L., J. Schrier, and K. B. Whaley. 2004. Electronic transport, structure, and energetics of endohedral Gd@C-82 metallofullerenes. *Nano Letters* 4:2073–2078.

36. Funasaka, H., K. Sakurai, Y. Oda, K. Yamamoto, and T. Takahashi. 1995. Magnetic-properties of Gd@C-82 metallofullerene. *Chemical Physics Letters* 232:273–277.

37. Huang, H. J., S. H. Yang, and X. X. Zhang. 2000. Magnetic properties of heavy rare-earth metallofullerenes M@C-82 (M = Gd, Tb, Dy, Ho, and Er). *Journal of Physical Chemistry B* 104:1473–1482.

38. MacKerell, A. D., D. Bashford, M. Bellott, R. L. Dunbrack, J. D. Evanseck, M. J. Field, S. Fischer et al. 1998. All-atom empirical potential for molecular modeling and dynamics studies of proteins. *Journal of Physical Chemistry B* 102:3586–3616.

39. Brooks, B. R., C. L. Brooks, 3rd, A. D. Mackerell, Jr., L. Nilsson, R. J. Petrella, B. Roux, Y. Won et al. 2009. CHARMM: The biomolecular simulation program. *Journal of Computational Chemistry* 30:1545–1614.

40. Jorgensen, W. L., J. Chandrasekhar, J. D. Madura, R. W. Impey, and M. L. Klein. 1983. Comparison of simple potential functions for simulating liquid water. *Journal of Chemical Physics* 79:926–935.

41. Darden, T., D. York, and L. Pedersen. 1993. Particle mesh Ewald: An N·log(N) method for Ewald sums in large systems. *Journal of Chemical Physics* 98:10089–10092.

42. Phillips, J. C., R. Braun, W. Wang, J. Gumbart, E. Tajkhorshid, E. Villa, C. Chipot, R. D. Skeel, L. Kale, and K. Schulten. 2005. Scalable molecular dynamics with NAMD. *Journal of Computational Chemistry* 26:1781–1802.

43. Kumar, S., C. Huang, G. Zheng, E. Bohm, A. Bhatele, J. C. Phillips, H. Yu, and L. V. Kale. 2008. Scalable molecular dynamics with NAMD on the IBM Blue Gene/L system. *IBM Journal of Research and Development* 52:177–188.

44. Tochowicz, A., K. Maskos, R. Huber, R. Oltenfreiter, V. Dive, A. Yiotakis, M. Zanda, T. Pourmotabbed, W. Bode, and P. Goettig. 2007. Crystal structures of MMP-9 complexes with five inhibitors: Contribution of the flexible Arg424 side-chain to selectivity. *Journal of Molecular Biology* 371:989–1006.

45. Orgel, J., T. Irving, A. Miller, and T. Wess. 2006. Microfibrillar structure of type I collagen in situ. *Proceedings of the National Academy of Sciences of the United States of America* 103:9001–9005.

46. Pires, J. R., F. Taha-Nejad, F. Toepert, T. Ast, U. Hoffmuller, J. Schneider-Mergener, R. Kuhne, M. J. Macias, and H. Oschkinat. 2001. Solution structures of the YAP65 WW domain and the variant L30 K in complex with the peptides GTPPPPYTVG, N-(n-octyl)-GPPPY and PLPPY and the application of peptide libraries reveal a minimal binding epitope. *Journal of Molecular Biology* 314:1147–1156.

47. Zuo, G., Q. Huang, G. Wei, R. Zhou, and H. Fang. 2010. Plugging into proteins: Poisoning protein function by a hydrophobic nanoparticle. *ACS Nano* 4:7508–7514.

48. Ibragimova, G. T. and R. C. Wade. 1999. Stability of the beta-sheet of the WW domain: A molecular dynamics simulation study. *Biophysical Journal* 77:2191–2198.

49. Ozkan, S. B., G. A. Wu, J. D. Chodera, and K. A. Dill. 2007. Protein folding by zipping and assembly. *Proceedings of the National Academy of Sciences of the United States of America* 104:11987–11992.

50. Macias, M. J., M. Hyvonen, E. Baraldi, J. Schultz, M. Sudol, M. Saraste, and H. Oschkinat. 1996. Structure of the WW domain of a kinase-associated protein complexed with a proline-rich peptide. *Nature* 382:646–649.

51. Toepert, F., J. R. Pires, C. Landgraf, H. Oschkinat, and J. Schneider-Mergener. 2001. Synthesis of an array comprising 837 variants of the hYAP WW protein domain. *Angewandte Chemie International Edition* 40:897–900.

52. Wu, X., B. Knudsen, S. M. Feller, J. Zheng, A. Sali, D. Cowburn, H. Hanafusa, and J. Kuriyan. 1995. Structural basis for the specific interaction of lysine-containing proline-rich peptides with the N-terminal SH3 domain of c-Crk. *Structure* 3:215–226.

53. Chawla, P., V. Chawla, R. Maheshwari, S. A. Saraf, and S. K. Saraf. 2010. Fullerenes: From carbon to nanomedicine. *Mini-Reviews in Medicinal Chemistry* 10:662–677.

54. Bakry, R., R. M. Vallant, M. Najam-ul-Haq, M. Rainer, Z. Szabo, C. W. Huck, and G. K. Bonn. 2007. Medicinal applications of fullerenes. *International Journal of Nanomedicine* 2:639–649.

55. Yamada, M., T. Akasaka, and S. Nagase. 2010. Endohedral metal atoms in pristine and functionalized fullerene cages. *Accounts of Chemical Research* 43:92–102.

56. Anilkumar, P., F. Lu, L. Cao, P. G. Luo, J. H. Liu, S. Sahu, K. N. Tackett, Y. Wang, and Y. P. Sun. 2011. Fullerenes for applications in biology and medicine. *Current Medicinal Chemistry* 18:2045–2059.

57. Kato, H., Y. Kanazawa, M. Okumura, A. Taninaka, T. Yokawa, and H. Shinohara. 2003. Lanthanoid endohedral metallofullerenols for MRI contrast agents. *Journal of the American Chemical Society* 125:4391–4397.

58. Mikawa, M., H. Kato, M. Okumura, M. Narazaki, Y. Kanazawa, N. Miwa, and H. Shinohara. 2001. Paramagnetic water-soluble metallofullerenes having the highest relaxivity for MRI contrast agents. *Bioconjugate Chemistry* 12:510–514.

59. Chen, C., G. Xing, J. Wang, Y. Zhao, B. Li, J. Tang, G. Jia et al. 2005. Multihydroxylated [Gd@C_{82}(OH)$_{22}$]$_n$ nanoparticles: Antineoplastic activity of high efficiency and low toxicity. *Nano Letters* 5:2050–2057.

60. Meng, H., G. Xing, E. Blanco, Y. Song, L. Zhao, B. Sun, X. Li et al. 2012. Gadolinium metallofullerenol nanoparticles inhibit cancer metastasis through matrix metalloproteinase inhibition: Imprisoning instead of poisoning cancer cells. *Nanomedicine* 8:136–146.

61. Yang, D., Y. Zhao, H. Guo, Y. Li, P. Tewary, G. Xing, W. Hou, J. J. Oppenheim, and N. Zhang. 2010. [Gd@C_{82}(OH)$_{22}$]$_n$ nanoparticles induce dendritic cell maturation and activate Th1 immune responses. *ACS Nano* 4:1178–1186.

62. Yin, J. J., F. Lao, J. Meng, P. P. Fu, Y. Zhao, G. Xing, X. Gao et al. 2008. Inhibition of tumor growth by endohedral metallofullerenol nanoparticles optimized as reactive oxygen species scavenger. *Molecular Pharmacology* 74:1132–1140.

63. Meng, H., G. Xing, B. Sun, F. Zhao, H. Lei, W. Li, Y. Song et al. 2010. Potent angiogenesis inhibition by the particulate form of fullerene derivatives. *ACS Nano* 4:2773–2783.

64. Tallant, C., A. Marrero, and F. X. Gomis-Ruth. 2010. Matrix metalloproteinases: Fold and function of their catalytic domains. *Biochimica et Biophysica Acta* 1803:20–28.

65. Deryugina, E. I. and J. P. Quigley. 2010. Pleiotropic roles of matrix metalloproteinases in tumor angiogenesis: Contrasting, overlapping and compensatory functions. *Biochimica et Biophysica Acta* 1803:103–120.
66. Ge, C., J. Du, L. Zhao, L. Wang, Y. Liu, D. Li, Y. Yang et al. 2011. Binding of blood proteins to carbon nanotubes reduces cytotoxicity. *Proceedings of the National Academy of Sciences of the United States of America* 108:16968–16973.
67. Zuo, G., X. Zhou, Q. Huang, H. P. Fang, and R. H. Zhou. 2011. Adsorption of villin headpiece onto graphene, carbon nanotube, and C60: Effect of contacting surface curvatures on binding affinity. *Journal of Physical Chemistry C* 115:23323–23328.
68. Overall, C. M. and C. Lopez-Otin. 2002. Strategies for MMP inhibition in cancer: Innovations for the post-trial era. *Nature Reviews Cancer* 2:657–672.
69. Overall, C. M. and O. Kleifeld. 2006. Towards third generation matrix metalloproteinase inhibitors for cancer therapy. *British Journal of Cancer* 94:941–946.
70. Martens, E., A. Leyssen, I. Van Aelst, P. Fiten, H. Piccard, J. Hu, F. J. Descamps et al. 2007. A monoclonal antibody inhibits gelatinase B/MMP-9 by selective binding to part of the catalytic domain and not to the fibronectin or zinc binding domains. *Biochimica et Biophysica Acta* 1770:178–186.
71. Zhou, R., B. J. Berne, and R. Germain. 2001. The free energy landscape for beta hairpin folding in explicit water. *Proceedings of the National Academy of Sciences of the United States of America* 98:14931–14936.
72. Baker, N. A., D. Sept, S. Joseph, M. J. Holst, and J. A. McCammon. 2001. Electrostatics of nanosystems: Application to microtubules and the ribosome. *Proceedings of the National Academy of Sciences of the United States of America* 98:10037–10041.
73. Kadler, K., D. Holmes, J. Trotter, and J. Chapman. 1996. Collagen fibril formation. *Biochemical Journal* 316:1–11.
74. Myllyharju, J. and K. Kivirikko. 2001. Collagens and collagen-related diseases. *Annals of Medicine* 33:7–21.
75. Grabowska, M. 1959. Collagen content of normal connective tissue, of tissue surrounding a tumour and of growing rat sarcoma. *Nature* 183:1186–1187.
76. Veld, P. and M. Stevens. 2008. Simulation of the mechanical strength of a single collagen molecule. *Biophysical Journal* 95:33–39.
77. Dai, N., X. Wang, and F. Etzkorn. 2008. The effect of a trans-locked Gly-Pro alkene isostere on collagen triple helix stability. *Journal of the American Chemical Society* 130:5396–5397.
78. Bachmann, A., T. Kiefhaber, S. Boudko, J. Engel, and H. Bachinger. 2005. Collagen triple-helix formation in all-trans chains proceeds by a nucleation/growth mechanism with a purely entropic barrier. *Proceedings of the National Academy of Sciences of the United States of America* 102:13897–13902.
79. Engel, J. and H. Bachinger. 2005. Structure, stability and folding of the collagen triple helix. *Topics in Current Chemistry* 247:7–33.
80. Gurry, T., P. Nerenberg, and C. Stultz. 2010. The contribution of interchain salt bridges to triple-helical stability in collagen. *Biophysical Journal* 98:2634–2643.
81. Raman, S., R. Gopalakrishnan, R. Wade, and V. Subramanian. 2011. Structural basis for the varying propensities of different amino acids to adopt the collagen conformation. *Journal of Physical Chemistry B* 115:2593–2607.

82. Freudenberg, U., S. H. Behrens, P. B. Welzel, M. Müller, M. Grimmer, K. Salchert, T. Taeger, K. Schmidt, W. Pompe, and C. Werner. 2007. Electrostatic interactions modulate the conformation of collagen I. *Biophysical Journal* 92:2108–2119.

83. Pawson, T. and P. Nash. 2003. Assembly of cell regulatory systems through protein interaction domains. *Science* 300:445–452.

84. Pawson, T. 2004. Specificity in signal transduction: From phosphotyrosine-SH2 domain interactions to complex cellular systems. *Cell* 116:191–203.

85. Pawson, T. and P. Nash. 2000. Protein–protein interactions define specificity in signal transduction. *Genes and Development* 14:1027–1047.

86. Sudol, M., K. Sliwa, and T. Russo. 2001. Functions of WW domains in the nucleus. *FEBS Letters* 490:190–195.

87. Ingham, R. J., K. Colwill, C. Howard, S. Dettwiler, C. S. Lim, J. Yu, K. Hersi et al. 2005. WW domains provide a platform for the assembly of multiprotein networks. *Molecular and Cellular Biology* 25:7092–7106.

88. Macias, M. J., S. Wiesner, and M. Sudol. 2002. WW and SH3 domains, two different scaffolds to recognize proline-rich ligands. *FEBS Letters* 513:30–37.

More Applications in Molecular Biology

Chapter 11

Modeling of DNA Sequencing with Solid-State Nanopores

Binquan Luan

CONTENTS

11.1 INTRODUCTION

Driven by the benefits of human genome sequencing to medical science, many research projects have been carried out to achieve the low-cost and high-throughput sequencing. When the sequencing of human genome is affordable in medical treatments, it will result in significant improvements in human healthcare, such as fast and accurate diagnostics or personalized medicine. To promote the research, NIH has launched the $1000 genome project. The goal is to develop a novel sequencing method whose cost for an entire human genome is less than $1000. The XPRIZE foundation also announced $10,000,000 award to the winning team that can sequence the whole genome of 100 subjects within 1 month for $10,000 or less per genome.

Till now, commercialized DNA sequencing technologies are based on the sequencing-by-synthesis method (Sanger's method [1]). Since the ground-breaking invention by Frederick Sanger (the Nobel Prize winner in 1980), many sequencing-by-synthesis methods (such as fluorescent in situ sequencing [2] and pyrosequencing [3]) were developed in the last 30 years. The cost of sequencing a whole human genome has been reduced dramatically from about $3 billion to about tens of thousands of dollars. Further reducing the cost could be very challenging. Recently, nanopore-based DNA sequencing becomes more and more promising as a potential low-cost method [4,5]. In the last decade, the nanopore-based DNA sequencing technology has been significantly improved. However, before being commercialized, several technical problems (see Sections 11.2 through 11.4) should be solved.

11.1.1 Sanger's Method

Sanger's method is also known as the chain-termination method. The essence of this method is to obtain positions of the same type of nucleotides in an ssDNA molecule in one measurement. Four independent measurements are required to sequencing ssDNA because of four different types of bases (adenine, thymine, cytosine, and guanine). In each measurement, dideoxynucleotides (such as dideoxyadenosine triphosphate [ddATP]) of the same type are added so that chain replication processes terminate at the same specific base (such as thymine if ddATPs are added). Different lengths of replicated ssDNA molecule, determined by gel electrophoresis, indicate relative positions of the same type of bases to the first one in ssDNA. Sanger's method has been dramatically improved since its invention. For example, in an automated sequencing, different kinds of dideoxynucleotides are labeled with different color dyes, allowing an easier and faster readout of DNA sequencing. Many recently developed DNA sequencing technologies, such as the ion torrent, were still based on the sequencing-by-synthesis method. Till now, for all available sequencing technologies, the cost of sequencing an entire human genome is still very high. A new method is required to make the sequencing affordable.

11.1.2 Next-Generation Sequencing Method

Nanopore-based DNA sequencing [4] could be the next-generation method. The fundamental idea is to use a nanopore to confine DNA in a linear conformation when the DNA is driven electrically through the pore. The stretched conformation of DNA in a nanopore can facilitate the electrical measurement of each DNA base. Ideally, this method requires little to no sample preparation; therefore, the cost is low. In experiment, the translocation speed of DNA in a nanopore is proven to be fast, about a few nanoseconds to a microsecond per nucleotide [6]. Currently, nanopore-based DNA sequencing holds the promise to be low cost and high throughput.

11.1.2.1 Protein Pores

The possibility of sequencing DNA in a nanopore was first demonstrated for a bacterial protein pore α-hemolysin [7]. To mimic a cell membrane artificially, a lipid bilayer is self-assembled across a hydrophobic orifice in a solid membrane. After that, α-hemolysin proteins can be spontaneously inserted into the lipid membrane, yielding channels connecting fluidic chambers above and below the membrane. This experimental setup has been widely used for investigating the transport of ssDNA through the protein pore, in a biasing electric field across the membrane. Because of the small constriction (~1.3 nm) of the pore, ssDNA can be electrically driven through the pore in a single file manner. Recent experiment shows that it is possible to differentiate four types of mononucleotides in the protein pore, by measuring signals of ionic current blockage [8]. Another protein pore that becomes more attractive for DNA sequencing is the MspA protein [9]. Since the MspA pore has a shorter channel and a narrower constriction site, even nucleotides in a short ssDNA molecule can be called from signals of an ionic current.

A major difficulty for nanopore-based DNA sequencing is that ssDNA threads quickly through a nanopore in an electric field. Ideally, each nucleotide in ssDNA should stop at the constriction site of a protein pore for a sufficient amount of time to allow the measurement of current signals. This has been achieved recently by adding a DNA polymerase on the top of the MspA protein pore [10], so that one strand of a dsDNA molecule can nucleotide by nucleotide pass through the pore constriction. With the controlled stepwise motion of ssDNA in a protein pore and the base-sensitive current recording, DNA sequencing using a protein pore could be realized in the near future.

The benefits to use a protein pore include the following: (1) both α-hemolysin and MspA pores (and their engineered ones) are structurally rigid in the membrane; (2) the atomic structure of a protein pore is known and same experimental results can be reproduced in different pores; and (3) ionic current through a protein (or engineered ones) pore is stable, that is, a good signal-to-noise ratio. The disadvantage is that the lipid membrane is fragile in an electric field. One solution is to reduce the size of an orifice to tens of nanometers. In an extreme case, a protein pore can be inserted directly into a nanopore (or nanometer-sized orifice) in a solid membrane [11].

11.1.2.2 Solid-State Nanopores

A solid-state nanopore is a nanometer-sized hole in a thin (30–100 nm) membrane that could be made of silicon dioxide (SiO_2) or silicon nitride (Si_3N_4). A typical diameter of a nanopore ranges from a few to about 30 nm. In a nanopore experiment (Figure 11.1), the membrane separates the *cis* and *trans* fluidic chambers. These two chambers are only connected through a nanopore in the membrane. In a biasing electric field (perpendicular to the membrane surface) applied across the two chambers, charged molecules (such as DNA) as well as ions are electrically driven through the nanopore. Because of the confined geometry of a nanopore, the ionic conductance of a nanopore is very sensitive to a transported molecule inside the pore. The ionic current resulting from electrically driven motion of ions can be altered during the translocation of a molecule.

The amount of change in current and the duration of such change provide estimations for physical properties (such as the size and the charge of a molecule) of a transported molecule. Interestingly, depending on the pore size and the concentration of an electrolyte, the pore current could be reduced (i.e., current blockage) or be enhanced [12]. In the case of dsDNA in a nanopore of a few nanometers in diameter, the current-blockage signal is normally observed in an experiment if the ion concentration is about 1 M; the current-enhancement

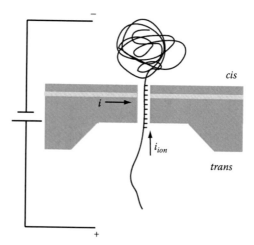

Figure 11.1   Sketch of the nanopore-based DNA sequencing device. DNA is driven through a nanopore by a biasing electric field. The translocation process of DNA is monitored by the transmembrane ionic current i_{ion}. A transversal tunneling current through a nucleotide that bridges a pair of nanoelectrodes is measured to call the type of the nucleotide.

signal occurs if the ion concentration is about 0.1 M. A theoretical study of signals of ionic current for DNA translocation is described in Ref. 13.

Although current signals are suitable for the detection of DNA translocation, it is difficult to sequence DNA using those signals. The constriction site of a typical solid-state nanopore is much larger than that of a protein pore, resulting in a larger structural fluctuation of each nucleotide. Thus, current signals cannot be sensitive to four different types of DNA bases. Although it is likely to reduce the size of a pore constriction to even 2 nm, unfavorable interaction between a DNA base and an untreated solid surface may prevent DNA from transiting the pore. Thus, for a solid-state nanopore, a different type of sensor is needed for sequencing DNA. Recent theoretical [14] and experimental [15] works show that it is possible to sequence DNA by measuring the transversal tunneling current through a pair of nanoelectrodes (bridged by a nucleotide in ssDNA) built inside a nanopore (Figure 11.1).

Therefore, to enable DNA sequencing in a solid-state nanopore, it is necessary to realize the following conditions. Firstly, the surface of a nanopore should be chemically treated [16,17] so that ssDNA does not stick to a pore surface. Secondly, the motion of ssDNA should be controlled, such as the base-by-base ratcheting motion of ssDNA in a nanopore [18]. Lastly, a new electric sensor, not measuring ionic currents, should be integrated inside a nanopore.

Till now, extensive experimental studies on DNA in a solid-state nanopore have been carried out and provided the understanding of translocation phenomena. Meanwhile, the problem is extremely difficult and is far from being solved till now. This is partly due to the fact that each solid-state nanopore is unique. Although sizes of different nanopores are close to each other, their shapes, surface roughness, and surface charge densities can be very different from each other, yielding different translocation behaviors of DNA. Limited by existing experimental approaches, it is hard to determine the dynamics of DNA translocation through a nanopore. Such dynamics can only be inferred from the ion current recording that is not informative enough. How does DNA interact hydrodynamically and electrically with a pore? What is the radial distribution of DNA translocation speed inside a pore? Answers to these questions are important to understand the translocation of DNA in a nanopore but cannot be obtained experimentally.

Recently, all-atom MD simulations provide an alternative way to attack those difficult problems. With an atomic level of detail, the MD method can be used as microscope to *image* the translocation of DNA in a pore or as an atomic force microscope (AFM) to study mechanical/electrical forces acting on DNA. With the fast development of supercomputers (such as the IBM Blue Gene), a simulated system can mimic a real experimental setup, and the time scale of MD simulation approaches the experimental one. Therefore, MD simulations can be applied to verify, understand, and even predict experimental results. The rest of this chapter is devoted to computational efforts on (1) investigating

DNA translocation through a solid-state nanopore, (2) studying a controlled motion of DNA in a solid-state nanopore, and (3) suggesting a new nanopore-based sequencing method.

11.2 SIMULATING THE NANOPORE-BASED DNA SEQUENCING

11.2.1 Simulation Methods

Molecular dynamics: All MD simulations described in this chapter were carried out using the program nanoscale molecular dynamics (NAMD) [19]. Standard force fields were applied for ions [20], water molecules (the transferable intermolecular potential 3P (TIP3P) model [21]), and DNA (the Amber force field [21]). The silica force field [22] was used to model interaction between a SiO_2 solid and water. After applying periodic boundary conditions, long-range Coulomb interactions were computed using particle mesh Ewald (PME) full electrostatics over a meshed grid. The mesh size was about 1 Å. The temperature was kept constant at 300 K, by applying the Langevin thermostat [23] to atoms in the SiO_2 solid. The simulation time step was 1 fs. The van der Waals' interactions between atoms were calculated using a smooth (10–12 Å) cutoff. A built system, such as the one shown in Figure 11.2a, was minimized for 10 ps and was followed by 5 ns equilibration in the constant particle number, volume, and temperature (NVT) ensemble at 300 K, before any production run.

Nanopores: A crystalline SiO_2 solid was melted and quenched in silico to obtain an amorphous SiO_2 solid. The force field used for simulating melting and quenching processes of a SiO_2 solid was developed by van Beest, Kramer, and van Santen (BKS) [24]. During the quenching process, atoms inside the cylinder (radius R) whose symmetry axis coincides with the z-axis were driven out, forming a cylindrical channel. The entire melted system was equilibrated at 5000 K and was gradually quenched to 300 K. For a Si_3N_4 nanopore [25], a cylindrical channel with the radius R was cut from a solid membrane that was built from a crystalline Si_3N_4 solid. Charges of surface atoms can be adjusted to allow a neutral or a charged nanopore.

DNA and electrolytes confined inside a nanopore: A dsDNA or ssDNA molecule was placed and harmonically confined on the symmetry axis of the channel. This simplified treatment was based on the fact that in a nanopore coated with a hydrophobic self-assembled monolayer, DNA does not stick to the surface and is almost always near the symmetry axis of the pore [17]. To focus on the electrohydrodynamics of DNA translocation inside a nanopore, a DNA molecule was covalently

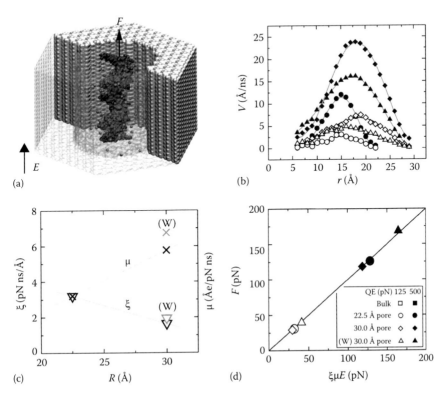

Figure 11.2 DNA electrophoresis through nanopores. (a) The MD system of DNA electrophoresis in a solid-state nanopore. A uniform external electric field E applies to all atoms, whereas a harmonic spring force F applies to DNA. The two strands of DNA are in light and dark gray. Ions are shown as vdW spheres; water is shown as a transparent surface. (b) Water flow profile in the three nanopores. Symbols used are defined in panel (d). (c) Friction coefficients (left ordinate) and electrophoretic mobility (right ordinate) of DNA in the three nanopores. (d) Stall force as a function of $\xi\mu E$. The line is the theoretical prediction, $F = \xi\mu E$. The error bars for the stall force are less than the symbols. (Adapted from Luan, B.Q. and Aksimentiev, A., *Phys. Rev. E*, 78, 021912, 2008.)

linked to itself through the periodic boundary of a simulated system. Thus, the DNA molecule as well as the nanopore/nanochannel was effectively infinite. Thus, the entropy force exerted on DNA fragment outside the pore was not modeled. The average spacing d between neighboring nucleotides in the same strand was about 3.4 Å for dsDNA or 7.4 Å for ssDNA (in a stretched conformation). DNA was solvated with an electrolyte whose ion (K^+ and Cl^-) concentration varied from 0.1 to 1 M. In a simulated nanopore system, the pressure inside the pore was approximately 1 bar.

ssDNA in the DNA transistor: The DNA transistor (Figure 11.3a), which was proposed to control the motion of ssDNA in a nanopore [26], contains a sandwich-like multilayered structure in which metal and dielectric layers are alternatively stacked on top of each other. To model the DNA transistor, the thicknesses of each metal and each dielectric layer are $2d$ and $2.5d$ (where d denotes the spacing between neighboring phosphate groups in ssDNA), respectively. As a typical voltage setup in the DNA transistor, two outer metal layers are grounded and the voltage on the middle metal layer is 2 V. To mimic electric fields in the DNA transistor, $\pm E_t$ ($E_t = 108$ mV/Å) was applied (using a grid force approach [27]) to all atoms in the dielectric layers. Ideally, ssDNA in these electric fields (provided by the DNA transistor) can be trapped [18].

11.2.2 Electrohydrodynamics of DNA Translocation through a Nanopore

When DNA transits a solid-state nanopore in a biasing electric field, DNA is under simultaneous actions of electric and hydrodynamic forces. Additionally, due to the charged surface of DNA, an electroosmotic flow is present in the biasing electric field. Therefore, DNA is screened by counterions both electrically and hydrodynamically. To understand the effective driving force on DNA (after the electrohydrodynamic screening) at a molecular level, MD simulations of DNA electrophoresis in Si_3N_4 nanopores were carried out [28]. In these simulations, three pores were considered: two atomically smooth, cylindrical pores of 22.5 and 30 Å radii and one corrugated pore of a 30 Å mean radius. The surface of the corrugated pore can be described by the equation $r(z) = 30 + 2 \cos (3\pi z/16)$ where r is the radius and z is the position along the long symmetry axis of the pore. To measure the effective electric driving force on DNA in the electric field E, a harmonic spring was used in simulation to stall DNA. One end of the spring was attached to the center of mass of DNA and the other end of the spring was fixed. In an electric field, negatively charged DNA moved opposite the field direction, and in the meanwhile, the spring attached to DNA was stretched. After about 5 ns of simulation time, the force of the spring balanced the effective electric driving force on DNA and consequently DNA stalls.

The measured stall force of the spring is much less than QE, where Q is the charge of bare DNA. Interestingly, the screening of DNA is not purely electric, as described in the counterion condensation theory. Because of motion of screening ions in the external electric field, an electroosmotic flow was present in each of the three nanopores, as shown in Figure 11.2b. Due to the hydrophilic pore surface (a nonslip boundary), the water velocity vanished at the pore surface. Since DNA was stalled, the water velocity on the DNA surface also approached zero. However, the water velocity was nonzero inside the major groove of DNA. When in the same electric field but in different pores, the mean velocity of the flow decreases with the nanopore radius. For the two 30 Å radius pores, the flow was slower in the pore with a rough surface, which was caused by higher interfacial friction forces.

To understand the dynamic screening process of counterions, two sets of independent simulations were performed: (1) without the electric field, DNA was pulled, by a harmonic spring, through the nanopore and (2) without the spring, DNA was electrophoretically driven, by the electric filed, through the nanopore. The first simulated process is characterized by $F = \xi v$ where v is the velocity of DNA. The second simulated process is characterized by $v = \mu E$, where μ is the electrophoretic mobility of DNA. From these simulations, electrophoretic mobilities and friction coefficients of DNA can be obtained for all three pores, as shown in Figure 11.2c. The friction coefficient ξ for DNA permeation is about 20% higher in the 30 Å radius pore with a rough surface than in the 30 Å radius pore with a smooth surface. The friction coefficient ξ for DNA is even larger for the pore with a smaller radius, as shown in Figure 11.2c. The electrophoretic mobility of DNA is larger for big pores, depending on both hydrodynamic friction and ionic screening.

Figure 11.2d shows that the simulated stall force linearly depends on the applied electric field. This simulation result can be understood theoretically by decomposing simultaneous actions of hydrodynamic and electric forces. Assume that the stalled DNA can be viewed as the result of two independent motions: (1) electrophoretic motion of DNA in an electric field at a velocity of v and (2) spring-pulling motion of DNA at a velocity of $-v$. Therefore, one can obtain $F = \xi v = \xi \mu E$, where the relation $v = \mu E$ is used. Therefore, the effective charge of DNA can be defined as $\xi \mu$, and the effective driving force on DNA in an electric field is $\xi \mu E$.

Note that the effective charge $\xi \mu$ of DNA cannot be explained by the counterion condensation theory. Here, the effective charge $\xi \mu$ of DNA depends on properties of a nanopore, such as the pore radius and the surface roughness, as shown in Figure 11.2d. But the charge of DNA after counterion condensation is predicted to be independent of the pore confinement. Since monovalent ions cannot be bound to the DNA surface due to the nonspecific interaction, it was concluded that the reduction of the electric driving force QE was mainly from the hydrodynamic drag of the electroosmotic flow around the DNA surface [28].

11.2.3 DNA Transiting a Nanopore: High Throughput but Not Too Fast

One essential requirement for nanopore-based DNA sequencing is to have controlled motion of an ssDNA molecule. Although the nanopore sequencing promises to be high throughput, limited by existing DNA sensing technologies (see the next section), the translocation speed of ssDNA has to be reduced for accurately sensing each DNA base along the ssDNA. Additionally, at a short time scale, ssDNA can diffuse backward or forward, resulting in some nucleotides being missed or being read multiple times. Ideally, ssDNA should be trapped each time when a nucleotide is read, and ssDNA moves forward one nucleotide spacing thereafter for the sensing of the next nucleotide. This ratchet-like motion of DNA might be achievable by exerting additional electric trapping force on ssDNA inside a nanopore. One possibility is to use built-in nanoelectrodes inside a nanopore to provide electric fields for DNA trapping, such as the DNA transistor [26]. In this section, the focus is to introduce simulation efforts on how the DNA transistor can be applied to control the motion of ssDNA in a nanopore.

The DNA transistor is a solid-state nanopore dressed with electric fields. Those electric fields are generated by built-in nanoelectrodes, shown as a metal/dielectric/metal sandwich-like structure in Figure 11.3a. When two outer electrodes are grounded and the voltage on the middle one is positive, electric fields in opposite directions are present in two dielectric regions (Figure 11.3b). When thicknesses of each metal and each dielectric layer are commensurate with the spacing d between neighboring nucleotides in a stretched ssDNA (see Section 11.2.1), ssDNA can be trapped in the DNA transistor [26,30]. Figure 11.3b shows how the setup of electric fields as well as thicknesses of layers can provide a resistant force when the ssDNA molecule moves away from a trapped state. Relative to the position of trapped ssDNA (middle one illustrated in Figure 11.3b), slightly displacing ssDNA downward or upward results in unbalanced numbers of nucleotides (or DNA charges) in both field regions (yellow in Figure 11.3b), yielding a net force that drags ssDNA back to the trapped state.

Figure 11.3c shows MD simulation of ssDNA trapping in the DNA transistor solvated with either water or glycerol. When the ssDNA molecule was in water and the trapping voltage on the middle electrode was 4 V, the center of mass of the ssDNA molecule (ssDNA position) stayed constant, indicating that the ssDNA molecule was electrically trapped. When decreasing the trapping voltage to 2 V, the ssDNA molecule can be thermally activated and diffuse away from a trapped state to the next trapped state as shown in Figure 11.3c (the orange line). Note that each time when the ssDNA molecule was thermally activated, it moved either forward or backward exactly by a distance of d and was in a new trapped state thereafter. Durations of these trapped states vary from a few nanoseconds to 15 ns. When the ssDNA molecule was in glycerol, it was always trapped in the 50 ns of simulation time even though the trapping voltage was 2 V. Therefore, the ssDNA molecule can be trapped for a longer time in a more viscous solvent. When the trapping voltage is even higher (4 V), the ssDNA molecule in glycerol is better trapped and the fluctuation of the ssDNA

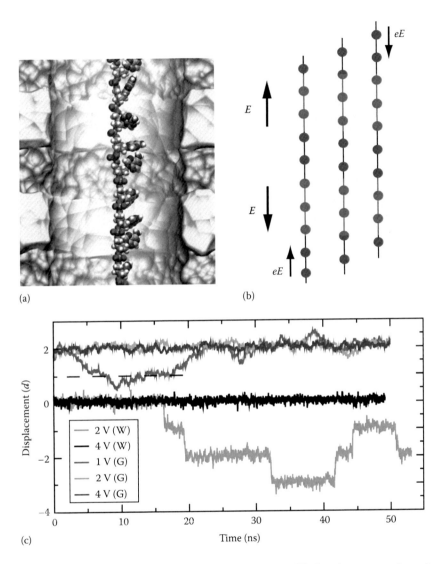

(a)

(b)

(c)

Figure 11.3 Trapping of ssDNA in the DNA transistor filled with aqueous glycerol. (a) The cross-sectional view of the DNA transistor region in the whole channel. Metal layers are in gray and dielectric layers are in light gray. (b) Illustration of the trapping of ssDNA in the DNA transistor. Each gray dot represents a nucleotide whose charge is e. The trapping electric field is E. The field regions are shown in light gray. (c) Motion of ssDNA in the DNA transistor solvated with water (denoted with "W") or glycerol (denoted with "G"). The trapping voltages are 1, 2, and 4 V. Results for ssDNA in the glycerol solvent are shifted by $2d$. (Adapted from Luan, B. et al., *Nanotechnology*, 23(45), 455102, 2012.)

position is less, as shown in Figure 11.3c (the blue line). However, if the trapping voltage is reduced to 1 V, even in glycerol, the ssDNA molecule can diffuse away from a trapped state. Figure 11.3c shows that at 1 V, ssDNA can move between two trapped states (at $1d$ and $2d$). Due to the low trapping voltage (1 V), the potential well of ssDNA trapping becomes shallow, and the thermal fluctuation allows ssDNA to move away from a trapped state. At an even lower trapping voltage, the ssDNA molecule can move more frequently from a trapping position to neighboring ones and the motion becomes random.

The potential of mean force for the ssDNA molecule in the DNA transistor can be approximated as [18], $F(z) = E_T[1 - \cos(2\pi z/d)]$, where the estimated trapping energy E_T is about $4.8k_BT$ and $9.6k_BT$ [30] for trapping voltages of 2 and 4 V, respectively. Thus, based on the reaction rate theory, the mean trapping time for the ssDNA molecule in water is about 10 ns when V_{trap} is 2 V. This agrees well with the simulation result shown in Figure 11.3c (the orange line). The friction coefficient for the ssDNA molecule in glycerol is approximately 18 times larger than in water [18]. Thus, the mean trapping times computed from the reaction rate theory are 18 ns, 0.2 μs, and 24 μs when V_{trap} = 1, 2, and 4 V, respectively. The computed mean trapping time is consistent with the simulation results (Figure 11.3c). Increasing the concentration of glycerol in the complex solvent can further increase the ssDNA friction coefficient and consequently increase the trapping time of the ssDNA molecule.

When ssDNA is in a trapped state, the type of the nucleotide before a sensor can be called. After sensing the nucleotide, the voltage on the middle electrode can be adjusted (e.g., at a smaller voltage) to allow ssDNA to move forward (in a biasing electric field) by one nucleotide spacing. Thereafter, the voltage on the middle electrode is raised again to trap ssDNA. This whole process is repeated for each nucleotide in ssDNA. Therefore, ssDNA could be sequenced while being ratcheted through the DNA transistor.

11.2.4 Sequencing DNA via Transversal Tunneling Currents

With the controlled motion (either ratcheting or moving at a constant velocity) of ssDNA in a solid-state nanopore, it is possible to design a sensor to detect each nucleotide in ssDNA. Additionally, the ssDNA should be in a linear and stretched conformation so that nucleotides in ssDNA can pass by a sensor one by one. A successful sensor should provide distinguishable signals for different types of nucleotides and a different baseline signal when there is no nucleotide in front of the sensor. One of the most promising methods is to measure the tunneling current between a pair of nanoelectrodes that are bridged by a nucleotide [14]. As shown in the inset of Figure 11.4, two pairs of nanoelectrodes are built inside the solid membrane; one end of each nanoelectrode is present inside the nanopore. Thus, tunneling currents in two perpendicular directions

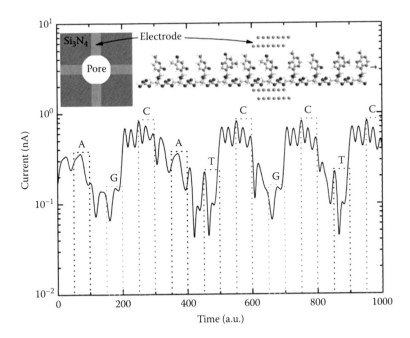

Figure 11.4 Transverse current versus time (in arbitrary units) of a highly idealized single strand of DNA translocating through a nanopore with a constant motion. The sequence of the single strand is AGCATCGCTC. The left inset shows a top view schematic of the pore cross section with four electrodes (represented by light gray rectangles). The right inset shows an atomistic side view of the idealized single strand of DNA and one set of electrodes across which electrical current is calculated. The boxes show half the time each nucleotide spends in the junction. Within each box, a unique signal from each of the bases can be seen. (Adapted from Lagerqvist, J. et al., *Nano Lett.*, 6, 779, 2006.)

are simultaneously monitored. Ideally, one pair of such nanoelectrodes might be enough for sensing nucleotides.

In an ideal case, if the stretched ssDNA (in a linear conformation) passes by the sensor at a constant velocity, the sensor indeed can provide different signals for four types of nucleotides, as shown in Figure 11.4. Here, the electric signal (tunneling currents through a pair of electrodes) was computed using the quantum mechanics approach with the tight-binding approximation. However, the structural fluctuation of ssDNA is inevitable and could yield significant noises to measured signals. Therefore, a tunneling current from a single measurement cannot tell what a nucleotide is. One possible solution is to make multiple measurements for the same nucleotide and obtain the distribution of tunneling currents [14]. MD simulation was used to sample possible configurations of a nucleotide in the gap between a pair of nanoelectrodes.

After that, the tunneling current was computed for each configuration of a nucleotide. Importantly, current distributions of all four types of nucleotides are very distinguishable. Therefore, if ssDNA can be stopped (such as that described in the process of ssDNA ratcheting in the DNA transistor) for sufficient time for multiple independent measurements, the type of a nucleotide close to the sensor can be determined. Potentially, this sequencing method is high throughput, that is, orders of magnitude faster than many other methods.

Recent experimental work [31] confirms that single mononucleotides could be statistically identified by measuring tunneling currents through two nanoelectrodes at the nanopore edge. In the experiment, single mononucleotides instead of ones in ssDNA were electrically driven through a nanopore. In the *cis* chamber, only mononucleotides of the same type were added for experiment. Each time when a mononucleotide arrived at the gap between two nanoelectrodes, the tunneling current of that single nucleotide was measured. Due to the interaction between a mononucleotide and nanoelectrodes as well as structural fluctuation of the mononucleotide, about 500 independent measurements of the tunneling current were required to produce a distinguishable distribution of the current for each type of mononucleotides. Note that each mononucleotide was not trapped in the gap between two nanoelectrodes. Therefore, it is difficult to make multiple measurements of the current for the same nucleotide. To sequence an ssDNA strand, the controlled motion of ssDNA (see Section 11.2.3) is required. Nevertheless, this experiment shows that such electric sensor is very promising for nanopore-based DNA sequencing. To improve electric signals of tunneling currents, another experimental study showed that the nanoelectrode could be functionalized with a *reader* molecule [32]. Because of hydrogen bonds between a DNA base and a *reader* molecule, the structural fluctuation of a nucleotide is greatly reduced. Thus, ideally, it is possible to tell the type of a nucleotide by one single measurement.

11.3   OUTLOOK OF THE NANOPORE-BASED DNA SEQUENCING: FROM THE POINT OF VIEW OF SIMULATION

Not directly measurable or observable in experiment, dynamics of DNA translocation through a solid-state nanopore can be revealed from realistic all-atom MD simulation. It was found that DNA could interact strongly with a nanopore surface, resulting in uncontrolled motion of DNA. Therefore, a treatment or modification of pore surface is essential to the success of DNA sequencing using a solid-state nanopore. Equally important, inside a nanopore, the surface of a metal electrode for DNA sensing should also be chemically treated to yield good electric tunneling signals. One possible way to reduce the interaction between DNA and the pore surface is to use a carbon nanotube (CNT) buried in a solid substrate as a nanopore/nanochannel. Recent experiment [33] shows that ssDNA can move through a CNT nanopore without friction, indicating weak DNA–surface interaction. Alternatively, the surface of a nanopore could be coated with a self-assembled monolayer [16,17]. If the functional end group of a coated molecule is negatively charged or hydrophobic, effectively, DNA is repelled away from the pore surface. This treatment of pore surface also prevents DNA from sticking to the surface and consequently clogging the pore.

With the chemically treated surface, motion control of DNA, and a built-in electric sensor, it is foreseeable that sequencing DNA in a solid-state nanopore will be realized in the near future. Till now, each simulation has been focused on only one aspect (such as translocation or sensing). To guide or facilitate experimental efforts on DNA sequencing, simulating an entire process of DNA sequencing inside a nanopore may be required. New simulation protocols and methods (such as quantum mechanics/molecular mechanics) should be developed to efficiently study both an ionic current for DNA translocation and an electron tunneling current for DNA sensing.

ACKNOWLEDGMENTS

The author gratefully acknowledges useful discussions with IBMers in the team for the DNA sequencing project. This work is partially supported by a grant from the National Institutes of Health (R01-HG05110-01).

REFERENCES

1. F. Sanger, S. Nicklen, and A.R. Coulson. DNA sequencing with chain-terminating inhibitors. *Proc. Natl. Acad. Sci. USA*, 74(12):5463–5467, 1977.
2. R.D. Mitra, J. Shendure, J. Olejnik, Edyta-Krzymanska-Olejnik, G.M. Church. Fluorescent in situ sequencing on polymerase colonies. *Anal. Biochem.*, 320(1):55–65, 2003.
3. M. Ronaghi. Pyrosequencing sheds light on DNA sequencing. *Genome Res.*, 11(1):3, 2001.
4. D. Branton, D.W. Deamer, A. Marziali, H. Bayley, S.A. Benner, T. Butler, M. Di Ventra et al. The potential and challenges of nanopore sequencing. *Nat. Biotechnol.*, 26:1146–1153, 2008.
5. J.J. Kasianowicz, J.W. Robertson, E.R. Chan, J.E. Reiner, and V.M. Stanford. Nanoscopic porous sensors. *Annu. Rev. Anal. Chem.*, 1:737–766, 2008.
6. C. Dekker. Solid-state nanopores. *Nat. Nanotechnol.*, 2:209–215, 2007.
7. J.J. Kasianowicz, E. Brandin, D. Branton, and D.W. Deamer. Characterization of individual polynucleotide molecules using a membrane channel. *Proc. Natl. Acad. Sci. USA*, 93:13770–13773, 1996.

8. J. Clarke, H.C. Wu, L. Jayasinghe, A. Patel, S. Reid, and H. Bayley. Continuous base identification for single-molecule nanopore DNA sequencing. *Nat. Nanotechnol.*, 4(4):265–270, 2009.

9. I.M. Derrington, T.Z. Butler, M.D. Collins, E. Manrao, M. Pavlenok, M. Niederweis, and J.H. Gundlach. Nanopore DNA sequencing with MspA. *Proc. Natl. Acad. Sci. USA*, 107(37):16060, 2010.

10. E.A. Manrao, I.M. Derrington, A.H. Laszlo, K.W. Langford, M.K. Hopper, N. Gillgren, M. Pavlenok, M. Niederweis, and J.H. Gundlach. Reading DNA at single-nucleotide resolution with a mutant MspA nanopore and phi29 DNA polymerase. *Nat. Biotechnol.*, 30(4):349–353, 2012.

11. A.R. Hall, A. Scott, D. Rotem, K.K. Mehta, H. Bayley, and C. Dekker. Hybrid pore formation by directed insertion of α-haemolysin into solid-state nanopores. *Nat. Nanotechnol.*, 5(12):874–877, 2010.

12. R.M.M. Smeets, U.F. Keyser, D. Krapf, M.Y. Wu, N.H. Dekker, and C. Dekker. Salt dependence of ion transport and DNA translocation through solid-state nanopores. *Nano Lett.*, 6:89–95, 2006.

13. B. Luan and G. Stolovitzky. An electro-hydrodynamics model on current signals of DNA transiting a solid-state nanopore. *Nanotechnology*, 24(19):195702S, 2013.

14. J. Lagerqvist, M. Zwolak, and M. Di Ventra. Fast DNA sequencing via transverse electronic transport. *Nano Lett.*, 6:779–782, 2006.

15. A.P. Ivanov, E. Instuli, C.M. McGilvery, G. Baldwin, D.W. McComb, T. Albrecht, and J.B. Edel. DNA tunneling detector embedded in a nanopore. *Nano Lett.*, 11(1):279, 2011.

16. M. Wanunu and A. Meller. Chemically modified solid-state nanopores. *Nano Lett.*, 7(6):1580–1585, 2007.

17. B. Luan, A. Afzali, S. Harrer, H. Peng, P. Waggoner, S. Polonsky, G. Stolovitzky, and G. Martyna. Tribological effects on DNA translocation in a nanochannel coated with a self-assembled monolayer. *J. Phys. Chem. B*, 114(51):17172–17176, 2010.

18. B. Luan, H. Peng, S. Polonsky, S. Rossnagel, G. Stolovitzky, and G. Martyna. Base-by-base ratcheting of single stranded DNA through a solid-state nanopore. *Phys. Rev. Lett.*, 104(23):238103, 2010.

19. J.C. Phillips, R. Braun, W. Wang, J. Gumbart, E. Tajkhorshid, E. Villa, C. Chipot, R.D. Skeel, L. Kale, and K. Schulten. Scalable molecular dynamics with NAMD. *J. Comput. Chem.*, 26:1781–1802, 2005.

20. D. Beglov and B. Roux. Finite representation of an infinite bulk system: Solvent boundary potential for computer simulations. *J. Chem. Phys.*, 100:9050–9063, 1994.

21. W.L. Jorgensen, J. Chandrasekhar, J.D. Madura, R.W. Impey, and M.L. Klein. Comparison of simple potential functions for simulating liquid water. *J. Chem. Phys.*, 79:926–935, 1983.

22. E.R. Cruz-Chu, A. Aksimentiev, and K. Schulten. Water-silica force field for simulating nanodevices. *J. Phys. Chem. B*, 110:21497–21508, 2006.

23. M.P. Allen and D.J. Tildesley. *Computer Simulation of Liquids*. Oxford University Press, New York, 1987.

24. B.W.H. van Beest, G.J. Kramer, and R.A. van Santen. Force fields for silicas and aluminophosphates based on ab initio calculations. *Phys. Rev. Lett.*, 64(16):1955–1958, 1990.

25. A. Aksimentiev, J.B. Heng, G. Timp, and K. Schulten. Microscopic kinetics of DNA translocation through synthetic nanopores. *Biophys. J.*, 87:2086–2097, 2004.
26. S. Polonsky, S. Rossnagel, and G. Stolovitzky. Nanopore in metal-dielectric sandwich for DNA position control. *Appl. Phys. Lett.*, 91:153103, 2007.
27. D.B. Wells, V. Abramkina, and A. Aksimentiev. Exploring transmembrane transport through alpha-hemolysin with grid-based steered molecular dynamics. *J. Chem. Phys.*, 127:125101, 2007.
28. B.Q. Luan and A. Aksimentiev. Electro-osmotic screening of the DNA charge in a nanopore. *Phys. Rev. E*, 78:021912, 2008.
29. B. Luan, D. Wang, R. Zhou, S. Harrer, H. Peng, and G. Stolovitzky. Dynamics of DNA translocation in a solid-state nanopore immersed in aqueous glycerol. *Nanotechnology*, 23(45):455102, 2012.
30. B. Luan and A. Aksimentiev. Electric and electrophoretic inversion of the DNA charge in multivalent electrolytes. *Soft Mater.*, 6(2):243–246, 2010.
31. M. Tsutsui, M. Taniguchi, K. Yokota, and T. Kawai. Identifying single nucleotides by tunnelling current. *Nat. Nanotechnol.*, 5:286–290, 2010.
32. S. Chang, J. He, A. Kibel, M. Lee, O. Sankey, P. Zhang, and S. Lindsay. Tunneling readout of hydrogen-bonding-based recognition. *Nat. Nanotechnol.*, 4:297–301, 2009.
33. V. Lulevich, S. Kim, C.P. Grigoropoulos, and A. Noy. Frictionless sliding of single-stranded DNA in a carbon nanotube pore observed by single molecule force spectroscopy. *Nano Lett.*, 11(3):1171, 2011.

Chapter 12

Biological Water under Confinement

Nanoscale Dewetting

Ruhong Zhou

CONTENTS

12.1 INTRODUCTION

Water is not only the most unique molecules known to mankind but also one of the most important molecules in biological systems. In fact, water might be the most crucial *biological molecule*, even though it is often ignored by biologists simply as *background*. Hydrophobicity, or simply *water fearing*, a much related physical property of molecule, is essential for sustainable life as we all know. It is associated with various unusual properties of aqueous solutions of polar or nonpolar solutes and contributes significantly to a wide variety of biological phenomena such as protein folding, DNA/RNA folding, and self-assembly of lipid membranes [1–7]. Biological water, particularly those under nanoscale confinement, plays a key role in so many important biological processes that considerable effort has been devoted to understanding it over the past decades [8–19]. Not long ago, Ball [13] presents an intriguing review and measured historical discussion of the contentious issues along with recent developments and how they correspond to other general questions about the role of water in biological systems. Although many of the controversies involve quantitative questions, they cannot be resolved by taking a rigid and simplistic view based on generally useful qualitative concepts, whether of *broken hydrogen bonds*, *dry interfaces*, or *hydrogen bond networks*. Since water is a subtle and flexible medium with a fluctuating and elusive balance between many nearly degenerate local configurations, it can respond to different perturbations in surprising ways. Therefore, we must be equally flexible in our attempts to understand its solvation properties with both experimental and theoretical approaches.

Hydrophobicity is now generally recognized to manifest itself differently on small and large length scales [10,11,13,20–29]. For example, with small hydrophobic molecules, such as methane, they can fit into the water hydrogen bond network without destroying any hydrogen bonds during the hydrophobic hydration [22]. Since no hydrogen bonds are broken, the enthalpy of solution (ΔH) is small. The formation of small cavities in the solvent to accommodate the solute is an entropically dominated process, and the presence of the solute constrains the freedom of movement of the neighboring water molecules. Thus, the entropy of solvation (ΔS) is negative (e.g., $\Delta S_{Ar}^{o} = -30.2$ cal/mol K) and is proportional to the molar excluded volume of the solute. This means that the dissolution of small nonpolar solutes is entropically driven where the free energy of solution (ΔG) is positive and increases with both the temperature (around biological temperatures) and excluded volume of the solute.

In contrast, as suggested by Stillinger [22], and confirmed by molecular dynamics (MD) simulations [30], *large* hydrophobic solutes such as proteins behave very differently. For these large hydrophobic solutes, they must break some hydrogen bonds at the interface. For each interfacial water molecule, one hydroxyl group tends to point into the hydrophobic surface so that the orientational ordering of water molecules at the solute–liquid interface resembles the ordering at a liquid–vapor interface. The *missing* interfacial hydrogen bonds produce a large positive enthalpy of solvation and corresponds to a free-energy change that is proportional to the solute's *surface area, A,* as opposed to being proportional to the *volume* for small hydrophobes. Stillinger [22] argued that if the solute–water attraction is sufficiently weak, a large smooth hydrophobic solute might *be immediately surrounded by a microscopically thin film of water vapor*, with its hydration free energy dominated by a term like $\gamma_{lv}A$, where γ_{lv} is the vapor–liquid surface tension and A is the surface area of the solute.

The hydrophobic interaction of large-scale hydrophobic solutes is also different from small-scale ones. When two large hydrophobes are brought into contact, fewer water hydrogen bonds will have to be broken down while they are apart, so that there will be a negative enthalpy change. Since the free-energy change is dominated by the enthalpy change, it too will be negative with a thermodynamic driving force toward aggregation. In other words, large-scale hydrophobicity is mainly enthalpically driven whereas small-scale hydrophobicity is expected to be entropically driven. For complex heterogeneous molecules such as proteins, the interaction becomes more complicated, with water playing many interesting and diversified roles near the interfaces or within the nanoconfinements.

One particular important effect not sufficiently appreciated, until fairly recently [23,24,31,32], is the nanoscale dewetting (drying). It occurs when two large-scale strongly hydrophobic solutes approach each other closer than a critical separation, which triggers a large-scale drying transition. Once triggered, the interfacial region, although large enough to accommodate a significant number of water molecules, desolvates and leads to hydrophobic collapse. This phenomenon can be correlated with the contact angle of water droplets on macroscopic hydrophobic surfaces [24,32–36]. However, even in idealized models, the critical separation is a decreasing function of the strength of the attractive interaction between the solute and water that can disappear entirely for small solutes. Surprisingly, we and others have observed this nanoscale dewetting transition in the folding, binding, or aggregation of protein complexes [36–38]. We will review recent MD simulations examining the possible relevance of such macroscopic dewetting transitions in the folding of heterogeneous globular proteins [36–38], the collapse of multidomain proteins [34], ligand–receptor binding [39–42], as well as ion-channel gating [43].

The focus of this chapter is on the large-scale hydrophobicity, particularly the dewetting (drying) of water molecules under nanoscale confinement within biological molecules such as proteins. We will try to distinguish a normal dehydration [44] versus more extreme dewetting transition [34,36,37] in the following sections. As for the nanoscale dewetting, we will first describe a simple macroscopic theory and then use several examples to illustrate this fascinating phenomenon in various complex biological systems and finally try to link this nanoscale dewetting transition with potential biological functions such as ion-channel gating.

12.2 SIMPLE MACROSCOPIC THEORY

In this section, we focus on a simple macroscopic theory describing the potential nanoscale dewetting transition between two large hydrophobic solutes, which can be directly tested and validated in MD simulations. As mentioned earlier, nanoscale drying can sometimes occur between two large hydrophobic particles as a function of separation. To understand some of the key features of this process, a simple macroscopic thermodynamic model [24,32,33,45,46] based on Young's equation [47] was generalized for the study. The model considers the change in the grand potential $\Delta\Omega(D)$ for the reaction in which the interplate region (the gap) between two fixed macroscopic coaxial cylindrical plates, with a separation D, transforms from being completely wet to dry, with the vapor occupying a cylindrical region between them. Ignoring the edge effects, the change in the grand potential for this reaction is

$$\Delta\Omega(D) = \left[(P - P_v)A_w + \gamma_{lv}C_w\right](D - D_c), \tag{12.1}$$

and the critical separation D_c below which vapor between the disks is stable is

$$D_c = \frac{2\Delta\gamma}{\left[(P - P_v) + 2\gamma_{lv}/R_m\right]}. \tag{12.2}$$

Here

D is the separation distance

P is the pressure on the liquid

P_v is the pressure of the vapor between the plates

$A_w = \pi R_m^2$ is the area of the plate face

γ_{lv} is the liquid–vapor surface tension

$C_w = 2\pi R_m$ is the circumference of the cylindrical plate

R_m is the radius of the plate

$\Delta\gamma \equiv \gamma_{wl} - \gamma_{wv} = -\gamma_{lv}\cos\theta_c$ is Young's equation with θ_c being the contact angle for the liquid in contact with the wall of the plate

γ_{wl} and γ_{wv} are the wall–vapor and wall–liquid surface tensions, respectively

For a hydrophobic surface, the contact angle is obtuse so that both $\Delta\gamma$ and D_c become positive and drying should occur. However, for hydrophilic plates, the contact angle is acute; therefore, $\Delta\Omega$ will be positive for all separations and the interplate region will be wet.

For small plate sizes (small R_m), the $(P - P_v)$ term usually can be neglected compared to the surface tension term. Consequently, $D_c = -R_m\cos\theta_c$ should grow linearly with the plate radius. However, for sufficiently large plates, the surface tension term in the denominator can be neglected. Hence, D_c should become independent of the plate size, and drying should occur for very large separations $D_c = 2\Delta\gamma/(P - P_v)$.

Since the critical distance for drying decreases as the strength of the plate–water attractive interaction increases, we expect that purely repulsive plates will be the most prone to dry, whereas substances like graphite with contact angles near $\pi/2$ will not. For the same reason, most proteins or protein complexes should not display drying transition during folding (and that is exactly why it is so surprising to observe such nanoscale dewetting transition in protein complexes that is counterintuitive). These simple thermodynamic arguments show that the critical distance D_c that depends on the geometry and thermodynamic properties is indeed a key factor that decides when macroscopic hydrophobic objects dissolved in water should expel the solvent trapped between them. However, the free-energy barrier to cross from the wet to the dry state during the drying transitions could be very large, effectively freezing the system in a metastable state [19,24,25,48,49]. This free-energy barrier is responsible for the phenomenon of cavitation-induced hysteresis observed [50,51] experimentally in surface-force measurements.

12.3 DEHYDRATION IN PROTEIN FOLDING

It is widely accepted that a necessary structural consequence of the protein folding reaction is the exclusion of water from its side and main chains that become buried in the native state. The thermodynamic consequence of the dehydration reaction reflects the overall gain in free energy realized by freeing water during folding. As one would expect, most proteins involve dehydration during their folding [52–57], but probably not the extreme nanoscale dewetting (drying) transition, which is the subject of the following section.

The role of water in protein folding reactions has been examined by both experimental and computational approaches. By replacing nonpolar side chains with isosteric polar side-chain analogs, mutational analyses have shown that water is selectively shed prior to the appearance of the native state to enable the formation of critical cores of stability in early intermediates [58] or transition state ensembles [59,60]. Experimental result from time-resolved infrared (IR) analysis revealed dehydration of the main-chain amides in the final step of folding from the alkaline-denatured states of both α-helical [61] and β-sheet proteins [62]. Another experimental approach toward examining the role of water in folding monitors the protection of main-chain amide hydrogens against exchange in deuterated water in partially folded states [63,64] and folding intermediates [65–67]. In particular, when hydrogen exchange (HX) techniques were applied to a pair of $(\beta\alpha)_8$ TIM barrel proteins [67,68], protection against exchange in folding intermediates was found to be selectively associated with clusters of branched aliphatic side chains, isoleucine, leucine, and valine (ILV cluster). This behavior was attributed to the preferential partitioning of side-chain analogs of saturated hydrocarbon moieties into the vapor phase, relative to their aromatic, sulfur, or polar-containing counterparts that spontaneously dissolve in water [69]. It was proposed that clusters of ILV side chains play crucial roles in stabilizing folding intermediates in TIM barrel proteins by selectively excluding water from their interiors [70,71].

Building on the results of the previous experiments and MD simulations, Zhou, Matthews, and coworkers adopted a combined experimental and computational approach [44] to test the conjecture that large ILV-rich clusters in TIM barrel proteins are prone to undergo dewetting from their interiors. As a target, they chose the alpha subunit of Trp synthase, αTS, a ~28 kDa TIM barrel $(\beta\alpha)_8$ protein that is a component of the $\alpha_2\beta_2$ tetrameric tryptophan synthase complex. αTS has been known to offer strong and selective protection against HX in an on-pathway intermediate associated with a large N-terminal ILV cluster [64]. To simulate the polarity introduced by water, two buried leucines in the N-terminal cluster and a single leucine in the C-terminal cluster were individually replaced with the isosteric and polar asparagine. The effects of these mutations on the water density within the clusters were predicted by MD simulations of artificially displaced versions of their preformed β-sheet and α-helical components. These predictions were then compared with the effects of the mutations on the experimentally determined stabilities and structures of the native state and the folding intermediate. The combined results support the conclusion that the drying of the large N-terminal ILV cluster is crucial to the stability and structure of the native state and of a productive folding intermediate in a TIM barrel protein.

12.3.1   Hydration in TIM Barrel ILV Cluster

A ribbon diagram of αTS and the locations of its three ILV clusters are shown in Figure 12.1. Cluster 1 (blue in Figure 12.1a), which contains 31 ILVs, is the focus of the current illustration that forms the interface between the exterior of the β-barrel and the interior of the α-helical shell. For the MD simulation study, a cavity filled with water molecules inside Cluster 1, with an estimated volume of ~1300 Å^3, was created by pulling the α1 and α2 helices away from the β1, β2, and β3 strands by a separation distance d varying from 4 to 6 Å. As shown in the following, previous works have reported nanoscopic dewetting transitions in proteins with cavity volumes of a similar order [37,72]. During the 16 ns simulation time, the water molecules were free to move, but the protein heavy atoms remained constrained. The cavity undergoes intermittent transitions between wet and dry states at the separation distance of 4 Å; however, no drying transition was observed at $d = 6$ Å. To check the convergence of the results, additional simulations were carried out with two different initial states, one wet and one dry, both with $d = 4$ Å. Within the first 1–4 ns, the cavity underwent wetting/dewetting transitions in which the normalized water density inside the cavity fluctuated between a maximum of 0.8 (wet) and a minimum of 0.2 (dry) from both initial states. The normalized water density is obtained by dividing the number of water molecules with maximum number of water molecules inside the cavity. Snapshots of the cavity in both states

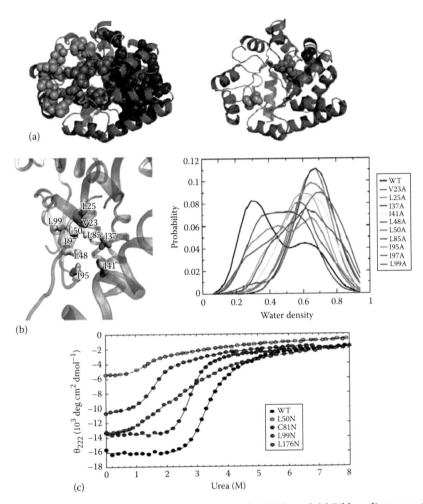

Figure 12.1   (See color insert.) Dehydration inside TIM barrel. (a) Ribbon diagrams of αTS (a) highlighting the three hydrophobic clusters formed by the ILV residues: Cluster 1 (blue), Cluster 2 (orange), and Cluster 3 (green) obtained using a 4.2 Å cutoff distance between pairs of ILV side chains and (b) showing the locations of hydration mutations in the crystal structure, Leu50 in β2 (purple), Cys81 in α2 (blue), Leu99 in β3 (green), and Leu176 in β6 (orange). Coordinates of αTS from *Salmonella typhimurium* were used to generate the figure from a refined version of PDB file 1BKS. (b) Ribbon diagram of the interior of Cluster 1, portraying the ILV side chains selected for alanine-scanning mutagenesis, in space filling format (left). Histograms of water density inside Cluster 1 for the wild-type protein and the 10 alanine variants (right). (c) Experimental urea-induced equilibrium unfolding profiles for wild type and the hydration variants, L50N, C81N, L99N, and L176N. The continuous lines represent the fit of the data to a three-state model. (Reproduced from Das, P. et al., *J. Am. Chem. Soc.*, 135, 1882, 2013. With permission.)

suggested that the water density was lowest near the center of the cavity, as a vapor bubble was frequently formed in this region and was stable for several nanoseconds. The two termini of the β-strand triplet remained relatively wet with the N-terminus being drier than the C-terminus. The latter results are consistent with the stronger protection against amide HX with solvent in this region observed in native-state HX experiments [68].

To provide insight into the role of individual ILV side chains to the dehydration observed in Cluster 1, 10 of its constituent members were individually substituted with alanines for MD simulations [44]. The residues selected (Figure 12.1), V23, L25, I37, I41, L48, L50, L85, I95, I97, and L99, have previously been shown to eliminate an early kinetic trap in folding when replaced by alanine, and almost all of them significantly destabilize the on-pathway equilibrium intermediate [71]. Figure 12.1b plots the histograms showing the probability of the water density within the cavity of Cluster 1 for the wild-type protein and all 10 alanine mutants. The histogram for the wild-type protein shows a bimodal distribution, confirming that the cavity undergoes transitions between a dry (water density ~0.3) and wet state (water density ~0.65) with the dry state being more probable over the wet one. All mutated proteins except V23A, L50A, and L99A were found to experience complete or nearly complete loss of dewetting in the cavity (Figure 12.1b). The histograms of water density for all of these mutants are Gaussian in nature with peaks centered at a water density of 0.6–0.7, showing that the cavity remains wet most of the time upon the alanine substitution. These findings indicate that Cluster 1 in the wild-type protein is fairly "optimized" in terms of drying and stability. However, subtle changes in the surface topography and chemistry (e.g., single mutation I/L → A) can tip the balance of the cavity to a wetter state, potentially lowering the stability of the cluster and protein. In contrast, alanine replacement to residues V23, L50, and L99 resulted in partial loss of dewetting, with the L99A mutation being the most resistant to wetting. The histograms of these three mutants show a considerable population of the low water density states (Figure 12.1b). These β-strand residues are centrally located in the cluster, facing the helical shell, and are surrounded by neighboring ILV residues indicating dewetting depends strongly on the local environment (Figure 12.1). In order to further *wet* the cavity inside Cluster 1, the authors performed a more radical perturbation on the hydrophobic surface by substituting L50 and L99 with their isosteric and polar counterpart, asparagine. The water density histograms of L50N and L99N mutant proteins illustrate that the introduction of a polar side chain at these two positions results in a complete or significant loss of dewetting. The histograms for those two asparagine mutants have a water density peak centered at 0.6–0.7. As confirmed experimentally (Figure 12.1c), these results led to the expectation that native states and the folding intermediates for the L50N and L99N variants of αTS would be substantially destabilized relative to their wild-type counterpart.

Inspection of the simulations for wild-type αTS showed residual water density near C81, which is adjacent to Cluster 1 in helix α2 and near the C-termini of strands β3 and β4. To investigate the effect of side chains proximal to Cluster 1 on hydration, C81 was substituted with isoleucine, valine, and asparagine, respectively. The simulated water density distributions of C81I and C81V mutants showed a cavity that fluctuates between wet and dry states with C81V making the cavity noticeably drier compared to the wild-type protein. These results suggest that C81V mutation would be the best candidate to further dewet. In contrast, the cavity in the C81N variant favors the wet state with the maximum of water density probability around 0.6. These findings illustrate the sensitivity of water probability inside the cavity to the local environment.

12.3.2   Experimental Spectra Confirm the Theoretical Prediction

The effects of the asparagine hydration mutations on the structural properties of αTS were determined by circular dichroism (CD) spectroscopy. The reductions in the ellipticities at 195 and 222 nm for the variants, L50N, C81N, L99N, and L176N, indicate that the introduction of a polar side chain at all four positions disrupts the secondary structure to varying degrees. The near-UV CD spectra, which provide insight into the chiral packing of aromatic side chains, reveal that all of the hydration variants have altered tertiary structures. The effects of the mutations on the thermodynamic properties were determined by monitoring the far-UV CD spectrum as the proteins were denatured with urea. The equilibrium unfolding transitions, as illustrated by the changes in ellipticity at 222 nm, are shown in Figure 12.1c [44]. All of the variants display a nearly urea-independent baseline indicative of a thermodynamically stable state in the absence of denaturant. As previously observed for wild-type αTS [71], the equilibrium unfolding reactions of the four hydration variants are well described by a three-state model, $N \leftrightarrow I \leftrightarrow U$, to fit the CD data. Because the amplitude is proportional to the fraction of the native state at the initial urea concentration, the fitting of the amplitude to a two-state model yields the desired thermodynamic parameters. The free-energy differences for the $N \rightarrow I$ and $I \rightarrow U$ transitions for the variants can then be compared. The stability of the N state versus the I state is substantially decreased for the L50N, L99N, and L176N mutations; however, the C81N mutation leaves the stability virtually unchanged [44].

12.3.3   Interplay between ILV Cluster Drying and Protein Stability

Theoretical study using MD simulations [44] reveal that either intermittent or strong water density fluctuations can occur inside the cavities of the large hydrophobic ILV clusters of αTS, depending on the size and composition of the cluster. The largest ILV cluster (Cluster 1) was found to be optimized in terms of dehydration, but it proved to be difficult to design a drier cavity. Selected ILV residues mutated to alanine was found to weaken or completely diminish dewetting, which strongly depends on the local environment of the mutation site. These simulations also showed that the replacement of buried leucines with asparagines is sufficient to completely wet their cavities.

Experimental analysis of leucine to asparagine mutations in the N-terminal ILV cluster of αTS not only demonstrated dehydration in both native and intermediate states but also revealed that the presence of polarity substantially decreased the stabilities and had a profound effect on the structures of both states. The substitution of an acetamide group for an isobutyl group at L50 and L99 reduced the secondary structure of the native state by 40%–70% and appeared to mobilize the tyrosine side chains [44]. The secondary structures of the corresponding intermediate states were also greatly diminished for these variants. The results are consistent with the prediction from MD simulations that the interior of this cluster strongly prefers to dewet in a TIM barrel configuration and the conclusion that this configuration also exists for the intermediate state. What is very surprising, however, is that these mutations do not simply destabilize the TIM barrel fold or its folding intermediate, but their associated polar side chains lead to distinct high-energy thermodynamic states in the native basin on the TIM barrel folding free-energy surface. In the case of Cluster 1, there appears to be sufficient driving force from the need to sequester the remaining 30 aliphatic side chains from solvent to populate these alternative states. The substantial decrease in the CD signal at ~222 nm [44] could reflect a highly dynamic α-helical shell that enables the partial exposure of the asparagine side chains at positions 50 and 99 to water while retaining the buried surface area.

The fact that the MD simulations for the entire set of 10 ILV → A mutations in Cluster 1 resulted in the wetting of the cavity was indeed very surprising. One might have expected that the removal of 2–3 carbons from a cluster of 31 branched aliphatic side chains would have little effect on the propensity of water to occupy the exposed nonpolar volume. However, the sensitivity of drying to the composition and/or structure of the cavity may explain the previous experimental observation that these same alanine replacements substantially reduce the stability of the intermediate in αTS [71]. The simulations suggest that the enhanced propensity of the alanine variants in Cluster 1 to wet (i.e., disfavoring a well-folded state) is accompanied by the loss of packing interactions, a mechanism for destabilizing the intermediate [44].

In short, the results of this combined experimental–simulation study demonstrate the critical role of dehydration in a large hydrophobic ILV cluster in determining the stability and structure of a TIM barrel fold and a critical folding intermediate. Even though no nanoscale drying transition was identified in the final collapse of the TIM barrel ILV clusters, the sharp transitions between the dry and wet states did indicate that it is probably *on the edge* of the dewetting transition. It should be noted that the nanoscale dewetting [14,23,32] transitions between hydrophobic surfaces have long been of interest for both physical [8,32,73–76] and biological systems [34,36,37,42,74]. As described in the following, previous MD simulation studies have identified several proteins or peptides in which a dewetting transition was observed prior to the docking of preformed elements of the secondary structure. For example, a remarkable dewetting transition was observed within the nanoscale channel between the four melittin α-helices each of whose hydrophobic interface is comprised of three isoleucines, four leucines, one tryptophan, and two valines [36]. A subsequent study on a variety of protein complexes (dimers, tetramers, and two-domain proteins) found that dewetting required large complementary hydrophobic surfaces with significant contributions from ILVs [37]. In contrast, a marked decrease in water density was not detected at the domain interface in the two-domain 2,3-dihydroxy-biphenyl dioxygenase (BphC) [34]. The domain interface in BphC is relatively heterogeneous in nonpolar side chains.

12.4 NANOSCALE DEWETTING IN PROTEIN COMPLEXES

12.4.1 Dewetting in Protein Folding

Proteins are often characterized by solvent-accessible surfaces containing extended nonpolar regions that form the hydrophobic parts. The aggregation and subsequent removal of water molecules between these hydrophobic surfaces is believed to be crucial to their self-organization.

However, trying to analyze the role of small- and large-scale hydrophobicity and especially dewetting in folding and aggregation of proteins can be a big challenge, not only because the amino acid residues range from being strongly hydrophobic to strongly hydrophilic (charged residues) but also because the carbonyl and amide groups along the peptide backbone are polar and capable of participating in hydrogen bonds. The nonpolar confinements of water in nanopores and cavities have also been heavily studied recently [8,13].

The phenomenon of large-scale capillary drying between approaching strongly hydrophobic plates suggests that fluctuations of water, in particular, bubble creation and annihilation, should be somehow reflected in the kinetics of plate association. The implications of this during a hydrophobic chain folding were poignantly suggested by ten Wolde and Chandler [14] in their study of the collapse from an extended coil state to a compact globule state of a string of 12 purely hydrophobic beads, with a volume similar to those of amino acids in solution, interacting with a coarse-grained lattice-gas model of water through purely repulsive forces. They found that the evaporation of water in the vicinity of the polymer is accompanied with the chain collapse. Once a sufficiently large cluster of beads formed, a Stillinger-like vapor interface formed and seemed to embrace the growing cluster until collapse was complete. They concluded that length-scale-dependent hydrophobic dewetting is the rate-limiting step in the hydrophobic collapse of the model chain system. Chandler and coworkers recently performed MD simulations on a similar chain, with larger beads, interacting repulsively with an all-atom model of water and found very similar results, although in this latter study the vapor seems to appear in hydrophobically enclosed regions rather than by surrounding a cluster as before [77]. This behavior is reminiscent of drying in the region between plates and in other enclosed regions as predicted by the macroscopic theory. Then, Chandler and coworkers posed the following question:

> Our findings would seem also pertinent to the mechanism of biological assembly, such as protein folding, but to demonstrate so with simulation will require an analogous simulation study of a protein-like chain.

At present, it is still not easy to answer this question even with large-scale simulations using supercomputers. A typical protein folds in milliseconds, which is still beyond the routine reach of modern computational resources [78]. Alternatively, it is possible to study the role of water in the collapse of multi-domain proteins, like the BphC enzyme, or aggregation of proteins, like the formation of tetramers of melittin, since one can investigate how water behaves as the domains or monomers (which are prefolded) are brought together. The first such study by Zhou et al. involved bringing the two domains of the BphC enzyme together to see if there is a critical distance for drying between them [34]. It was found that when combination of the van der Waals and electrostatic interactions between the protein and water were turned off at the region between the domains, it exhibited a drying transition. But when the full force field was turned on, no complete drying transition was found. Based on this study, it was at first thought that dewetting would not occur in proteins because the van der Waals and electrostatic attractions would be too strong. However, it is expected that this will be sensitive to temperature (and pressure). For example, in recent studies of the assembly of Alzheimer amyloid-beta Aβ16–22 protofilaments, no drying was observed at room temperature but was observed at a slightly higher temperature [55]. When the strength of the $1/r^6$ terms in the protein–water LJ potential was decreased by 10%, the protofilaments exhibited a drying transition.

A good place to look for dewetting in protein folding is probably in the protein complex formation during the final stages of folding. In a recent work by Zhou, Berne, and coworkers [36], the authors studied the water drying transition inside a protein complex—the collapse of the melittin tetramer. The two dimers of the melittin tetramer were separated by a distance to create a *nanoscale channel*, then solvated in a water box. The authors found that there is a clear dewetting transition in which essentially all water molecules are expelled from the nanoscale channel in a few hundred picoseconds (Figure 12.2). Simulations with many other separation distances show that the critical distance for the melittin tetramer system is approximately 5.5–7.0 Å, which is equivalent to two to three water molecule diameters. A water contact angle up to ~113° was calculated around its large hydrophobic patch at the interface [79]. This critical distance also agrees well with the prediction from the simple macroscopic theory with a simplified model for the protein using the water contact angle [35]. The reasons that the melittin tetramer channel exhibits a drying transition that cannot be observed in the previously studied two-domain protein BphC [34] appear to be twofold. First, the melittin channel is more enclosed, like a 1D tube [36], and it is less costly with respect to free energy to disrupt the hydrogen bonds in a tubelike channel. On the contrary, the interdomain region in the two-domain protein is more like a 2D interplate region between two plates [34]. Second, the unique surface topology springing from the isoleucine residues in melittin disrupts the water hydrogen bonds in the channel, thus destabilizing the wet state. When certain isoleucines are mutated to alanine or even to valine, the channel wets. Later, these authors examined the entire protein data bank (PDB) for potential nanoscale dewetting in protein complexes, such as dimers, trimers, and tetramers, and found a handful of similar cases as the melittin tetramer [37].

As shown in the previous section, the ILV cluster (Cluster 1) in TIM barrel experiences frequent transitions between a wet and dry state. This study with melittin tetramer also shows that small perturbations, such as single amino acid substitutions, can shift the protein from a dry state to a wet state [36]. For example, a single I2A or I2V mutation can tip the protein melittin tetramer channel from dry to wet [36]. Debenedetti, Rossky, and coworkers [38] have also recently found similar results with slightly modified melittin dimers. Along this line, recent simulation studies by Garde and coworkers also show that water near protein surfaces can be sensitive to subtle changes in surface conformation, topology, and chemistry, and small changes can tip the balance from dry to wet or vice versa. That is, the protein can be *sitting at the edge* of the dewetting transition [12]. For example, melittin is closer to the dry side of a dewetting transition, while another protein BphC is biased to the wet side. It is possible to tip the balance to the opposite side for both melittin and BphC proteins by introducing additional perturbations, for example, point mutations. Taken together, these findings suggest that biomolecules often sit at the edge of dewetting transitions and are sensitive to perturbations [81].

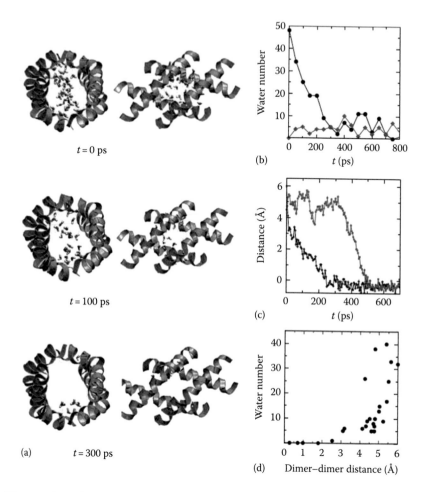

Figure 12.2   (See color insert.) Dewetting in melittin. (a) Snapshots of water molecules inside the gap of the melittin tetramer (the protein is shown as ribbons and water as sticks). Only water molecules near the center of the channel are plotted, which is defined as the region with a spherical radius of 10 Å from the center of the enlarged tetramer. (b) Plot of the number of water molecules inside the channel against the MD time for *dewetting* (black) and *wetting* (red) simulations. (c) The kinetics of the *folding* simulation starting from two different initial separations, $d = 6$ Å (black) and $d = 4.5$ Å (red). (d) Plot of the number of water molecules inside the channel against the dimer–dimer distance for one folding trajectory starting at an initial separation of 6 Å, which indicates a drying-induced collapse. (Reproduced from Liu, P. et al., *Nature*, 437, 159, 2005. With permission.)

12.4.2 Dewetting in Protein–Ligand Binding

During protein–ligand binding, the water solvating the active site is expelled into the bulk fluid, making enthalpic and entropic contributions to the binding free energy of the complex. The less energetically or entropically favorable to the expelled water, the more favorable its contributions to the binding free energy. The active sites of proteins provide very diverse environments for solvating water. For example, water solvating narrow hydrophobic enclosures such as the Cox-2 binding cavity is energetically unfavorable because it cannot form a full complement of hydrogen bonds. Similarly, water molecules solvating enclosed protein hydrogen bonding sites are entropically unfavorable since the number of configurations they can adapt while simultaneously forming hydrogen bonds with the protein and their water neighbors is severely reduced [39–42,80]. The expulsion of water from such enclosed regions has been shown to lead to enhancements in ligand–receptor binding affinity [39–42,80]. From these observations, Abel et al. determined that a computationally derived map of the thermodynamic properties of the active site solvent could be used to rank the relative binding affinities of certain classes of congeneric compounds [39,40].

Young et al. have recently studied the water drying inside the protein–ligand binding active sites using MD simulations [41] to investigate the binding cases of streptavidin, antibody DB3, and Cox-2. In all three cases, water seems to be simply eager to get out of the protein binding cavity. For example, in the streptavidin case, water molecules in the cavity do manage to form some hydrogen bonds with each other and with the protein residues. However, they are largely trapped on the top and bottom of the hydrophobic surroundings, forming a five-membered ringlike structure, which in turn makes them less stable both enthalpically and entropically (Figure 12.3). On the other hand, the Cox-2 active site was found to contain no persistent hydration sites and is in fact entirely devoid of solvent in 80% of the simulation, despite the cavity being sterically able to accommodate approximately seven water molecules [41]. The high excess chemical potential of the binding-cavity solvent is due to the inability of the water molecules to make hydrogen bonds among themselves, and with the surrounding hydrophobic protein residues, which results in an extreme enthalpic perturbation (≥ 8 kcal/mol) that drives the dewetting of the cavity. In the simulation, the active site water molecules of an artificially hydrated Cox-2 structure were found to evacuate within 100 ps. This is because the active site of Cox-2 is predominantly a narrow paraffin-like tube and is therefore in line with other studies of hydrophobically induced dewetting [15,36]. A similar effect has been reported for bovine β-lactoglobulin (BLG) [81]. Halle and coauthors show that an extreme case of a completely dehydrated free binding site is realized for the large nonpolar

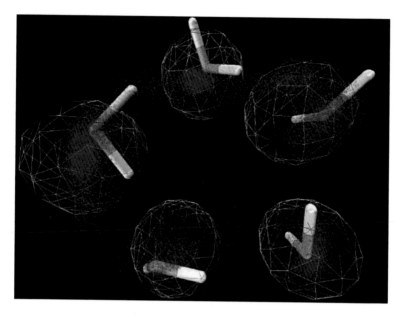

Figure 12.3 Five-membered ringlike solvating water configuration in the binding cavity of streptavidin. The conformations of water are shown with the solvent density averaged over all simulations in gray dots. The hydration sites determined by the clustering of this density are shown in wireframe. Note the near absence of water molecules in the inner part of the five-membered ring and in between the ring sites. (Reproduced from Young, T. et al., *Proc. Natl. Acad. Sci. USA*, 104, 808, 2007. With permission.)

binding cavity in the protein BLG. They use a combination of water ^{17}O and ^2H magnetic relaxation dispersion (MRD), ^{13}C NMR spectroscopy, MD simulations, and free-energy calculations to empty out the water from the binding cavity, a 315 Å binding pore, to a completely dry state. The apo protein is thus poised for efficient binding of fatty acids and other nonpolar ligands [81].

Following the earlier work [40], Abel et al. then applied the inhomogeneous solvation approach to study ligand binding in the factor Xa, an important drug target in the thrombosis pathway, several inhibitors of which are currently in Phase III clinical trials [39,40]. They build a map of water occupancy in the factor Xa active site using a clustering technique and assign chemical potentials to the water sites using the inhomogeneous solvation theory [82]. They then construct a semiempirical extension of the model that enables computation of free-energy differences ($\Delta\Delta G$ values) for pairs of congeneric factor Xa ligands that differ by deletions of atoms and thus will displace different portions of the solvent. These free-energy differences are shown to correlate exceptionally well with experimental data without complicated adjustable parameters. These findings indicate that contributions to the binding free energy of adding a *complementary chemical group* that makes hydrogen bonds where appropriate and hydrophobic contacts otherwise to a given ligand scaffold could largely be understood by the analysis of the solvent alone.

12.4.3 Dewetting in Ion-Channel Gating

Pentameric ligand-gated ion channels [83] (pLGICs) form a large family of membrane proteins that transmit external signals—the binding of a ligand—through the opening of their ion-conducting pores. With this important biological role in nerve signal transduction, pLGICs have recently become major drug targets. Accordingly, the gating mechanism, as an essential element of their function, has been a hot subject of extensive study. pLGICs are composed of five subunits that assemble into an extracellular domain (ECD) and a transmembrane domain (TMD) whose ion conductivity is gated by signals (i.e., ligand binding) from the ECD. Zhu and Hummer [43] recently studied this interesting ion-channel gating in an isolated TMD of *Gloeobacter violaceus* channel (GLIC) [84,85] (Figure 12.4). Unlike most other pLGICs, GLIC is a pH-gated channel,

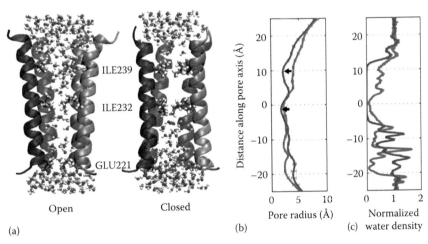

Figure 12.4 Structure and hydration of open (light gray lines) and closed (dark gray lines) channels. (a) Snapshots of the M2 helices in the open and closed channels at the transition endpoints, with the extracellular side up. The subunit in the front is removed to reveal the water molecules inside the pore. The side chains of three residues forming constrictions are also shown. (b) Pore radius along the axis, calculated using the program HOLE [33] and averaged over all frames of the respective simulations. The two narrowest parts of the closed conformation are indicated by arrows. (c) Distribution of the water density within a 5 Å cylinder around the pore axis, relative to the bulk density. (Reproduced from Zhu, F. and Hummer, G., *Proc. Natl. Acad. Sci. USA*, 107, 19814, 2010. With permission.)

a characteristic reflected in the unique features of its ECD. The TMD of GLIC, in contrast, fully reflects the common features of the membrane protein family.

In pLGICs, the transmembrane helices M2 from each of the five subunits together form the ion-conducting pore (Figure 12.4). Zhu and Hummer [43] found that along the pathway, the most prominent conformational change is a tilting motion of the M2 helices, with the two ends undergoing much larger displacements than the center of the helix. The helix tilting axes do not intersect the pore axis. Consequently, the tilting results in an iris-like motion of the helix atoms, altering both their distance from the symmetry axis and their polar angles. The gate in the GLIC channel is formed by the rings of residues Ile232 and Ile239 (Figure 12.4). Their hydrophobic and bulky side chains constrict the pore at its middle and near the extracellular entrance, respectively. In the open state, the region between the isoleucine rings is fully hydrated; upon channel closure, surprisingly, the ~15 Å long segment of the central pore becomes completely dry—a nanoscale dewetting transition similar to what was observed in the melittin tetramer channel [36]. Drying of the pore is induced by remarkably subtle changes in the pore width near the hydrophobic constriction. The authors defined a *gate radius* for each quintet of Ile residues by fitting a circle through the five centers of mass of their side chains in the xy plane parallel to the membrane. They found that the gate radius at Ile239 undergoes a larger change during the transition than that at Ile232, indicating that Ile239 might play a more significant role in controlling the state of the pore. Meanwhile, the water occupancy of the pore changes more abruptly along the pathway. Beyond a reduced path length of $L = 0.2$ (in reduced unit), the average occupancy is low and the hydrophobic region is rarely hydrated during the simulations, indicating an early drying and closure of the pore during the open-to-closed transition. The water in the hydrophobic constriction of the channel exhibits cooperative wetting and dewetting transitions, with channel closure shifting the equilibrium toward the dry state. The alternation between wetting and dewetting occurs intermittently on the nanosecond time scale, associated with sharp transitions in the number of water molecules. Interestingly, the gate radii remain nearly constant during the simulation.

Zhu and Hummer [43] also show that for an isolated TMD, the closed state is thermodynamically more stable based on the simulations, and closing is driven by a large free-energy gradient. By integrating the restraining forces along the string connecting the open and closed configurations, they calculated the free-energy profile for the open-to-closed transition. Interestingly, the open conformation does not appear to be in a local free-energy minimum. Instead, starting from the open conformation, the free energy decreases monotonically by a total of ~35 kcal/mol during the first half of the transition. The free energy then levels off after the midpoint of the transition. With the ECD removed, the TMD of GLIC alone does not actually have a metastable open state, as the earlier simulations indicate a spontaneous collapse of the pore into a narrow and dry hydrophobic gate.

This strong bias toward a closed state in isolated TMDs has possible physiological relevance. Although the exact function of GLIC has not been clearly established, its eukaryotic homolog, nAChR, triggers the contraction of muscle cells. After the nAChRs are opened, the resulting small change in the membrane potentially can activate nearby voltage-gated ion channels, which then quickly depolarize the membrane. Once the latter channels are also opened, the contraction of the muscle cell is inevitable even if the nAChRs return to the closed or inactivated state. Therefore, the risk of accidental damage to the muscle would impose evolutionary pressure to have a low opening probability of TMDs uncoupled from their ECDs, as found here in the case of GLIC. Interestingly, for voltage-gated K^+ channels, recent MD simulations [86] revealed that with the regulatory domain (the voltage sensor) removed, the pore domain also spontaneously underwent hydrophobic closure. It thus appears to be a common feature of gated ion channels that the pore domains alone would remain closed and can be opened only by a proper coupling to the regulatory domains.

12.5 SUMMARY

Water plays a wide variety of roles in various biological processes including shaping the complex biomolecular structures, activation and modulation of protein dynamics, facilitating ligand–receptor binding, and acting as a switchable ion channel for gating across membranes. Many of these properties do seem to depend, to a greater or lesser degree, on the special attributes of the water molecule, especially its ability to accommodate various solutes with different hydrophobicity. This chapter on biological water under confinement and its related hydrophobicity, in particular the nanoscale dewetting, is by no means complete and has at best touched on only a few interesting topics that have captivated a vast array of researchers with many different perspectives such as normal dehydration during protein folding, nanoscale dewetting in protein complex aggregation, dewetting in ligand–receptor binding, and dewetting in ion-channel gating. This diversity has also produced its share of contentious issues, but more importantly, it has illuminated the problem from many different directions. Because of the subtle nature of the hydrophobic effect itself, and its interplay with an array of other forces in the realistic biological environments, much more work is still needed before we can claim with any confidence that we have a thorough understanding on this important subject.

REFERENCES

1. Dinner, A. R., A. Sali, L. J. Smith, C. M. Dobson, and M. Karplus. 2000. Understanding protein folding via free-energy surfaces from theory and experiment. *Trends Biochem. Sci.* 25:331–339.

2. Dobson, C. M. and M. Karplus. 1999. The fundamentals of protein folding: Bringing together theory and experiment. *Curr. Opin. Struct. Biol.* 9:92–101.

3. Brooks, C. L., 3rd, M. Gruebele, J. N. Onuchic, and P. G. Wolynes. 1998. Chemical physics of protein folding. *Proc. Natl. Acad. Sci. USA* 95:11037–11038.

4. Hummer, G., S. Garde, A. E. Garcia, M. E. Paulaitis, and L. R. Pratt. 1998. The pressure dependence of hydrophobic interactions is consistent with the observed pressure denaturation of proteins. *Proc. Natl. Acad. Sci. USA* 95:1552–1555.

5. Klein-Seetharaman, J., M. Oikawa, S. B. Grimshaw, J. Wirmer, E. Duchardt, T. Ueda, T. Imoto, L. J. Smith, C. M. Dobson, and H. Schwalbe. 2002. Long-range interactions within a nonnative protein. *Science* 295:1719–1722.

6. Plaxco, K. W., J. I. Guijarro, C. J. Morton, M. Pitkeathly, I. D. Campbell, and C. M. Dobson. 1998. The folding kinetics and thermodynamics of the Fyn-SH3 domain. *Biochemistry* 37:2529–2537.

7. Dill, K. A., Bromberg, S., Yue, K., Chan, H. S., Ftebig, K. M., Yee, D. P., and Thomas, P. D. 1995. Principles of protein folding—A perspective from simple exact models. *Protein Sci.* 4:561–602.

8. Hummer, G., J. C. Rasaiah, and J. P. Noworyta. 2001. Water conduction through the hydrophobic channel of a carbon nanotube. *Nature* 414:188–190.

9. Chandler, D. 2005. Interfaces and the driving force of hydrophobic assembly. *Nature* 437:640–647.

10. Athawale, M. V., G. Goel, T. Ghosh, T. M. Truskett, and S. Garde. 2007. Effects of lengthscales and attractions on the collapse of hydrophobic polymers in water. *Proc. Natl. Acad. Sci. USA* 104:733–738.

11. Berne, B. J., J. D. Weeks, and R. Zhou. 2009. Dewetting and hydrophobic interaction in physical and biological systems. *Annu. Rev. Phys. Chem.* 60:85–103.

12. Patel, A. J., P. Varilly, S. N. Jamadagni, M. F. Hagan, D. Chandler, and S. Garde. 2012. Sitting at the edge: How biomolecules use hydrophobicity to tune their interactions and function. *J. Phys. Chem. B* 116:2498–2503.

13. Ball, P. 2008. Water as an active constituent in cell biology. *Chem. Rev.* 108:74–108.

14. ten Wolde, P. R. and D. Chandler. 2002. Drying-induced hydrophobic polymer collapse. *Proc. Natl. Acad. Sci. USA* 99:6539–6543.

15. Collins, M., G. Hummer, M. Quillin, B. Matthews, and S. Gruner. 2005. Cooperative water filling of a nonpolar protein cavity observed by high-pressure crystallography and simulation. *Proc. Natl. Acad. Sci. USA* 102:16668–16673.

16. Kalra, A., S. Garde, and G. Hummer. 2003. Osmotic water transport through carbon nanotube membranes. *Proc. Natl. Acad. Sci. USA* 100:10175–10180.

17. Arkin, I. T., H. Xu, M. O. Jensen, E. Arbely, E. R. Bennett, K. J. Bowers, E. Chow et al. 2007. Mechanism of Na^+/H^+ antiporting. *Science* 317:799–803.

18. Jensen, M. O., R. O. Dror, H. Xu, D. W. Borhani, I. T. Arkin, M. P. Eastwood, and D. E. Shaw. 2008. Dynamic control of slow water transport by aquaporin 0: Implications for hydration and junction stability in the eye lens. *Proc. Natl. Acad. Sci. USA* 105:14430–14435.

19. Sarupria, S. and S. Garde. 2009. Quantifying water density fluctuations and compressibility of hydration shells of hydrophobic solutes and proteins. *Phys. Rev. Lett.* 103:037803.

20. Patel, B. A., P. G. Debenedetti, F. H. Stillinger, and P. J. Rossky. 2007. A water-explicit lattice model of heat-, cold-, and pressure-induced protein unfolding. *Biophys. J.* 93:4116–4127.

21. Pratt, L. R. and H. S. Ashbaugh. 2003. Self-consistent molecular field theory for packing in classical liquids. *Phys. Rev. E: Stat. Nonlinear Soft Matter Phys.* 68:021505.

22. Stillinger, F. H. 1973. Structure in aqueous solutions of nonpolar solutes from the standpoint of scaled-particle theory. *J. Solut. Chem.* 2:141–158.

23. Lum, K., D. Chandler, and J. D. Weeks. 1999. Hydrophobicity at small and large length scales. *J. Phys. Chem. B* 103:4570–4577.

24. Lum, K. and A. Luzar. 1997. Pathway to surface-induced phase transition of a confined fluid. *Phys. Rev. E: Stat. Nonlin. Soft Matter Phys.* 56:R6283–R6286.

25. Luzar, A. and K. Leung. 2000. Dynamics of capillary evaporation. I. Effect of morphology of hydrophobic surfaces. *J. Chem. Phys.* 113:5836–5844.

26. Soper, A. K., E. W. Castner, and A. Luzar. 2003. Impact of urea on water structure: A clue to its properties as a denaturant? *Biophys. Chem.* 105:649–666.

27. Webb, A. E. and K. M. Weeks. 2001. A collapsed state functions to self-chaperone RNA folding into a native ribonucleoprotein complex. *Nat. Struct. Mol. Biol.* 8:135–140.

28. Davis, J. G., K. P. Gierszal, P. Wang, and D. Ben-Amotz. 2012. Water structural transformation at molecular hydrophobic interfaces. *Nature* 491:582–585.

29. Davis, J. G., B. M. Rankin, K. P. Gierszal, and D. Ben-Amotz. 2013. On the cooperative formation of non-hydrogen-bonded water at molecular hydrophobic interfaces. *Nat. Chem.* 5:796–802.

30. Lee, C. Y., J. A. McCammon, and P. J. Rossky. 1984. The structure of liquid water at an extended hydrophobic surface. *J. Chem. Phys.* 80:4448–4455.

31. Wallqvist, A. and B. J. Berne. 1995. Computer simulation of hydrophobic hydration forces on stacked plates at short-range. *J. Chem. Phys.* 99:2893.

32. Huang, X., C. J. Margulis, and B. J. Berne. 2003. Dewetting-induced collapse of hydrophobic particles. *Proc. Natl. Acad. Sci. USA* 100:11953–11958.

33. Parker, J., P. Claesson, and P. Attard. 1994. Bubbles, cavities, and the long-ranged attraction between hydrophobic surfaces. *J. Phys. Chem.* 98:8468–8480.

34. Zhou, R., X. Huang, C. J. Margulis, and B. J. Berne. 2004. Hydrophobic collapse in multidomain protein folding. *Science* 305:1605–1609.

35. Huang, X., R. Zhou, and B. J. Berne. 2005. Drying and hydrophobic collapse of paraffin plates. *J. Phys. Chem. B* 109:3546–3552.

36. Liu, P., X. Huang, R. Zhou, and B. J. Berne. 2005. Observation of a dewetting transition in the collapse of the melittin tetramer. *Nature* 437:159–162.

37. Hua, L., X. Huang, P. Liu, R. Zhou, and B. J. Berne. 2007. Nanoscale dewetting transition in protein complex folding. *J. Phys. Chem. B* 111:9069–9077.

38. Giovambattista, N., C. F. Lopez, P. J. Rossky, and P. G. Debenedetti. 2008. Hydrophobicity of protein surfaces: Separating geometry from chemistry. *Proc. Natl. Acad. Sci. USA* 105:2274–2279.

39. Abel, R., L. Wang, R. A. Friesner, and B. J. Berne. 2010. A displaced-solvent functional analysis of model hydrophobic enclosures. *J. Chem. Theory Comput.* 6:2924–2934.

40. Abel, R., T. Young, R. Farid, B. J. Berne, and R. A. Friesner. 2008. Role of the active-site solvent in the thermodynamics of factor Xa ligand binding. *J. Am. Chem. Soc.* 130:2817–2831.

41. Young, T., R. Abel, B. Kim, B. J. Berne, and R. A. Friesner. 2007. Motifs for molecular recognition exploiting hydrophobic enclosure in protein-ligand binding. *Proc. Natl. Acad. Sci. USA* 104:808–813.

42. Young, T., L. Hua, X. Huang, R. Abel, R. Friesner, and B. J. Berne. 2010. Dewetting transitions in protein cavities. *Proteins* 78:1856–1869.

43. Zhu, F. and G. Hummer. 2010. Pore opening and closing of a pentameric ligand-gated ion channel. *Proc. Natl. Acad. Sci. USA* 107:19814–19819.

44. Das, P., D. Kapoor, K. T. Halloran, R. Zhou, and C. R. Matthews. 2013. Interplay between drying and stability of a TIM barrel protein: A combined simulation-experimental study. *J. Am. Chem. Soc.* 135:1882–1890.

45. Yaminsky, V. V. and B. W. Ninham. 1993. Hydrophobic force: Lateral enhancement of subcritical fluctuations. *Langmuir* 9:3618–3624.

46. Yaminsky, V. V., V. S. Yusshchenko, E. A. Amelina, and E. D. Shchukin. 1983. Cavity formation due to a contact between particles in a nonwetting liquid. *J. Colloid Interface Sci.* 96:301–306.

47. Rowlinson, J. S. and B. Widom. 1989. *Molecular Theory of Capillarity.* Clarendon Press, Oxford, U.K.

48. Lum, K. and D. Chandler. 1998. Phase diagram and free energies of vapor films and tubes for a confined fluid. *Int. J. Thermophys.* 19:845–855.

49. Bratko, D., R. A. Curtis, H. W. Blanch, and J. M. Prausnitz. 2001. Interaction between hydrophobic surfaces with metastable intervening liquid. *J. Chem. Phys.* 115:3873–3877.

50. Christenson, H. K. and P. M. Claesson. 1988. Cavitation and the interaction between macroscopic hydrophobic surfaces. *Science* 239:390–392.

51. Claesson, P. M. and H. K. Christenson. 1988. Very long range attractive forces between uncharged hydrocarbon and fluorocarbon surfaces in water. *J. Phys. Chem.* 92:1650–1655.

52. Zhou, R. and B. J. Berne. 2002. Can a continuum solvent model reproduce the free energy landscape of a beta-hairpin folding in water? *Proc. Natl. Acad. Sci. USA* 99:12777–12782.

53. Zhou, R. 2003. Trp-cage: Folding free energy landscape in explicit water. *Proc. Natl. Acad. Sci. USA* 100:13280–13285.

54. Hua, L., X. Huang, R. Zhou, and B. J. Berne. 2006. Dynamics of water confined in the interdomain region of a multidomain protein. *J. Phys. Chem. B* 110:3704–3711.

55. Krone, M. G., L. Hua, P. Soto, R. Zhou, B. J. Berne, and J. E. Shea. 2008. Role of water in mediating the assembly of Alzheimer amyloid-beta Abeta16–22 protofilaments. *J. Am. Chem. Soc.* 130:11066–11072.

56. Chang, C. W., L. Guo, Y. T. Kao, J. Li, C. Tan, T. Li, C. Saxena et al. 2010. Ultrafast solvation dynamics at binding and active sites of photolyases. *Proc. Natl. Acad. Sci. USA* 107:2914–2919.

57. Zhang, L., L. Wang, Y. T. Kao, W. Qiu, Y. Yang, O. Okobiah, and D. Zhong. 2007. Mapping hydration dynamics around a protein surface. *Proc. Natl. Acad. Sci. USA* 104:18461–18466.

58. Bartlett, A. I. and S. E. Radford. 2010. Desolvation and development of specific hydrophobic core packing during Im7 folding. *J. Mol. Biol.* 396:1329–1345.
59. Brun, L., D. G. Isom, P. Velu, B. Garcia-Moreno, and C. A. Royer. 2006. Hydration of the folding transition state ensemble of a protein. *Biochemistry* 45:3473–3480.
60. Fernandez-Escamilla, A. M., M. S. Cheung, M. C. Vega, M. Wilmanns, J. N. Onuchic, and L. Serrano. 2004. Solvation in protein folding analysis: Combination of theoretical and experimental approaches. *Proc. Natl. Acad. Sci. USA* 101:2834–2839.
61. Nishiguchi, S., Y. Goto, and S. Takahashi. 2007. Solvation and desolvation dynamics in apomyoglobin folding monitored by time-resolved infrared spectroscopy. *J. Mol. Biol.* 373:491–502.
62. Kimura, T., A. Maeda, S. Nishiguchi, K. Ishimori, I. Morishima, T. Konno, Y. Goto, and S. Takahashi. 2008. Dehydration of main-chain amides in the final folding step of single-chain monellin revealed by time-resolved infrared spectroscopy. *Proc. Natl. Acad. Sci. USA* 105:13391–13396.
63. Englander, S. W. 2000. Protein folding intermediates and pathways studied by hydrogen exchange. *Annu. Rev. Biophys. Biomol. Struct.* 29:213–238.
64. Wintrode, P. L., T. Rojsajjakul, R. Vadrevu, C. R. Matthews, and D. L. Smith. 2005. An obligatory intermediate controls the folding of the alpha-subunit of tryptophan synthase, a TIM barrel protein. *J. Mol. Biol.* 347:911–919.
65. Miranker, A., C. V. Robinson, S. E. Radford, R. T. Aplin, and C. M. Dobson. 1993. Detection of transient protein folding populations by mass spectrometry. *Science* 262:896–900.
66. Jones, B. E. and C. R. Matthews. 1995. Early intermediates in the folding of dihydrofolate reductase from *Escherichia coli* detected by hydrogen exchange and NMR. *Protein Sci.* 4:167–177.
67. Gu, Z., M. K. Rao, W. R. Forsyth, J. M. Finke, and C. R. Matthews. 2007. Structural analysis of kinetic folding intermediates for a TIM barrel protein, indole-3-glycerol phosphate synthase, by hydrogen exchange mass spectrometry and Go model simulation. *J. Mol. Biol.* 374:528–546.
68. Vadrevu, R., Y. Wu, and C. R. Matthews. 2008. NMR analysis of partially folded states and persistent structure in the alpha subunit of tryptophan synthase: Implications for the equilibrium folding mechanism of a 29-kDa TIM barrel protein. *J. Mol. Biol.* 377:294–306.
69. Radzicka, A. and R. Wolfenden. 1988. Comparing the polarities of the amino acids: Side-chain distribution coefficients between the vapor phase, cyclohexane, 1-octanol, and neutral aqueous solution. *Biochemistry* 27:1664–1670.
70. Kathuria, S. V., I. J. Day, L. A. Wallace, and C. R. Matthews. 2008. Kinetic traps in the folding of beta alpha-repeat proteins: CheY initially misfolds before accessing the native conformation. *J. Mol. Biol.* 382:467–484.
71. Wu, Y., R. Vadrevu, S. Kathuria, X. Yang, and C. R. Matthews. 2007. A tightly packed hydrophobic cluster directs the formation of an off-pathway sub-millisecond folding intermediate in the alpha subunit of tryptophan synthase, a TIM barrel protein. *J. Mol. Biol.* 366:1624–1638.
72. Yu, N. and M. F. Hagan. 2012. Simulations of HIV capsid protein dimerization reveal the effect of chemistry and topography on the mechanism of hydrophobic protein association. *Biophys. J.* 103:1363–1369.

73. Zhang, F., H.-N. Du, Z.-X. Zhang, L.-N. Ji, H.-T. Li, L. Tang, H.-B. Wang et al. 2006. Epitaxial growth of peptide nanofilaments on inorganic surfaces: Effects of interfacial hydrophobicity/hydrophilicity. *Angew. Chem. Int. Ed.* 45:3611–3613.

74. Hua, L., R. Zangi, and B. J. Berne. 2009. Hydrophobic interactions and dewetting between plates with hydrophobic and hydrophilic domains. *J. Phys. Chem. C* 113:5244–5253.

75. Li, J., T. Liu, X. Li, L. Ye, H. Chen, H. Fang, Z. Wu, and R. Zhou. 2005. Hydration and dewetting near graphite–CH(3) and graphite–COOH plates. *J. Phys. Chem. B* 109:13639–13648.

76. Li, X., J. Li, M. Eleftheriou, and R. Zhou. 2006. Hydration and dewetting near fluorinated superhydrophobic plates. *J. Am. Chem. Soc.* 128:12439–12447.

77. Miller, T. F., E. Vanden-Eijnden, and D. Chandler. 2007. Solvent coarse-graining and the string method applied to the hydrophobic collapse of a hydrated chain. *Proc. Natl. Acad. Sci. USA* 104:14559–14564.

78. Lindorff-Larsen, K., S. Piana, R. O. Dror, and D. E. Shaw. 2011. How fast-folding proteins fold. *Science* 334:517–520.

79. Wang, J., D. Bratko, and A. Luzar. 2011. Probing surface tension additivity on chemically heterogeneous surfaces by a molecular approach. *Proc. Natl. Acad. Sci. USA* 108:6374–6379.

80. Wang, L., R. Abel, R. A. Friesner, and B. J. Berne. 2009. Thermodynamic properties of liquid water: An application of a nonparametric approach to computing the entropy of a neat fluid. *J. Chem. Theory Comput.* 5:1462–1473.

81. Qvist, J., M. Davidovic, D. Hamelberg, and B. Halle. 2008. A dry ligand-binding cavity in a solvated protein. *Proc. Natl. Acad. Sci. USA* 105:6296–6301.

82. Lazarids, T. 1998. Inhomogeneous fluid approach to solvation thermodynamics. 1. Theory. *J. Phys. Chem. B* 102:3531–3541.

83. Sine, E. M. and A. G. Engel. 2006. Recent advances in Cys-loop receptor structure and function. *Nature* 440:448–455.

84. Hilf, R. J. and R. Dutzler. 2008. X-ray structure of a prokaryotic pentameric ligand-gated ion channel. *Nature* 452:375–379.

85. Hilf, R. J. and R. Dutzler. 2009. Structure of a potentially open state of a proton-activated pentameric ligand-gated ion channel. *Nature* 457:111–114.

86. Jensen, M. O., D. W. Borhani, K. Lindorff-Larsen, P. Maragakis, V. Jogini, M. P. Eastwood, R. O. Dror, and D. E. Shaw. 2010. Principles of conduction and hydrophobic gating in K⁺ channels. *Proc. Natl. Acad. Sci. USA* 107:5833–5838.

Index